# Springer Series in Optical Sciences     Volume 25

Edited by David L. MacAdam

W0225789

# Springer Series in Optical Sciences

Edited by David L. MacAdam

Editorial Board:  J. M. Enoch    D. L. MacAdam    A. L. Schawlow    T. Tamir

David C. Brown

# High-Peak-Power
# Nd:Glass Laser Systems

With 135 Figures

Springer-Verlag Berlin Heidelberg GmbH 1981

DAVID C. BROWN, Ph. D.
University of Rochester, College of Engineering and Applied Science,
Laboratory for Laser Energetics, 250 East River Road,
Rochester, NY 14623, USA

ISBN 978-3-662-13516-7    ISBN 978-3-540-38508-0 (eBook)
DOI 10.1007/978-3-540-38508-0

Library of Congress Cataloging in Publication Data. Brown, D. C. (David C.), 1942-. High peak power Nd. (Springer series in optical sciences ; v. 25). Bibliography: p. Includes index. 1. Neodymium glass lasers. I. Title. II. Series. TA1705.B76  621.36'63  81-4286 AACR2

2153/3130-543210

# Preface

During the 1970s we have witnessed the development of the Nd: glass laser from a laboratory tool operating typically in the gigawatt-power regime to large operating systems such as Argus, Shiva, GDL, and OMEGA with power outputs typically in the terawatt beamline range. The major impetus for this development has been the growth of inertial confinement of laser fusion as well as the weapons-effects simulation capabilities of such systems used in the national laboratories. Recently, high-power systems have begun to be used for other diverse applications, such as the development of the X-ray laser, time-resolved X-ray diffraction studies, studies of astrophysics and equations of state, materials studies, and the reconversion of spent nuclear fuel rods from fissile reactors.

   Two years ago while reading Dr. Koechner's book "Solid State Laser Engineering" (Springer Series in Optical Sciences, Vol.1), it occurred to me that there was a real need for a book devoted specifically to high-peak-power Nd:glass lasers, primarily because most of the physics of such systems had been worked out and was well understood, and because much of the literature in this field is found in laboratory reports, government reports, or conference proceedings. While high-peak power lasers are treated exclusively, since the preparation of this book began the emphasis has shifted from that regime to rather-longer-pulse lasers (typically 1-3 ns) where the dominant effects are saturation and damage phenomena. Nevertheless, the reader will find in this book all of the relevant physics, regardless of the emphasis on short-pulse operation.

   My initial plans concerning this volume were to have a number of authors contribute. Dr. Lotsch, however, encouraged me to write the entire book myself, a difficult task, but one which has led to a consistency of notation, well-planned-out contents, and lack of duplication or overlap. The subject matter has been chosen to cover those major areas of interest to the design and operation of Nd:glass laser systems. The author does not apologize for

the lack of comprehensiveness in this book; such completeness would have resulted in a volume two to three times as large. Instead, what has been striven
for is to cover all of the major developments in the field and to present them
here in a coherent and, I hope, understandable fashion.

This book will be useful to the graduate student or engineer just beginning
in this field and could also be used as a text in a graduate course in solid-
state laser engineering. I also hope that it will be useful to many colleagues
working in this or related fields to have all of the physics drawn together
in a single volume.

## *Acknowledgement*

It is difficult, if not impossible, to acknowledge the large number of persons
who have contributed to this book in a variety of ways. I would like to thank
Dr. Moshe Lubin, director of the Laboratory for Laser Energetics, who has pro-
vided the encouragement, time, and support, even in the difficult periods,
Dr. David L. MacAdam for excellent editing of the text in a timely and efficie
fashion, Mrs. Edna Hughes for accurately typing the manuscript, and Dr. Helmut
Lotsch for the encouragement and patience displayed during its completion. Als
I am indebted to my colleagues at the Laboratory for Laser Energetics, particu
ly J. Kelly, S. Jacobs, J. Abate, K. Lee, J. Hoose, J. Eastman, L. Lund, J.
Soures, W. Seka, J. Rinefierd, J. Bunkenburg, R. Sampath, J. Wilson, R. Hop-
kins, J. Boles and others who have all contributed to this volume, and to my
good friend Farrés Mattar for encouragement and advice during its writing.
Finally, special thanks to my wife Joan and my two daughters Jennifer and
Vanessa who have always been there through the good times and the bad.
The author gratefully acknowledges the partial support by the following
sponsors: Exxon Research and Engineering Company, General Electric Company,
Northeast Utilities Service Company, New York State Energy Research and De-
velopment Authority, The Standard Oil Company (Ohio), The University of
Rochester, and Empire State Electric Energy Research Corporation. Such sup-
port does not imply endorsement of the content by any of the above parties.

Rochester, January, 1981                                    *David C. Brown*

# Contents

# 1. Glass Laser Physics

This book is concerned with the physics of high-power Nd:glass lasers. To understand the design and operation of such devices we first explore the fundamental physics of the trivalent rare-earth Nd ion in various glass and crystalline hosts. It is worthwhile pausing here to understand the reasons for Nd:glass becoming the predominant optical material to be used today. An incomplete listing would include the following:

1) The absorption spectrum of $Nd^{3+}$ in typical optical glasses extends from $\simeq 350\,nm$ in the ultraviolet to $\simeq 900\,nm$ in the infrared, thus overlapping well with high-brightness pump sources such as Xe flashlamps.

2) The output wavelength in the vicinity of 1.06 $\mu m$ is of interest in laser inertial confinement experiments and, although a shorter wavelength is desirable, leads to reasonable plasma-coupling efficiencies.

3) Owing to a combination of favorable energy-level dynamics, the laser transition of most interest at 1.06 $\mu m$ is capable of large energy storage.

4) The stimulated-emission cross section for the $Nd^{3+}$ laser transition is in the intermediate regime, large enough to provide gain in reasonably sized amplifiers, but not so large as to make the problem of amplified spontaneous emission (ASE) severe. ASE is a common problem to most lasers and will be investigated in Chap.4.

5) The spectroscopy of $Nd^{3+}$ is well understood. This makes predictions of optical and nonlinear properties possible. Also, the technology of forming rare-earth-doped glasses is advanced, and the composition of a glass can be varied to achieve optimum properties for a given application.

Although many other rare earths have displayed laser transitions in the near-ultraviolet, visible, and near-infrared regions of the electromagnetic spectrum, not one has been found to have all of the desirable properties mentioned above. Lest the reader get the impression, however, that Nd:glass is the ideal material for use in high-peak power lasers, we mention two problems that are at present obstacles to the use of such systems in any laser-fusion power plant. They are the efficiency and repetition rate. Typical Nd:glass amplifiers convert typically a maximum of only 1%-3% of the initial electrical energy to useful inversion or stored energy density for use in laser amplification. Present laser amplifiers are capable of being fired, typically, only once every 1/2-3 hours, depending upon the geometry, principally because of the great amount of heat generated in the pumping process as well as the low thermal conductivity of glass. It is likely that in the next few years significant progress will be made in addressing both problems.

## 1.1 Spectroscopy of $Nd^{3+}$

An understanding of $Nd^{3+}$ lasers necessitates familiarity with the details of the energy-level structure of $Nd^{3+}$. A more complete analysis for the reader interested in such details may be found in the classic work of DIEKE and CROSSWHITE [1.1]. Referring to Fig.1.1, the ground state is a $^4I_{9/2}$ configuration; the visible absorption spectrum occurs from the ground state to

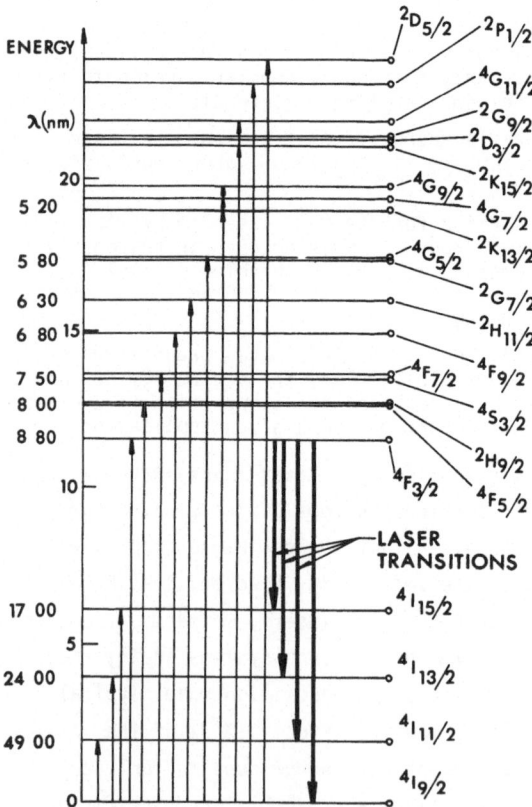

Fig. 1.1. Energy levels of $Nd^{3+}$

various excited-state levels. Atoms excited to one of the excited states re-
lax via radiative or multiphonon transitions to the upper laser level ($^4F_{3/2}$)
in a time short compared to its radiative lifetime (typically 300-600 $\mu s$). The
$^4F_{3/2}$ level decays radiatively through the four laser transitions [$^4F_{3/2} \rightarrow$
$^4I_{15/2}$ (1.80 $\mu m$), $^4I_{13/2}$ (1.35 $\mu m$), $^4I_{11/2}$ (1.06 $\mu m$), $^4I_{9/2}$ (0.88 $\mu m$)]. Note
that the transition ($^4F_{3/2} \rightarrow {}^4I_{9/2}$) is a resonant line in the sense that it
is both absorbed and emitted. In most common laser materials, the radiation
at 1.80 $\mu m$ and 1.35 $\mu m$ is not detected because of strong host background absorption in that spectral region. The relative intensities of the various
laser lines are determined by the branching ratios that are discussed in
Sect.1.4.

One of the fundamental parameters in Nd:glass laser design is the lifetime
of the $^4I_{11/2}$ transition because its value determines whether three- or four-
level laser operation is obtained (see Sect.1.3). Attempts to measure the
lifetime directly by absorption from the ($^4I_{11/2} \rightarrow {}^4I_{9/2}$) transition have been
fruitless because its wavelength at 5 $\mu m$ is strongly absorbed by the host. A
pulse-saturation technique was used by DUSTON [1.2], who found that the terminal-level lifetime varied with temperature and wt-% doping of the sample.
Owens-Illinois ED-2 silicate laser glass with 3 wt-% doping gave a lifetime
of $\approx$11 ns. Other measurements yielded values in the range of 10-50 ns [1.3-5].
More recently [1.6] for example, the $^4I_{11/2}$ lifetime was measured for ED-2
laser glass of unspecified doping, and was found to be $\approx$1.25 $\pm$ 0.2 ns. Terminal-level lifetime effects will be reviewed in detail in Sect.1.3.

In Fig.1.2 we show two typical absorption spectra for commercially available silicate (ED-2) and phosphate (LGH-5) laser glasses [1.7]. It should be
noted that the phosphate glass is clearer in the ultraviolet than the silicate, with a band centered at about 353 nm uncovered owing to the lack of
background absorption. The ultraviolet (UV) edge, which is so apparent in
the silicate glass, has been shifted considerably towards the blue in the
phosphate, leading to a significant pumping contribution ($\approx$7-10%) from the
353 nm band ($^4I_{9/2} \rightarrow {}^4D_{3/2}$) as well as alleviating severe thermal problems
associated with the absorption of UV from typical Xe flashlamps. The absorption bands in the phosphate host are typically more intense but narrower than
those in the silicate glasses, and slightly shifted toward the UV.

The location, intensity, and breadth of the absorption bands are determined by the interaction of the $Nd^{3+}$ ion with the local crystalline field as
we discuss in the next section. It is important to realize, however, that as
can be seen from Fig.1.2, each absorption band usually consists of a multi-
plicity of levels (a manifold) so that at $\approx$580 nm, for example, the transi-

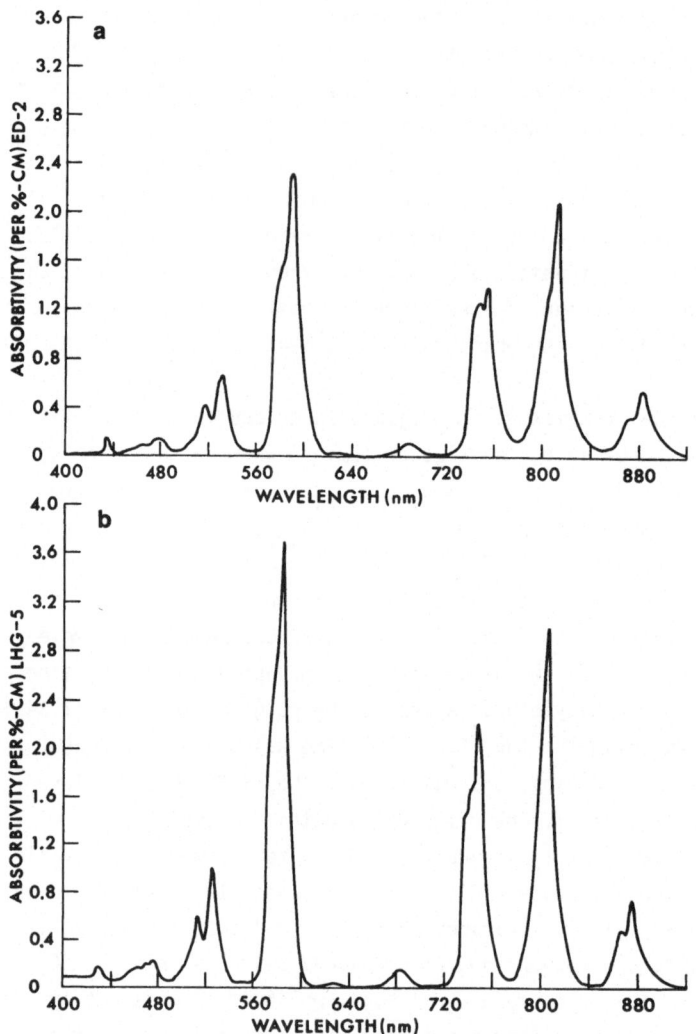

Fig. 1.2a,b. Absorption spectrum of $Nd^{3+}$ in ED-2 silicate (a) and LHG-6 phosphate (b) glasses

tions involved are $(^4I_{9/2} \rightarrow ^4G_{5/2}, ^2G_{7/2})$. Owing to the inhomogeneous broadening from local field interactions, adjacent energy levels often overlap and appear in the absorption spectrum as one. This behavior is to be contrasted to, say, Nd:YAG where the absorption spectrum is clearly resolved (see, for example, [Ref.1.8, p.57]). We now turn our attention to the local environment that a $Nd^{3+}$ ion sees in various hosts and the reasons for the broad-absorption-bands characteristic of Nd:glass.

## 1.2  Nd$^{3+}$ Environment in Glass

The breadth of Nd$^{3+}$ absorption lines is determined by the Stark broadening experienced by individual Nd$^{3+}$ ions in the host. Unlike the regular local crystal field experienced by Nd$^{3+}$ in hosts such as Nd:YAG, the sites in glass are randomly disordered, leading to a statistical distribution. Thus, macroscopic parameters such as stimulated-emission cross section and lifetimes are really statistical averages and are determined by superposition of contributions from ions in a myriad of local environments. An indication of this was the careful work performed at Lawrence Livermore Laboratory [1.9] which showed that the characteristic fluorescence decay at 1.06 µm ($^4F_{3/2} \rightarrow {}^4I_{11/2}$) was nonexponential. One experimental result, seen in Fig.1.3 [1.10], was obtained for ED-2 silicate laser glass with various doping levels and shows that even for low dopings where ion-ion interactions should be negligible, the decay is indeed nonexponential and can be explained only by site-to-site variations of the radiative decay rate. Also evident in Fig.1.3 is the so-called "concentration quenching" where the lifetime decreases with increasing Nd$^{3+}$ concentration due to ion-ion interactions. The average rate of decay of excited Nd$^{3+}$ ions (lifetime $\tau$) increases to approximately twice the zero concentration value (lifetime $\tau_0$) for $\simeq 6$ wt-% doping. The quantum efficiency $\eta$ defined by

$$\eta = \tau/\tau_0 \tag{1.1}$$

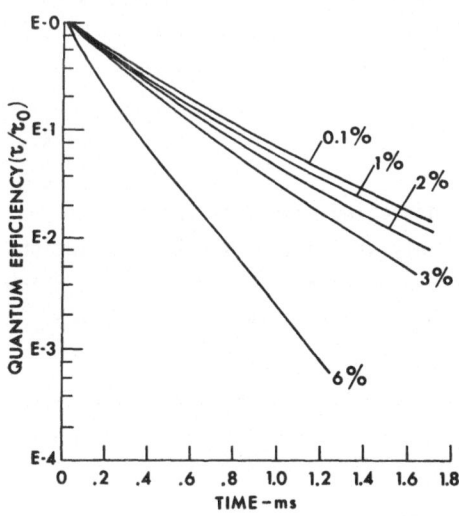

Fig. 1.3. Time dependence of Nd$^{3+}$ fluorescence decay in ED-2 glass for several Nd$^{3+}$ concentrations (in weight percent) [1.9]

thus decreases from a value close to 100% near zero wt-% doping to approximately 50% at the six-percent level.

In a detailed study of nonexponential decay rates in various laser glasses, LAYNE [1.11] summarized the results as follows:

1) Decay is nonexponential at low concentrations owing to site-to-site variations of the radiative decay rates. If the distribution of sites (as determined, for example, by fluorescence line narrowing (FLN) experiments to be discussed below) can be accurately determined, it should be possible to calculate the observed decay rate.

2) For concentrations larger than $\approx 1\%$, the decay consists of one fast initial part that is nonexponential, whereas for long times the decay is near exponential. The fast initial decay may be accurately fitted by

$$\Phi(t) = \exp(-kNt^{\frac{1}{2}}) \quad , \tag{1.2}$$

where k is a quenching constant, N the $Nd^{3+}$ density, and t the time.

3) For large concentrations, the long-term exponential decay rate was found to be proportional to $N^2$, where N is again the $Nd^{3+}$ density. Because of the $N^2$ dependence, energy diffusion was assumed to be operative.

At any given site, an initially excited ion may decay radiatively, by multi-phonon transitions, or by transferring its excitation to another nearby ion. Energy transfer between like ions is generally referred to as diffusion whereas that between unlike ions is called quenching. Theory has shown [1.12] that quenching is characterized by a nonexponential fluorescence decay with the rate $\Phi(t)$ approaching the purely radiative rate for long times. In particular, the following expression for $\Phi(t)$ can be derived

$$\Phi(t) = \exp\left(-t/\tau - k_1 N t^{\frac{1}{2}}\right) \quad , \tag{1.3}$$

where $\tau$ is the radiative lifetime, $k_1$ the quenching constant, and N the concentration of quenching ions. It is valid only when dipole-dipole interactions occur and has been found to agree well with experiment. Physically, ions which are near quenching centers decay quickly, whereas those that are not near quenching centers remain excited and decay at the radiative rate.

The theory of diffusion among like ions predicts a decay rate $\Phi(t)$ of the form

$$\Phi(t) = \exp\left(-t/\tau - k_2 N^2 t\right) \quad , \tag{1.4}$$

where $k_2$ is a constant and N the ion density. In agreement with experimental results, for short times after excitation, quenching is dominant according to (1.3) whereas at long times, diffusion is the predominant mechanism with a decay rate of $(1/\tau + k_2 N^2)$. Although quenching normally takes place between unlike-ion populations, it has been proposed that the $Nd^{3+}$ ions themselves are responsible [1.11]. Impurity ions have been ruled out and it is not clear whether ions in a particular local-field environment or any $Nd^{3+}$ ion may act as quencher while also taking part in diffusion.

Recently, investigation has shown that radiative decay in Nd pentaphosphate crystals ($Nd:P_5O_{14}$) is purely exponential and that concentration quenching is to a large degree absent [1.13]. For example, the radiative lifetime for near-zero Nd concentration is $\tau_0 \simeq 310$ μs, whereas for pure $Nd:P_5O_{14}$ (which corresponds to ≈40 wt-% doping) the radiative lifetime is $\tau \simeq 115$ μs. This corresponds to 35% quantum efficiency which is reached in commercial laser glasses for a 6-7 wt-% doping. Because $Nd:P_5O_{14}$ is a crystal and the local field seen by a $Nd^{3+}$ ion is regular, large site-to-site variations in radiative decay do not occur and the measured fluorescence is purely exponential. Unlike commercial laser glasses, it has also been found that the decay rate $\Phi(t)$ does not involve the square of N as in (1.4), but is linear.

From this, we may conclude that the local environment seen by a $Nd^{3+}$ ion strongly affects the macroscopic properties. Recently, FLN experiments have demonstrated wide variations of local sites [1.14]. The $Nd^{3+}$ ion is inhomo-geneously broadened, hence it is possible, by using a dye-laser source whose emission is narrow compared to the transition of interest, to excite only the ions within the source bandwidth that are representative of a particular site. By varying the source center wavelength, the entire set of local environments may be probed and eventually their distribution may be determined. BRECHER et al. [1.14] performed such an experiment on ED-2 silicate laser glass, exciting the $(^4I_{9/2} \rightarrow {}^2P_{1/2})$ transition using a dye laser operating in the vicinity of 430.8 nm. That transition was chosen because Stark splitting of the $^2P_{1/2}$ level does not occur in the local crystal field and, at low temperatures, only the lowest Stark level of the ground-state manifold is populated. Ions decay from the $^2P_{1/2}$ state to the $^4F_{3/2}$ upper laser level with a lifetime of ≈20 ns owing to multi-phonon emission. Detection was via the $(^4F_{3/2} \rightarrow {}^4I_{11/2})$ transition at 1.06 μm. In Figs.1.4 and 1.5 we show their results; in the former, it can be seen that there is a variation of fluorescence intensity greater than a factor of 2 across the absorption profile; the maximum intensity occurs at longer wavelengths. The ratio of the 1.06 μm to 0.88 μm fluorescence intensity (branching ratio) changes similarly, except that it is maximum at shorter wave-

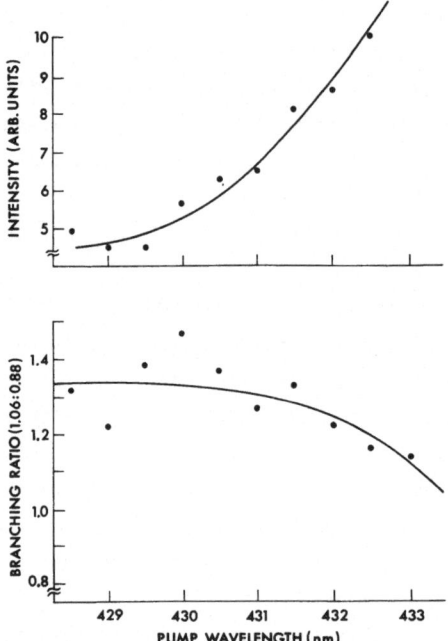

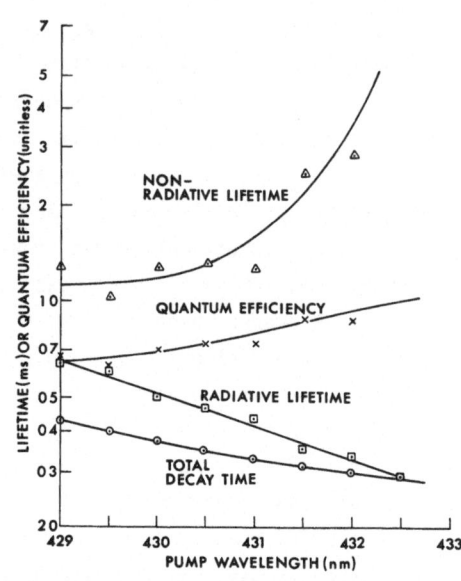

Fig. 1.4. 1.06 μm fluorescence intensity and branching ratio as a function
of pump wavelength, ED-2 laser glass
[1.14]

Fig. 1.5. Radiative and nonradiative
lifetime, total decay time, and relative quantum efficiency in ED-2 laser
glass as a function of pump wavelength
[1.14]

lengths. In Fig.1.5 the radiative, nonradiative, and total decay times are
shown along with the relative quantum efficiency. It is remarkable that the
total decay time varies from ≈440 μs at 429 nm to ≈300 μs at 432.5 nm, whereas the relative quantum efficiency varies from an assumed 100% at 432.5 nm
to ≈65% at 429 nm. Of further note here is the observed fact that the radiative decays can be described by pure exponential terms (confirming earlier
suspicions) mentioned previously, that under broadband excitation nonexponential decay was due to site-to-site variations in the $Nd^{3+}$ environment. Additionally, the experiment was performed at 15 K where ion cross relaxation was
assumed to be absent. The fact that quantum efficiency varies widely is similarly due to site-to-site variations of the strength of the ion-phonon coupling. The strength of that interaction is determined by the local field and
vibrational modes, both of which are site dependent. In a more recent publication, BRECHER et al. [1.15] have shown similar results in a FLN study of the
following glass types: silicate, phosphate, borate, fluoroberyllate, and
fluorophosphate. As in the foregoing results, intensity, branching ratio,

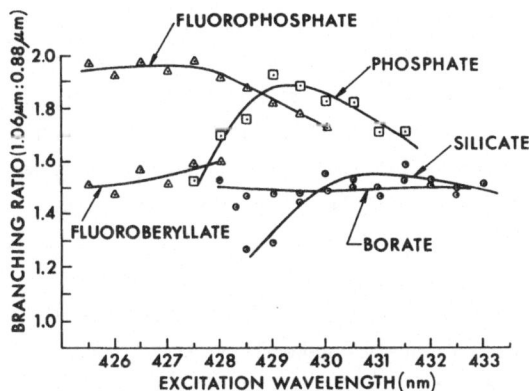

Fig. 1.6. Branching ratio (1.06/0.88 μm) in silicate, phosphate, fluoro-phosphate, fluoroberyllate, and borate glasses as a function of excitation wavelength [1.15]

relative quantum efficiency, and radiative and nonradiative decay rates were investigated. Although we cannot present the details here, of note are the differences of behavior exhibited by each of the five glasses, an example of which is shown in Fig.1.6. The branching ratio (1.06/0.88 μm) for each glass has been plotted as a function of excitation wavelength and is differ-ent for each. The silicate and borate glasses are shown to display similar behavior for the relative quantum efficiency and radiative decay rate. Both have large Stark splittings and low degrees of local order. Similarly, the phosphate and fluoroberyllate glasses have high degrees of local order, dis-play higher resolution of spectral components than the silicate or borate glasses, and show rather similar behaviors of relative quantum efficiencies. The other glass, fluorophosphate, contains both oxygen and fluorine anions and seems to exhibit behavior between the phosphate and fluoroberyllate. The authors propose that at shorter excitation wavelengths, emission is from $Nd^{3+}$ ions that are largely coordinated with fluorine, whereas the longer wavelength excitations are predominantly from oxygen coordinates sites. Thus, the two extremes involve chemically distinct species. Additionally, the fluorophos-phate results show that the largest changes of the obtained parameters are due to the coordinating species, superimposed upon the usual geometrical co-ordination changes.

Finally, similar measurements have been done on fluoroberyllate glasses, with substantially smaller variation of the observed quantities, which dem-onstrates that the distribution of sites is a property of the particular glass host.

This concludes our discussion of the $Nd^{3+}$ environment. We hope that the reader will have been made aware of the vast progress made in recent years in our understanding of the interaction of $Nd^{3+}$ ions with their local environment and that macroscopic quantities normally quoted to describe a given material, such as quantum efficiency and radiative lifetime, are really only averages over the statistical ensemble of available sites. Such studies can often have considerable practical utility. For example, LAYNE [1.11] has shown that in one pumping configuration, the gain obtained by use of the nonexponential-gain function for ED-2 laser glass with 3 wt-% $Nd_2O_3$ doping is $\approx 13\%$ less than that obtained by use of purely exponential decay. Such effects are now routinely incorporated into the computer design codes for laser amplifiers at the Lawrence Livermore Laboratory (LLL) and the Laboratory for Laser Energetics (LLE).

## 1.3  Amplification of Pulses in Nd:Glass

A good review of the amplification process in Nd:glass has been provided recently by KOECHNER [1.8]; it is not our intent to repeat that treatment here. In any real laser system, it is usually desirable to operate with a pulse width (FWHM) in the range from a few tens of picoseconds well into the nanosecond regime. Thus, terminal-level lifetime effects will be important, because in the former case the pulse width $\tau$ (FWHM) is substantially less than the reported terminal-level lifetime $\tau_T$ of silicate and phosphate glasses, whereas in the latter case $\tau \gtrsim \tau_T$. Here we review the level dynamics, rate equations, and the limiting three- and four-level systems, and show how a finite terminal-level lifetime may be included in the modeling of a real system. This section will also serve to establish basic notation that will be used throughout the remainder of the book.

As shown in Fig.1.1, it can be seen that $Nd^{3+}$ is normally a four-level system. Its spectroscopy can be simplified to a set of pump bands (level 3) whose relaxation rate is rapid compared with the fluorescence lifetime $\tau_F$ of the upper ($^4F_{3/2}$) metastable laser level (level 2). The lower laser level (level 1 or terminal level) ($^4I_{11/2}$) also relaxes with a lifetime $\tau_T << \tau_F$. The first level (level 0) is the ground state ($^4I_{9/2}$). If the terminal level emptied infinitely fast ($\tau_T \approx 0$) onto the ground state, it would be possible to describe $Nd^{3+}$ as a three-level system, but, of course, the threshold behavior would not be the same as, say, in a true three-level system. Equivalently, if the pulse width $\tau \gg \tau_T$, the situation is analogous to three-level operation whereas if $\tau \lesssim \tau_T$, four-level operation must be considered.

The general rate equations that govern the four-level system are

$$\frac{dN_2}{dt} = N_2\omega_{21} - W(N_2-N_1) + R_{32} + R_{02} \quad , \tag{1.5}$$

$$\frac{dN_1}{dt} = N_2\omega_{21} + W(N_2-N_1) + R_{01} - \omega_{10}N_1 + R_T \quad , \tag{1.6}$$

where the subscripts 2 and 1 refer to the metastable ($^4F_{3/2}$) and terminal ($^4I_{11/2}$) states, respectively; the $\omega_{ij}$ and $R_{ij}$ are, respectively, the relaxation and pumping rates from the $i^{th}$ to the $j^{th}$ level, and $R_T$ is the pumping of the terminal level to some possible excited state. The rate of stimulated emission W is given by

$$W = \frac{\sigma_p}{h\nu} P(t) \quad , \tag{1.7}$$

where $\sigma_p$ is the peak stimulated-emission coefficient, $h\nu$ the photon energy, and P(t) the power.

If we assume that, compared with $\omega_{10}$, the $\omega_{ij}$ and $R_{ij}$ as well as $R_T$ are unimportant (stationary), we have from (1.5,6)

$$\frac{dN_2}{dt} = -W(N_2-N_1) \quad , \tag{1.8}$$

$$\frac{dN_1}{dt} = W(N_2-N_1) - \omega_{10}N_1 \quad , \tag{1.9}$$

which represent the coupled equations to be solved when the terminal-level lifetime cannot be ignored.

We now consider two limiting cases in which the pulse width (FWHM) $\tau$ is either much shorter or longer than the terminal-level lifetime $\tau_T$.

*Case 1:* $\tau \ll \tau_T$ (Three-Level System)

Here we can assume $\omega_{10} = 0$ and (1.8,9) may be combined to give

$$\frac{d}{dt} (N_2-N_1) = -2W(N_2-N_1) \quad . \tag{1.10}$$

Defining the stored energy density

$$E_s = h\nu(N_2-N_1) \quad , \tag{1.11}$$

we have

$$\frac{dE_s}{dt} = -2WE_s \quad . \tag{1.12}$$

Using (1.7),

$$E_s = E_{s0} \exp\left[-2 \frac{\sigma_p}{h\nu} \int P(t)dt\right] \quad . \tag{1.13}$$

Letting $E = \int P(t)dt$, and defining

$$E_{SAT} = \frac{h\nu}{2\sigma_p} \quad , \tag{1.14}$$

we have

$$E_s = E_{s0} \exp(-E/E_{SAT}) \quad . \tag{1.15}$$

The significance of $E_{SAT}$ is that it is the energy density (fluence) at which the stored energy density has been reduced to $e^{-1}$ of its initial value; $E_{SAT}$ is called the saturated fluence. Because we have assumed that, during the pulse of duration $\tau$, the relaxation of the terminal level was not important, the system is a three-level one.

## Case 2: $\tau \gg \tau_T$ (Four-Level System)

If the pulse duration is much longer than the terminal-level lifetime, we can assume $N_1 = 0$. Hence, we have from (1.8,9)

$$\frac{dN_2}{dt} = -WN_2 \quad , \tag{1.16}$$

and because here

$$E_s = h\nu N_2 \quad , \tag{1.17}$$

$$\frac{dE_s}{dt} = -WE_s \quad . \tag{1.18}$$

Following the treatment in Case 1, the solution of (1.18) is identical to that of (1.12) with the saturation energy density $E_{SAT}$, now given by

$$E_{SAT} = \frac{h\nu}{\sigma_p} \quad . \tag{1.19}$$

As before, $E_{SAT}$ is the fluence required to deplete the $E_s$ or gain to $e^{-1}$ of its initial value, but the essential difference between the three- and four-level systems used for pulse amplification is that the three-level system requires twice the energy density before saturation occurs. This is an important consideration in the design of high-peak-power or energy-laser systems. Long pulses ($\tau \gg \tau_T$) will contain twice the energy before the gain saturates in an amplifier.

We have examined the gain at a specific location in a solid-state laser, and have seen how saturation affects gain amplification when terminal-level-lifetime effects are taken into account. Pulses that propagate in a real solid-state laser along a direction x obey the photon-transport equations given by [1.16]:

$$\frac{dE}{dx} = \frac{1}{2} E_s (1 - e^{-E/E_{SAT}}) - \alpha_L E \quad , \quad (\tau \ll \tau_T) \tag{1.20}$$

$$\frac{dE}{dx} = E_s (1 - e^{-E/E_{SAT}}) - \alpha_L E \quad . \quad (\tau \gg \tau_T) \tag{1.21}$$

Here E is the energy density at a location x and $\alpha_L$ is the passive loss coefficient which accounts for losses in solid-state media due to scattering, host absorption, etc.. The saturation fluence $E_{SAT}$ in (1.20,21) is $E_{SAT} = h\nu/2\sigma_p$ and $E_{SAT} = h\nu/\sigma_p$, respectively, which correspond to three- and four-level operation, respectively. We now examine a number of limiting cases of (1.20,21). If $E/E_{SAT} \ll 1$, both equations reduce to

$$\frac{dE}{dx} = (\alpha_0 E_s - \alpha_L)E = \alpha E \quad , \tag{1.22}$$

where the specific gain coefficient $\alpha_0$ is given by

$$\alpha_0 = \frac{\sigma_p}{h\nu} \quad , \tag{1.23}$$

and the gain per unit length is

$$\alpha = \alpha_0 E_s - \alpha_L \quad . \tag{1.24}$$

If we integrate (1.22) along our amplifier of length L and make the assumption that $\alpha$ is independent of x, we obtain the small-signal gain in a lossy amplifier

$$G_0 = \frac{E_{out}}{E_{in}} = \exp\left(\int_0^L \alpha dx\right) = \exp(\alpha_L) \quad . \tag{1.25}$$

In practice, $\alpha_0$ and L are accurately known and measurement of $G_0$ permits determination of the stored-energy density $E_s$ in an amplifier if the probe-beam-energy fluence is small compared to the saturation fluence. In the case $\alpha_L = 0$, (1.20,21) may be integrated to obtain the saturated gain

$$G = \frac{E_0}{E_{in}} = \frac{E_{SAT}}{E_{in}} \ln\left[G_0(e^{E_{in}/E_{SAT}}-1)+1\right] \tag{1.26}$$

as a function of input fluence $E_{in}$.

In the small-signal limit, at which $E_{in}/E_{SAT} \to 0$,

$$G \to G_0 = e^{\alpha_0 E_s L} = e^{\alpha L} \quad . \tag{1.27}$$

We recover (1.25) with $\alpha_L = 0$. Similarly, if the input fluence is large compared to the saturation fluence, the saturated gain G is given by

$$G \to 1 + \alpha L \frac{E_{SAT}}{E_I} \quad , \tag{1.28}$$

and all available energy is added to the input pulse.

In cases of practical interest, the energy fluence usually falls between the two limits represented by (1.27,28). Furthermore, the passive loss coefficient $\alpha_L$ is not zero; it usually falls in the range $(0.001 \leq \alpha_L \leq 0.005 \text{ cm}^{-1})$. Equations (1.20,21) are not integrable in that case and the laser designer is forced to utilize numerical techniques on high-speed digital computers.

Programs to model pulse propagation in large Nd:glass laser systems are now commonplace in most laboratories. They have been found to model observed effects very accurately.

Until recently, there was a real paucity of data on the terminal-level lifetime in glasses. The first definitive study was performed by DUSTON [1.2] who, using crossed probe and saturating beams measured $\tau_T$ in a number of early laser glasses. $\tau_T$ was found to vary both with temperature and wt-% Nd doping. A measurement of $\tau_T$ for 1 wt-% ED-2 silicate glass at room temperature yielded $\tau_T \simeq 11$ ns. Relaxation from the $^4I_{11/2}$ terminal level to the $^4I_{9/2}$ ground state is nonradiative and takes place via multi-phonon emission. Multi-phonon relaxation will be treated in detail in Section 1.5, but we mention here that recent work by LAYNE [1.11] predicted that $\tau_T \simeq 10$ ns, in good agreement with DUSTON [1.2]. More recently, MARTIN and MILAM [1.6] have measured the gain recovery in ED-2 glass using the crossed beam technique and found $\tau_T = 1.25 \pm 0.2$ ns, the most accurate experiment yet performed and in apparent disagreement with [1.11]. They also found a surprising polarization dependence

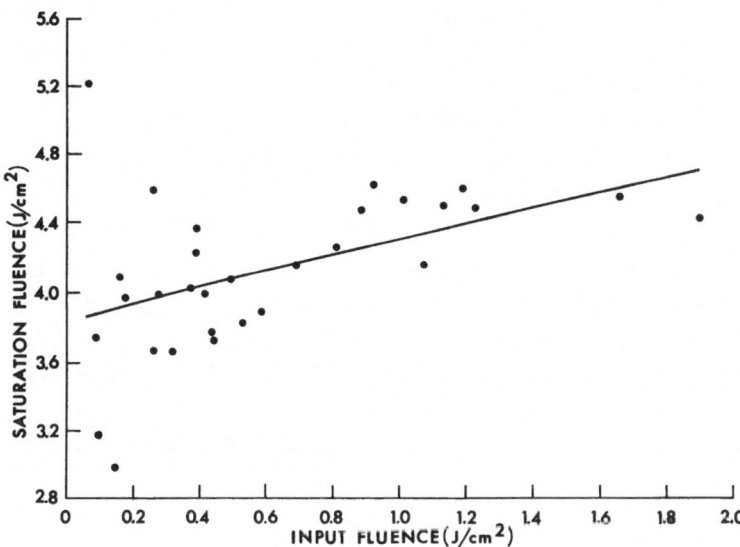

<u>Fig. 1.7.</u> Measured saturation fluence of LHG-8 laser glass as a function of incident fluence [1.19]

of gain recovery. If the probe beam was polarized parallel to the saturating pulse, the usual gain-recovery behavior was observed. If, however, a perpendicularly polarized probe was used, no gain-saturation recovery was found. No satisfactory explanation of the phenomenon has yet been offered; it is possible that the saturating pulse is polarizing the $Nd^{3+}$ ions so that interaction with perpendicular radiation is minimized, although the characteristic time for that process is apparently too long. One consequence of this experiment is that the saturation behaviors of amplifiers that are propagating linear or circular polarization should be substantially different. On the contrary, however, an experiment by LEWIS and SEKA [1.17] measured the gains in a phosphate rod amplifier and found that the saturation behaviors are the same, regardless of polarization state. As of this writing, the matter is not resolved.

In the past two years, the gain-recovery times of a number of laser glasses have been measured at LLE by MILAM and MARTIN [1.18] and coworkers. While the data are still unpublished, values in the range of 1-2 ns were obtained. The saturation behavior of several laser glasses has been measured by MARTIN and MILAM [1.19] in the past year and surprising results noted. The saturation fluence $E_{SAT}$, taken as a constant in (1.20,21), has been found to be a function of the incident fluence. An example of this is shown in Fig.1.7, where it can be seen that the saturation fluence for LHG-8 laser glass increases with the input fluence incident on the amplifier. The experimental points are unpublished data [1.19] at one and nine ns pulse duration; the curve is the result of a nonlinear least-squares fit to their data by the author. Most glasses seem to display a similar type of behavior; it is conjectured at present that it is due to the inhomogeneous nature of $Nd^{3+}$ ions in glass [1.20]. It is easy to calculate the saturation fluence for a three- or four-level LHG-8 glass system as 2.38 and 4.76 $J/cm^2$, respectively. For low incident fluence, it can be seen from Fig.1.7 that the glass operates like a four-level system. As we have seen previously, in four-level systems the terminal-level lifetime $\tau_T$ is shorter than the pulse width $\tau$. Since the experiments used one and 9 ns pulses and the data are virtually identical for both cases, a short terminal-level lifetime would be expected (<1 ns) contrary to gain-recovery experiments [1.18] but in agreement with multiphonon theory.

## 1.4  Judd-Ofelt Theory

The design of a high-peak-power solid-state-laser system involves consideration of a number of glass parameters. By far the two most important are the peak stimulated-emission cross section $\sigma_p$ and the nonlinear index of refraction $n_2$. The latter parameter will be treated in detail in Sect.1.6; here we are particularly interested in the optical properties of laser glass. In amplifiers that are pump limited, it is of interest to maximize $\sigma_p$ whereas in a parasitic-limited laser amplifier (Chaps.2 and 4) other properties, such as the nonlinear index $n_2$, may be more important. How $\sigma_p$ varies within a given glass composition (e.g. silicate, phosphate, fluorophosphate, etc.) has become the subject of intensive study because it is of interest to adjust that parameter to achieve the maximum performance from an amplifier or laser system at minimum cost. The method for predicting the value of $\sigma_p$ for a given composition has been developed by the application of the powerful Judd-Ofelt (J-O) approach which we review here. By using the J-O theory, data obtained by examination of small laboratory samples of glass compositions can be used to predict accurately the performance of an amplifier or system on the basis of scaled-up sizes.

The rare earths (or lanthanide series) are usually doped into solid-state hosts in the trivalent state, although many divalent systems have also been

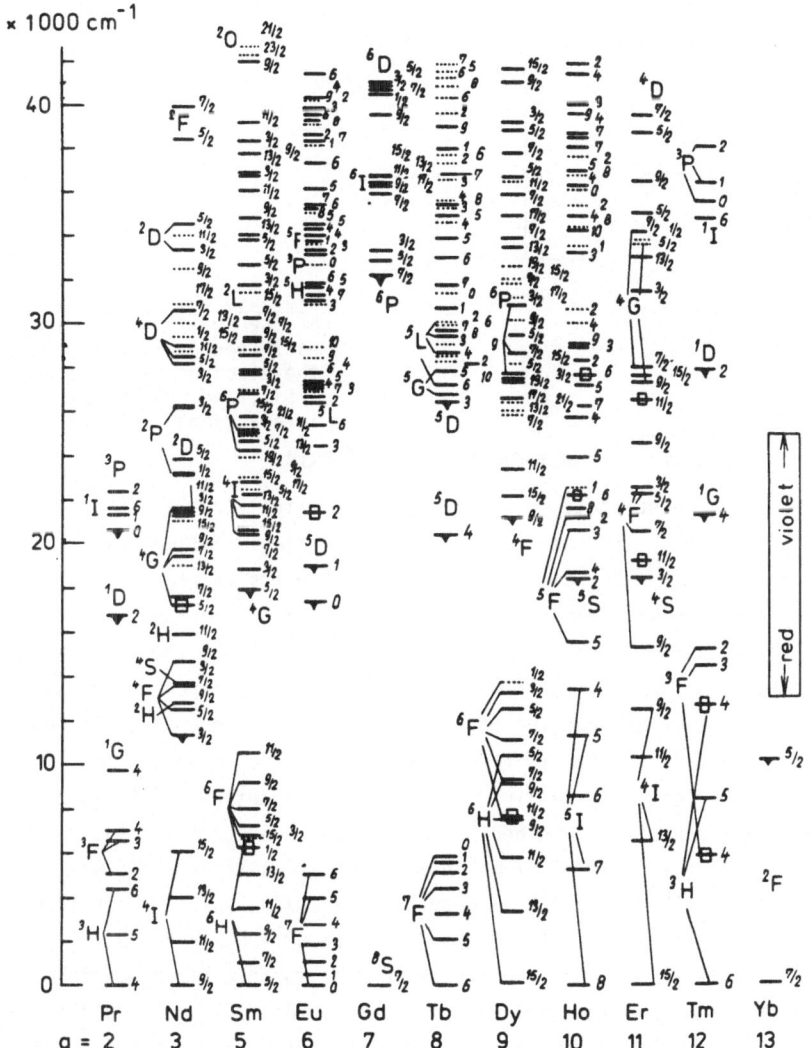

× 1000 cm⁻¹

**Fig. 1.8.** Energy levels (j indicated at the right hand) of the trivalent lanthanides (except cerium and promethium) in the unit 1000 cm⁻¹ as a function of the number q of 4f electrons. Excited levels frequently showing luminescence are indicated by a black triangle. The excited levels corresponding to hypersensitive transitions from the ground state are marked with a square. In cases where the quantum numbers S and L are reasonably well defined, the Russell-Saunders terms are given at the left. Calculated energy levels are shown as stippled lines. This figure is an extended and modified version of a diagram (known from a book by Dieke) that summarizes the work of a generation of rare-earth spectroscopists [1.21]

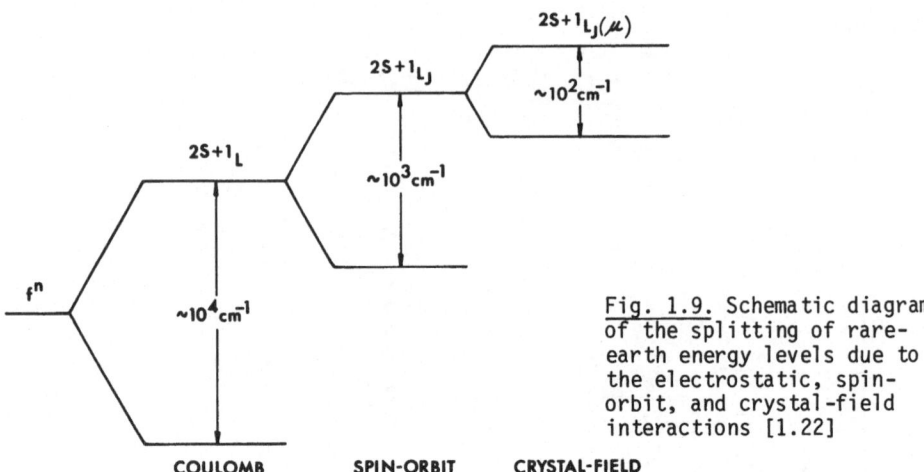

$2S+1_{L_J}(\mu)$

$\sim 10^2 cm^{-1}$

$2S+1_{L_J}$

$\sim 10^3 cm^{-1}$

$2S+1_L$

$f^n$

$\sim 10^4 cm^{-1}$

COULOMB        SPIN-ORBIT        CRYSTAL-FIELD

Fig. 1.9. Schematic diagram of the splitting of rare-earth energy levels due to the electrostatic, spin-orbit, and crystal-field interactions [1.22]

demonstrated. The ground-state configuration is Xe like ($1s^2$ $2s^2$ $2p^6$ $3s^2$ $3p^6$ $3d^{10}$ $4s^2$ $4d^{10}$ $5s^2$ $5p^6$), whereas the 4f shell is partially filled. Excited states are $4f^{N-1}5d$, $4f^{N-1}5g$, etc. The absorption spectra of rare earths in the optical region arises from transitions within the $4f^N$ configuration. In Fig.1.8 we show a recent compilation of the energy levels of the rare earths [1.21]. The location of the levels is determined by three effects. In the free state the Coulomb interaction of the electrons and the spin-orbit coupling are important. If the ion is incorporated into a solid host, then interaction with the local crystalline field is also important. RISEBERG and WEBER [1.22] have recently described these effects in a schematic fashion that is shown in Fig.1.9. The Coulomb interaction gives terms $2S+1_L$ with typical separation of $\geq 10^4$ cm$^{-1}$. The spin-orbit interaction splits the L terms into J states with separation of $\approx 10^3$ cm$^{-1}$. The crystal Stark field, if present, removes the (2J+1)-fold degeneracy which results in a Stark manifold a few hundred cm$^{-1}$ wide. In the rare earths, the crystal-field interaction with the 4f electrons is weak, owing to the shielding effect of the outer 5s and 5p-shell electrons. Hence, the crystal-field potential may be expanded in a spherical harmonic series

$$V_{CF} = \sum_{k,q,i} B_q^k (C_q^k)_i \quad , \tag{1.29}$$

where the coefficients $B_q^k$ are parameters that describe the strength of the crystal-field components, the $C_q^k$ are tensor-operator components and the summation is over the i electrons of the ion. For the f shell, k is restricted to $k \leq 6$. It is possible by use of group-theoretical arguments and the local point symmetry at the rare-earth site to find the type and number of terms in (1.29). In general, the calculation of the parameters $B_q^k$ is very difficult and has not met with much success.

Optical transitions in rare-earth ions have been found to be predominantly electric-dipole (ED) in origin. Although magnetic-dipole and electric-quadrupole transitions are allowed by the selection rules, their contributions to radiative decay are exceedingly small. For ED transitions, the well-known selection rules are $\Delta \ell = \pm 1$, $\Delta S = 0$, and $|\Delta L|$, $|\Delta J| \leq 2$ , where here $\ell = 3$. For transitions between $f^N$ electrons, no change of parity is involved and, by Laporte's rule, ED transitions are forbidden. That ED transitions are found experimentally and are indeed predominant is a consequence of the fact that the odd harmonics of the local crystal field can admix states of opposite parity into $4f^N$. One way in which this occurs is if the rare-earth ion is located at a site that has no inversion symmetry. According to RISEBERG and WEBER [1.22], the effect can occur for the point-group symmetries: $C_s$, $C_1$, $C_2$, $C_{2v}$, $C_3$, $C_{3v}$, $C_4$, $C_{4v}$, $C_6$, $C_{6v}$, $D_{2d}$, $D_3$, $D_{3h}$, $D_4$, $D_5$, $D_{5h}$, $S_4$, $T$, $T_d$, $O$. Opposite parity states have been mentioned previously and include 5d and 5g.

It can be shown by perturbation theory that consideration of the odd parity (k odd) terms in the crystal-field potential (1.29) gives, to first order, the $a^{th}$ level eigenstate for the $f^N$ configuration as

$$|\psi_a> = |\phi_a> - \sum_\beta (E_a - E_\beta)^{-1} <\phi_\beta | V_{CF}^{ODD} | \phi_a > | \phi_\beta > \quad . \tag{1.30}$$

Here $|\phi_a>$ is given by

$$|\phi_a> = |f^N [\gamma SL] J J_z> \quad , \tag{1.31}$$

where $\gamma$ includes any other quantum numbers besides the usual [SLJ] Russell-Saunders ones. The summation in (1.30) is over all $\beta$ opposite parity states including $4f^{N-1} n' \ell'$ where $n' \ell' = 5d$ or $5g$ as well as $3d^9 4f^{N+1}$. Calculation of the intensity or line strength for a transition involves consideration of the electric-dipole matrix elements of the operator P. They are given by

$$<\psi_a|P|\psi_b> = \sum_\beta \left[ (E_b - E_\beta)^{-1} <\phi_a|P|\phi_\beta> <\phi_\beta | V_{CF}^{ODD} | \phi_b > \right.$$
$$\left. + (E_a - E_\beta)^{-1} <\phi_a | V_{CF}^{ODD} | \phi_\beta > <\phi_\beta|P|\phi_b> \right] \quad , \tag{1.32}$$

where the $p^{th}$ component of P is

$$P_\rho = -e \sum_i r_i \left( c_\rho^{(1)} \right)_i \quad . \tag{1.33}$$

In addition to the values $B_q^k$ (k odd) in $V_{CF}$, which have not been calculated satisfactorily, the energy levels $E_\beta$ of the excited-state configurations are difficult to obtain. For their calculation the radial integrals $\int R(4f)R(n'\ell')$ $\cdot r^k dr$ must have good wavefunctions R, and the computation of the matrix elements given by (1.32) is tedious. It has been shown independently by JUDD [1.23] and OFELT [1.24] that it is possible to make the calculation of ED probabilities tractable by assuming certain approximations in (1.32). The first is replacement of $C_q^k |\phi_\beta> <\phi_\beta| C_\rho^{(1)}$ by a tensor-operator component $U_{\rho+q}^{(t)}$, where t is even. Second, the $4f^N$ and excited opposite-parity configurations are treated as degenerate with a single average energy separation. It is then possible to invoke closure over the summation in (1.32) and obtain

$$<\psi_a|P_\rho|\psi_b> = \sum_{q,t \text{ even}} Y(t,q,\rho) \left\langle f^N \gamma SLJJ_z |U_{\rho+q}^{(t)}| f^N \gamma'S'L'J'J_z' \right\rangle \quad . \quad (1.34)$$

We obtain an expression considerably easier to evaluate because the crystal-field parameters, the radial integrals, and the average energy separation have all been included in the phenomenological parameters $Y(t,q,\rho)$.

For intermediate coupling, the line strength

$$S = |<\psi_a|P|\psi_b>|^2 \quad\quad (1.35)$$

of an electric-dipole transition may be shown to be given by

$$S = \sum_{t=2,4,6} \Omega_t |<(S'',L')J'||U^{(t)}||(\bar{S},\bar{L})\bar{J}>|^2 \quad , \quad\quad (1.36)$$

where the terms $<||U^{(t)}||>$ are the doubly reduced unit-tensor operators calculated in the intermediate-coupling approximation. As above, the phenomenological parameters $\Omega_t$ contain the energy denominators, the odd-symmetry crystal-field terms, and the radial integrals. The line strength S can be related to the integrated absorbance of an ED transition $\int k(\lambda)d\lambda$ by use of

$$\int k(\lambda)d\lambda = \frac{8\pi^2 e^2 \bar{\lambda}\rho}{3ch(2J+1)} \frac{(n^2+2)^2}{9n} S \quad . \quad\quad (1.37)$$

Here, $k(\lambda)$ is the absorption coefficient at $\lambda$, $\rho$ the concentration of ions, e the electric charge, $n(\bar{\lambda})$ the index of refraction, c the speed of light, h Planck's constant, $\bar{\lambda}$ the mean wavelength of the band, and the factor $(n^2+2)^2/9n$ is the local field correction for a dielectric medium in the tight binding approximation. A number of radiative properties are usually of interest, including the spontaneous-emission probability A from an initial manifold $|(S',L')J'>$ to a final manifold $|(\bar{S},\bar{L})\bar{J}>$ given by

$$A[(S',L')J';(\bar{S},\bar{L})\bar{J}] = \frac{64\pi^4 e^2 n}{3h(2J'+1)\bar{\lambda}^3} \frac{(n^2+2)^2}{9}$$

$$\cdot \sum_{t=2,4,6} \Omega_t |<(S',L')J'||U^{(t)}||(\bar{S},\bar{L})\bar{J}>|^2 \quad , \qquad (1.38)$$

the branching ratio

$$\beta[(S',L')J';(\bar{S},\bar{L})\bar{J}] = \frac{A[(S',L')J';(\bar{S},\bar{L})\bar{J}]}{\sum_{\bar{S},\bar{L},\bar{J}} A[(S',L')J';(\bar{S},\bar{L})\bar{J}]} \quad , \qquad (1.39)$$

and the sum is over all terminal manifolds. The radiative lifetime of a level is defined by

$$\tau_R^C = \left\{ \sum_{\bar{S},\bar{L},\bar{J}} A[(S',L')J';(\bar{S},\bar{L})\bar{J}] \right\}^{-1} = A_{tot}^{-1} \quad , \qquad (1.40)$$

and the quantum efficiency $\eta$ by

$$\eta = \tau_f^M / \tau_R^C \quad . \qquad (1.41)$$

Here $\tau_f^M$ is the measured fluorescence lifetime of the $|(S',L')J'>$ manifold. Finally, the peak stimulated-emission cross section $\sigma_p$ for $Nd^{3+}$ may be determined by

$$\sigma_p(\lambda_p) = \frac{\lambda_p^4}{8\pi c n^2 \Delta\lambda_{eff}} A(^4F_{3/2};^4I_J) \qquad (1.42)$$

where $\lambda_p$ is the peak wavelength, $\Delta\lambda_{eff}$ the effective bandwidth of the transition (not the full width at half-maximum because such emission lines are asymmetrical) and $^4F_{3/2}$ the metastable level in $Nd^{3+}$.

The J-O approach has been responsible for the increased understanding of the spectra of rare earths in the liquid, solid, and gaseous phase. We restrict ourselves here to the application of the J-O approach to $Nd^{3+}$ in glass. The first report of such a study was by KRUPKE [1.25], who calculated emission cross sections, radiation lifetimes, and branching ratios for four commercial laser glasses (3669A, ED-2, LSG-91H, S-33). The essence of the method was use of (1.37). The line strength S was calculated from (1.36) by use of the unit-tensor operators evaluated for $Nd^{3+}$ [1.26]. Such operators have been shown to be almost invariant to the host composition [1.21]. Absorption spectra are obtained on small samples of each composition and the bands are numerically integrated. A computerized least-squares fitting routine is then employed that determines the values of $\Omega_2$, $\Omega_4$, and $\Omega_6$ by use of (1.36,37) to eliminate S. With the determination of $\Omega_2$, $\Omega_4$, and $\Omega_6$, it is a simple matter to determine the radiative probability, branching ratios, fluorescence lifetime, quantum efficiency, and peak induced-emission cross section from (1.38-42), respectively. Note that to obtain $\eta$ and $\sigma_p$ the calculation must be augmented with separate measurements of $\tau_f^M$, $\lambda_p$ and $\Delta\lambda_{eff}$.

From (1.38,42) it should be noted that the peak cross section $\sigma_p$ is dependent upon the $\Omega_i$ parameters and $\Delta\lambda_{eff}$, both of which are very composition dependent. The inhomogeneous line width $\Delta\lambda_{eff}$ results from Stark splitting of the J manifolds, the extent of which depends upon the nature of the local ligand field. That field is in turn determined by the particular composition. The way in which the $\Omega_i$ and all of the parameters deduced from them vary with composition has been the subject of a number of excellent studies at the LLL and will be discussed below. It has been shown [1.10] that the intensity of the $^4F_{3/2} \rightarrow {}^4I_J$, laser transitions in $Nd^{3+}$ are governed by the triangle rule ($|J-\bar{J}| \leq t \leq |J+\bar{J}|$) where J and $\bar{J}$ refer to the initial and final manifolds, respectively. Simple application of this rule to the above transitions gives the following results in Table 1.1. Thus, to obtain a large $\sigma_p$ for the usual transition of interest ($^4F_{3/2} \rightarrow {}^4I_{11/2}$), it is necessary to maximize both $\Omega_4$ and $\Omega_6$. To maximize the intensity, the branching ratio $\beta$ should also be maximized for the transition of interest. $\beta$ can be shown to depend only on the ratio $\Omega_4/\Omega_6$ [1.10]. In Fig.1.10, the branching ratio has been plotted as a function of $\Omega_4/\Omega_6$ for the four $Nd^{3+}$ laser transitions of interest. To maximize the intensity of the ($^4F_{3/2} \rightarrow {}^4I_{11/2}$) transition it is clearly necessary that $\Omega_6 \gg \Omega_4$, whereas to maximize ($^4F_{3/2} \rightarrow {}^4I_{9/2}$) the opposite is required. In practice, it has been determined that $\Omega_4/\Omega_6$ varies only in the relatively narrow range of 0.9-1.1 [1.10]. Figure 1.4 also shows that the intensity of

Table 1.1

| Transition | $\Omega_i$ | Wavelength [μm] |
|---|---|---|
| ($^4F_{3/2} \rightarrow {}^4I_{15/2}$) | $\Omega_6$ | 1.85 |
| ($^4F_{3/2} \rightarrow {}^4I_{13/2}$) | $\Omega_6$ | 1.33 |
| ($^4F_{3/2} \rightarrow {}^4I_{11/2}$) | $\Omega_4, \Omega_6$ | 1.06 |
| ($^4F_{3/2} \rightarrow {}^4I_{9/2}$) | $\Omega_4, \Omega_6$ | 0.88 |

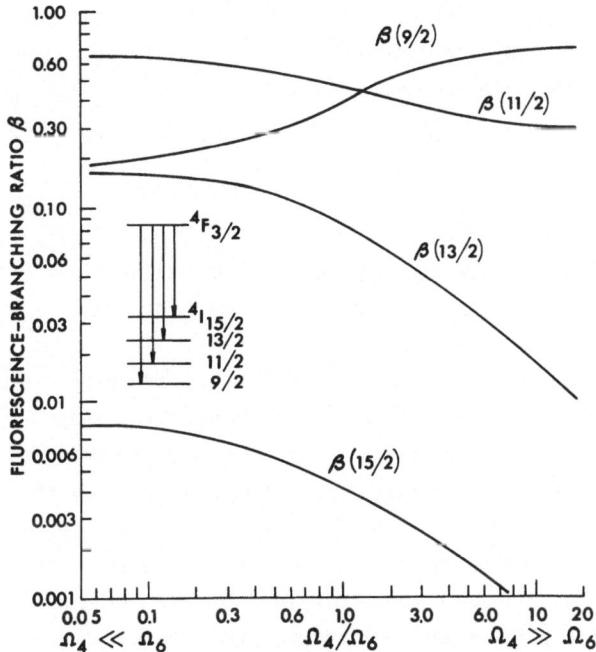

Fig. 1.10. Fluorescence branching ratio β for transition from the $^4F_{3/2}$ to the $^4I_J$ terminal level [1.10]

Table 1.2. Neodymium laser glasses

| Glass | $^4F_{3/2} \rightarrow {}^4I_{13/2}$ | | | $^4F_{3/2} \rightarrow {}^4I_{11/2}$ |
|---|---|---|---|---|
| | $\Delta\lambda_{eff}$[nm] | $\lambda_p$[m] | $\sigma_p[10^{-20}cm^2]$ | $\sigma_p[10^{-20}cm^2]$ |
| *Silicates* | | | | |
| 3669A | 53 | 1.326 | 0.30 | 1.2 |
| LSG-91H | 64 | 1.335 | 0.65 | 2.7 |
| LSG-95 | 58 | 1.335 | 0.78 | 2.8 |
| ED-2 | 67 | 1.335 | 0.72 | 2.9 |
| ED-8 | 63 | 1.334 | 0.82 | 3.1 |
| LG 650 | 52 | 1.325 | 0.31 | 1.1 |
| *Phosphates* | | | | |
| LHG-5 | 49 | 1.323 | 0.94 | 3.9 |
| P-25 | 45 | 1.324 | 1.1 | 4.7 |
| P-107 | 55 | 1.324 | 0.89 | 3.9 |
| L-41 | 54 | 1.320 | 0.50 | 1.9 |

the $(^4F_{3/2} \rightarrow {}^4I_{15/2})$ transition is small compared with the transition $(^4F_{3/2} \rightarrow {}^4I_{11/2})$. Nevertheless, its output, located at 1.33 μm, is important in the consideration of parasitic oscillations and amplified spontaneous emission in large-aperture amplifiers (Chap.4). JACOBS and WEBER [1.27] have utilized the J-O approach to evaluate the cross section $\sigma_p(^4F_{3/2} \rightarrow ^4I_{13/2})$ for a number of silicate and phosphate glasses. Their results, which list $\Delta\lambda_{eff}$, $\lambda_p$ and $\sigma_p$ for the $(^4F_{3/2} \rightarrow ^4I_{13/2})$ transition as well as $\sigma_p$ for the $(^4F_{3/2} \rightarrow ^4I_{11/2})$ transition, are reproduced in Table 1.2. They show the previously mentioned compositional dependence of $\sigma_p$, because the values obtained are generally larger in the phosphate than in the silicate ones. In all cases, $\sigma_p(^4F_{3/2} \rightarrow ^4I_{13/2}) < \sigma_p(^4F_{3/2} \rightarrow ^4I_{11/2})$. From (1.14,19) it can also be seen that the saturation fluence for the $(^4F_{3/2} \rightarrow ^4I_{13/2})$ transition is larger than that for the $(^4F_{3/2} \rightarrow ^4I_{11/2})$ one.

We now review results of compositional studies reported in a study by JACOBS and WEBER [1.28]. By use of the J-O theory as outlined above, they studied a series of five variations of composition and discussed the changes of $\beta$, $\sigma_p$, $\lambda_p$, $\tau_{rad}^e$, and $\Delta\lambda_{eff}$. The first series involved changes of only the glass-network formed in eight different glasses. The compositions considered, whose radiative parameters are shown in Table 1.3, include the currently important silicate, phosphate, and fluorophosphate glasses. The branching ratio for all transitions is relatively invariant with composition, whereas the total transition rate (or lifetime) varies significantly. For the $(^4F_{3/2} \rightarrow ^4I_{11/2})$ transition, $\lambda_p$ varies from 1055 nm in the phosphate (H14) to 1069 nm in the aluminate (L65) glass. The cross section $\sigma_p$ is maximum in the phosphate glass (H14) and minimum in the silicate (H11). The corresponding effective bandwidth is least in the phosphate glass (H14) and maximum in the aluminate (L65). The magnitude of the cross sections in the H14 and L65 glasses is partially explained by (1.42) where $\sigma_p$ varies inversely as $\Delta\lambda_{eff}$. It should also be noted from (1.38,32) that the spontaneous-emission probability A depends upon the host refractive index, as $\simeq n[(n^2+2)^2/9]$, and the peak cross section $\sigma_p$ as $n^{-1}[(n^2+2)^2/9]$. These factors arise from the local field correction, and can have a significant effect upon the values of A or $\sigma_p$. In Table 1.3, n varies in the range of $\simeq 1.5$ to $> 2$, resulting in values for $n[(n^2+2)^2/9]$ in the range 3-8. Thus for equal values of the $\Omega_i$, A can be significantly different for glasses that have different indices; the effect has been observed in tellurite glasses [1.28]. In the second series, only silicate glasses were considered and the alkali-modifying ion varied ($M^+$). The results are reproduced in Table 1.4 below, where M = Li, Na, K, and Rb. The J-O parameters $\Omega_i$ as well as $A_{TOTAL}$, $\beta$, $\tau_{rad}^c$, $\tau_f^M$, $\lambda_p$, $\sigma_s$ and $\Delta\lambda_{eff}$ are given for the four compositions. As has been explained, the cross section $\sigma_p$ is directly related to $\Omega_4$ and $\Omega_6$. Table 1.4 shows the trend that as $\Omega_4$ and $\Omega_6$ decrease, so does $\sigma_p$. $\Delta\lambda_{eff}$ changes only minimally in the compositions studied. Although the

Table 1.3. Summary of the $Nd^{3+}$, $^4F_{3/2}$ radiative properties for network former variations (series I) [1.28]

| Property | Phosphate (H14) | Borate (H13) | Germanate (H12) | Silicate (H11) | Tellurite (162M) | Aluminate (L65) | Titanate (L220) | Fluoro-phosphate (L-223) |
|---|---|---|---|---|---|---|---|---|
| *Composition* [mol. %] (all but the last sample contain 0.3-0.5% $Nd_2O_3$) | 67 $P_2O_5$ 18 BaO 15 $K_2O$ | 67 $B_2O_3$ 18 BaO 15 $K_2O$ | 67 $GeO_2$ 18 BaO 15 $K_2O$ | 67 $SiO_2$ 18 BaO 15 $K_2O$ | 80 $TeO_2$ 20 BaO | 11 $SiO_2$ 5 BaO 32 $Al_2O_3$ 52 CaO | 50 $SiO_2$ 25 $Na_2O$ 25 $TiO_2$ | 80 LiF 20 $Al(PO_3)_3$ (0.5% $NdF_3$) |
| **$^4F_{3/2}$ *Fluorescence:*** | | | | | | | | |
| $A_{Total}$ ($^4F_{3/2}$) [$s^{-1}$] | 2888 | 2386 | 2299 | 1602 | 4193 | 2660 | 3232 | 2695 |
| $\beta(^4F_{3/2}; ^4I_{9/2})$ | 0.43 | 0.42 | 0.46 | 0.45 | 0.45 | 0.45 | 0.43 | 0.42 |
| $\beta(^4F_{3/2}; ^4I_{11/2})$ | 0.48 | 0.49 | 0.45 | 0.46 | 0.46 | 0.46 | 0.47 | 0.48 |
| $\beta(^4F_{3/2}; ^4I_{13/2})$ | 0.09 | 0.09 | 0.08 | 0.08 | 0.08 | 0.08 | 0.09 | 0.09 |
| $\beta(^4F_{3/2}; ^4I_{15/2})$ | 0.005 | 0.005 | 0.004 | 0.004 | 0.004 | 0.004 | 0.004 | 0.005 |
| $\tau_{rad}^c$ ($^4F_{3/2}$) [µs] | 346 | 419 | 435 | 624 | 239 | 392 | 309 | 371 |
| **$^4F_{3/2} \rightarrow ^4I_{11/2}$ *Transition:*** | | | | | | | | |
| $\lambda_p$ [nm] | 1055 | 1061 | 1062 | 1060 | 1063 | 1069 | 1064 | 1054 |
| $\Delta\lambda_{eff}$ [nm] | 25.3 | 36.8 | 34.7 | 34.9 | 28.9 | 43.1 | 38.6 | 27.2 |
| $\sigma_p$ [$10^{-20}$ $cm^2$] | 4.1 | 2.2 | 1.9 | 1.5 | 2.9 | 1.8 | 2.5 | 3.5 |

Table 1.4. Summary of $\Omega_i$ values and $Nd^{3+}$, $^4F_{3/2}$ radiative properties for binary silicate glasses (series II) [1.28]

| Property | L215 | L213 | L217 | L235 |
|---|---|---|---|---|
| *Composition* [mol. %] (samples contain 66.5% $SiO_2$, 33% $M_2O$, 0.5% $Nd_2O_3$) | $Li_2O$ | $Na_2O$ | $K_2O$ | $Rb_2O$ |
| *J–O Parameters* [$10^{-20}$ $cm^2$]: | | | | |
| $\Omega_2$ | $3.4 \pm 0.2$ | $4.3 \pm 0.2$ | $5.1 \pm 0.1$ | $5.7 \pm 0.1$ |
| $\Omega_4$ | $4.5 \pm 0.2$ | $3.2 \pm 0.2$ | $2.4 \pm 0.2$ | $2.2 \pm 0.2$ |
| $\Omega_6$ | $4.6 \pm 0.1$ | $3.2 \pm 0.1$ | $2.0 \pm 0.1$ | $1.9 \pm 0.1$ |
| $^4F_{3/2}$ *Fluorescence:* | | | | |
| $A_{Total}$ $(^4F_{3/2})$ [$s^{-1}$] | 2523 | 1661 | 1100 | 1061 |
| $\beta(^4F_{3/2}; {}^4I_{9/2})$ | 0.43 | 0.43 | 0.45 | 0.45 |
| $\beta(^4F_{3/2}; {}^4I_{11/2})$ | 0.48 | 0.48 | 0.46 | 0.46 |
| $\beta(^4F_{3/2}; {}^4I_{13/2})$ | 0.09 | 0.09 | 0.08 | 0.08 |
| $\beta(^4F_{3/2}; {}^4I_{15/2})$ | 0.004 | 0.004 | 0.004 | 0.004 |
| $\tau_{rad}^c$ $(^4F_{3/2})$ [$\mu s$] | 396 | 602 | 909 | 942 |
| $\tau_f^M$ [s](for 0.05 mole % $Nd_2O_3$) | $370 \to 460$ | $450 \to 600$ | $600 \to 1000$ | * |
| $^4F_{3/2} \to {}^4I_{11/2}$ *Transition:* | | | | |
| $\lambda_p$ [nm] | 1061 | 1060 | 1059 | 1058 |
| $\Delta\lambda_{eff}$ [nm] | 33.8 | 33.4 | 35.6 | 37.7 |
| $\sigma_p$ [$10^{-20}$ $cm^2$] | 2.6 | 1.8 | 1.1 | 0.97 |

branching ratio for all transitions is again almost invariant with composition, the calculated radiative lifetime $\tau_{rad}^c$ increases and $A_{TOTAL}$ decreases as M is changed in the sequence Li, Na, K, Rb. Owing to the low $Nd_2O_3$ concentrations used (0.5% mol.), concentration quenching by ion-ion interactions was absent. The important fact to emerge from this work in agreement with earlier studies [1.29,30], is that, as the composition is changed, increasing the $M^+$ radius in the sequence Li $\to$ Na, $\to$ K, $\to$ Rb, $\sigma_p$ decreases, $A_{TOTAL}$ decreases, and $\tau_{rad}^c$ increases. There is a corresponding decrease of the value of $\Omega_i$ which can be related to those quantities. The way in which the various $\Omega_i$ change with $M^+$ is related to changes of the odd-parity terms in the local crystal-field expansion. The three other series represented changes of the

Table 1.5. $\Delta\lambda_{eff}$, $\lambda_p$, and $\sigma_p$ for the $^4F_{3/2}$ $^4I_{11/2}$ transition and $\tau_{rad}^c(^4F_{3/2})$ in selected Nd-doped silicate and phosphate laser glasses [1.28]

| Property | Glass→ (a) | Silicates | | | | | | Phosphates | | |
|---|---|---|---|---|---|---|---|---|---|---|
| | | LSG-91H | LSG-95 (a) | ED-2 | ED-8 | LG 650 | LHG-5 | P107 | EV 1 | L-41 |
| $\Delta\lambda_{eff}$ [nm] | | 33.5 | 35.1 | 34.0 | 36.1 | 33.5 | 25.5 | 25.5 | 21.7 | 32.5 |
| $\lambda_p$ [nm] | | 1062 | 1063 | 1062 | 1063 | 1059 | 1054 | 1054 | 1054 | 1054 |
| $\tau_{rad}^c(^4F_{3/2})$ [µs] | | 361 | 250 | 326 | 287 | 856 | 351 | 322 | 314 | 564 |
| $\sigma_p$ [$10^{-20}$ cm$^2$] | | 2.7 | 2.8 | 2.9 | 3.1 | 1.1 | 3.9 | 3.9 | 4.7 | 1.9 |

alkali concentration in one of the binary silicates (Na$_2$O) in Table 1.4, variation of alkali additive for fixed alkali-earth concentration in ternary silicates, and variation of the alkali-earth additives for fixed alkali concentration in ternary silicates. The interested reader should consult the original reference [1.28] for details. The J-0 approach was also applied to nine commercially available laser glasses, the results being shown in Table 1.5. In agreement with the above results, the phosphate glasses display smaller bandwidths, larger cross sections, and shorter center wavelengths than the silicates. $\sigma_p$ can vary by as much as 2.5-3.0 in silicate and phosphate hosts with different compositions. The relatively large cross sections of glasses such as ED-2 and ED-8 can be explained by the presence of lithium and calcium in those compositions, in agreement with the results of this study that small-radius ions (lithium) and small alkali earths (calcium) lead to larger cross sections. In LG 650, a glass with a small cross section, larger-size alkali and alkali earths such as potassium and barium are present.

We have attempted to indicate how the J-0 theory has rationalized our understanding of the way in which rare-earth ions interact with a somewhat bewildering variety of host materials. The variation of cross section, lifetime, branching ratio, and nonlinear index (Sect.1.6) with glass composition is now well understood; it is now possible to design a laser system using all of those variables as inputs. Indeed, in a single glass-laser system, glass is likely to be used at each stage, whose properties have been adjusted to be near optimal by use of the J-0 theory.

Recently, a voluminous amount of useful information has been published by LLL for a large number of commercially and scientifically important laser glasses [1.31]. An example is shown in Table 1.6, which includes all sample parameters, absorption data, the J-0 parameters, band strengths, branching ratios, fluorescence data, calculated and measured radiative lifetimes, stimulated-emission cross section, peak wavelength, effective bandwidth, linear refractive indices at various wavelengths, Abbe value, and nonlinear index.

<u>Table 1.6.</u> Sample identification: B101 fluoroberyllate. From [1.31]

*Sample Data*

Host: FL Beryllate     Dopant: ND

Concentrations:   6.19 wt-%  4.85600E + 20 ions/cm$^3$
Density:          2.621 G/cm$^3$
Sample thickness: 0.3735 cm

*Absorption Data*

| Band | Wavelength [nm] | Index | Absorptance |
|------|-----------------|---------|-------------|
| 2 | 866 | 1.34298 | 74.81 |
| 3 | 796 | 1.34344 | 249.14 |
| 4 | 741 | 1.3439 | 209.04 |
| 5 | 681 | 1.34454 | 12.19 |
| 6 | 624 | 1.34531 | 3.4 |
| 7 | 576 | 1.34617 | 148.13 |
| 8 | 520 | 1.34744 | 76.17 |
| 9 | 475 | 1.32883 | 17.76 |
| 10 | 426 | 1.35086 | 4.26 |

Total absorption (relative to ED2.2): 0.64

*Judd-Ofelt Parameters*

| Scale E-20 | [cm$^2$] | Standard deviation |
|------------|----------|--------------------|
| OMEGA- 2 | 0.226311 | 0.273754 |
| OMEGA- 4 | 3.9349 | 0.401255 |
| OMEGA- 6 | 4.61084 | 0.181478 |

Band Strengths:

| Band | Measured | Calculated | Difference |
|------|----------|------------|------------|
| 2 | 0.953 | 1.163 | -0.21 |
| 3 | 3.451 | 3.332 | 0.119 |
| 4 | 3.11 | 3.211 | -0.102 |
| 5 | 0.197 | 0.221 | -0.023 |
| 6 | 0.06 | 0.058 | 2.00000E-03 |
| 7 | 2.835 | 2.846 | -0.011 |
| 8 | 1.611 | 1.476 | 0.135 |
| 9 | 0.411 | 0.349 | 0.061 |
| 10 | 0.11 | 0.167 | -0.057 |

Goodness of fit (reduced CHI SQR):   0.0157474

*Fluorescence Data*

| Wavelength [nm] | Terminal state | Transition rate [s$^{-1}$] | Branching ratio |
|-----------------|----------------|----------------------------|-----------------|
| 1871.3 | 4I 15/2 | 7.521 | 5.00000E-03 |
| 1350.3 | 4I 13/2 | 154.263 | 0.094 |
| 1047.5 | 4I 11/2 | 826.049 | 0.505 |
| 888.1 | 4I 09/2 | 647.938 | 0.396 |

Radiative lifetime:                   611 µs
Measured lifetime E-folding [µs]: 271,345,367
Stimulated emission cross section:
  Sigma (4F 3/2 - 4I 11/2): 3.157 +/- 0.151 E-20 cm$^2$
  Peak wavelength [nm]:    1047.5
  Linewidth [FWHM, nm]:    19.35
  Effective linewidth [nm]: 23.19

*Refractive Indices*

Refractive index ($n_d$): 1.3459      Abbe number: 96
Nonlinear refractive index ($n_2$):   0.343 E-13 ESU
Nonlinear index coefficient (m$^2$/w): 1.07120E-29

Ordinary refractive indices:

| Wavelength [nm] | Index |
|-----------------|---------|
| 300 | 1.36143 |
| 400 | 1.35226 |
| 500 | 1.34801 |
| 600 | 1.34571 |
| 700 | 1.34432 |
| 800 | 1.34341 |
| 900 | 1.3428 |
| 1000 | 1.34235 |
| 1100 | 1.34203 |
| 1200 | 1.34178 |
| 1300 | 1.34158 |
| 1400 | 1.34143 |
| 1500 | 1.34131 |

Also included (not shown) are the absorption spectrum, fluorescence profile, measured fluorescence decays, and relative absorption. These data are very useful, if not necessary, to the physicist involved in the design of high-power Nd:glass laser systems.

As may be inferred from the foregoing, the application of the J-O theory to the study of new laser materials requires an extensive computational capability, particularly in determining the three $\Omega_i$ parameters by a nonlinear least-squares fitting program. DEUTSCHBEIN [1.32] has published a simplified method for determining peak induced-emission cross sections for Nd-doped glasses. In comparing the matrix elements $U^{(t)}$ [see (1.36)] of different Nd absorption transitions with the laser transitions, he noted a similarity between the latter and the Nd absorption band located at 750 nm, which corresponds to the transitions (unresolved) ($^4I_{9/2} \to {}^4F_{7/2}$, $^4I_{9/2} \to {}^4S_{3/2}$). The line strengths $S_{1.35}$ and $S_{1.06}$ of the ($^4F_{3/2} \to {}^4I_{13/2}$) and ($^4F_{3/2} \to {}^4I_{11/2}$) transitions, respectively, may be written in terms of the line strength $S_{0.750}$ of those absorption transitions and the ratio $X = \Omega_4/\Omega_6$

$$S_{1.06} = S_{0.75}\left[0.617 \frac{(1+0.349\,X)}{(1+0.068\,X)}\right] \quad , \tag{1.43}$$

$$S_{1.35} = S_{0.75}\left[\frac{0.321}{(1+0.068\,X)}\right] \quad . \tag{1.44}$$

Note that (1.43) differs from [Ref.1.32, Eq.(4a)] because of a mistake in the latter. The line strength of the two principal laser transitions of interest in an Nd-doped material can thus be determined by finding $S_{0.75}$ and X. The factor X can be determined by a second absorption line strength determination, or more easily by the ratio of the fluorescence branching ratios $\beta(^4F_{3/2} \rightarrow ^4I_{11/2})$ and $\beta(^4F_{3/2} \rightarrow ^4I_{9/2})$. By using (1.38,42), $\sigma_p$ can be shown to be

$$\sigma_p(1.06,1.33) = \frac{4\pi^2\lambda\alpha\eta}{(2J+1)\Delta\lambda_{eff}} S_{1.06,1.33} \quad , \tag{1.45}$$

where $\alpha$ is the fine-structure constant ($\alpha=2\pi e^2/hc$), $\lambda$, J, and $\Delta\lambda_{eff}$ are defined previously, and $\eta$ the local-field correlation factor given by $\eta = n^{-1}[(n^2+2)^2/9]$. To determine $\sigma_p$, it is then necessary to determine the integrated absorbance at 750 nm $S_{0.75}$, the effective bandwidth $\Delta\lambda_{eff}$, and X. SARKIES [1.33] has shown, as have JACOBS and WEBER [1.28], that X varies only in the range 0.9-1.1 in a wide variety of compositions. Taking X = 0.9 allows this simplified determination of $\sigma_p$ to be carried out with satisfactory accuracy. Then (1.43,44) reduce to

$$S_{1.06} = 0.766\,S_{0.75} \quad , \tag{1.46}$$

$$S_{1.35} = 0.301\,S_{0.75} \quad . \tag{1.47}$$

Calculations of $\sigma_p$ for the transition ($^4F_{3/2} \rightarrow ^4I_{11/2}$) using (1.45,46) were compared to published data obtained by use of the full J-O analysis (1.36), with the results shown in Fig.1.11. The straight line corresponds to the approximate method discussed here, and the dots, the complete calculation. The RMS error was $\approx 6\%$, or slightly greater than the 5% value reported by KRUPKE [1.25] in a complete calculation. By use of (1.45,47), $\sigma_p$ was also calculated for the ($^4F_{3/2} \rightarrow ^4I_{13/2}$) transition of 1.33 $\mu$m and compared with the values obtained by the complete analysis of JACOBS and WEBER [1.27] discussed previously. Table 1.7 shows that the agreement is excellent. This method,

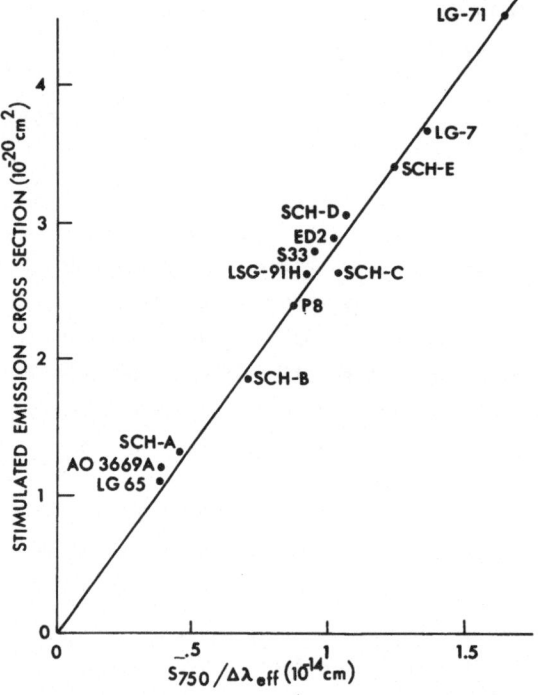

Fig. 1.11. Induced-emission cross sections: the straight line is obtained from (1.45, 46), the dots by the complete J-O calculation [1.32]

Table 1.7. Induced-emission cross sections for $\lambda = 1335$ nm (in units of $10^{-20}$ cm$^2$) [1.32]

| Glass | 3669A | LSG-91H | ED 2 |
|---|---|---|---|
| $\sigma_p$ [1.27] | 0.30 | 0.65 | 0.72 |
| $\sigma_p$ (1.45,47) | 0.31 | 0.65 | 0.70 |

because it involves only the determination of $S_{0.75}$ and $\Delta\lambda_{eff}$, is useful for the rapid evaluation of new laser materials.

## 1.5 Multiphonon Relaxation in Nd:Glass

The previous section has dealt with the essential features of the J-O theory, which has been responsible for the rationalization of radiative processes involving rare earths in various amorphous (and crystalline) hosts. As was observed very early, however, whether fluorescence is observed in a given rare-earth-host system depends upon the energy gap to the next level [1.11].

DIEKE [1.34], for instance, observed that an energy gap of 1000 cm$^{-1}$ was needed before radiative decay can be observed in the anhydrous rare-earth trichlorides. Because the phonon spectrum of LaCl$_3$ was known to extend only to about 250 cm$^{-1}$, it became apparent that a number of phonons must be involved in the nonradiative decay. Multi-phonon relaxation has become important in the study of the rare-earth amorphous system because of its obvious importance in determining the dynamics involved in Nd:glass lasers. Such parameters as the lifetimes of the $^4F_{3/2}$ metastable and $^4I_{11/2}$ terminal levels may be determined to first approximation by invoking the multi-phonon theory to be subsequently discussed. The quantum efficiency of the excited states and the $^4F_{3/2}$ level are also important in designing high-peak-power laser systems. We concentrate here only on results that have been obtained recently in amorphous systems. Although the multi-phonon theory was applied first to crystalline systems and later extended to amorphous materials, we cannot hope to discuss that work here; the interested reader is referred to the review article mentioned previously, by RISEBERG and WEBER [1.22].

When an excited rare-earth ion is produced by absorption of a photon or by energy transfer from another ion, relaxation to the ground-state configurations may take place via three distinct processes: radiative, with the emission of a photon, nonradiatively by the simultaneous emission of many phonons, or it may transfer its energy to another ion. Energy transfer has been discussed briefly in Sect.1.2 and is responsible for concentration quenching in Nd:glass. Radiative relaxation has been treated in Sect.1.4; here we concentrate solely on multiphonon relaxation processes. The theory we pursue here closely follows the treatment of RISEBERG and MOOS [1.35], LAYNE [1.11] and LAYNE et al. [1.36], and is based upon the perturbation-theory approach of KEIL [1.37]. Electrons of the trivalent rare earths are shielded from the local crystalline field by the outer-lying 5s, 5p and 6s orbits, resulting in only weak coupling to the lattice. Lattice vibrations are responsible for a modulation in the local crystal field experienced by any 4f electrons; transitions in a one-phonon process are induced between rare-earth electronic transitions whose energy separations are equal to that of the lattice mode. Much-less-probable, higher-order (p-phonon) processes may also occur if the energy separation exceeds that of the maximum energy of the host phonon spectrum. Owing to the insensitivity of the 4f electrons to the crystal field, the quantum mechanics perturbation-theory description may be used; the process is then treated as a perturbation of the free-ion states. L-S coupling is usually a good approximation, and the corresponding quantum numbers are L, S, J and $m_j$. As in the J-O theory, an exact treatment is extremely difficult, if not impossible; and for the theory we follow here, it is unnecessary to extract the essential results. We begin by expressing the system wave function $|\psi\rangle$ as a product of the electric wave function $|\psi_e\rangle$, and the vibrational wave function $|\psi_v\rangle$, the latter being expressed as a product of the individual harmonic-oscillator wave functions $|n_i\rangle$ characterized by the phonon occupation numbers $n_i$. Thus we have

$$|\psi\rangle = |\psi_e\rangle|\psi_v\rangle = |\psi_e\rangle \prod_i |n_i\rangle \quad . \tag{1.48}$$

The system Hamiltonian $\mathcal{H}$ consists of the sum of the free-ion Hamiltonian $H_0$ plus a perturbation V,

$$\mathcal{H} = H_0 + V \quad , \tag{1.49}$$

where

$$V = \sum_i V_i Q_i + \frac{1}{2} \sum_{i,j} V_{ij} Q_i Q_j + \cdots \quad . \tag{1.50}$$

Here the $V_i$, $V_{ij}$ are derivatives of the crystal field with respect to the normal modes $Q_i$,

$$V_i = \frac{\partial V}{\partial Q_i} \quad , \quad V_{ij} = \frac{\partial V}{\partial Q_i \partial Q_j} \quad . \tag{1.51}$$

Using Fermi's golden rule, we can calculate the transition rate $W_i$ for a one-phonon process involving matrix elements that couple to the ion through the first term in (1.50) between an initial electronic state $\psi_a$ and the final state $\psi_b$ as

$$W_1 = \frac{2\pi}{\hbar} \left| \sum_i \langle \psi_b | V_i | \psi_a \rangle \prod_j \langle n_j | Q_i | n_i \rangle \prod_{j \neq i} |n_j\rangle \right|^2 \quad . \tag{1.52}$$

It can also be shown that, in the case where the electronic states have an energy separation $\Delta E$ which is large compared to the maximum phonon energy in the glass, the perturbation calculation must be carried to a higher order to allow for the emission of two or more phonons. The rate of emission $W_p$ in a $p^{th}$-order process across an energy gap $\Delta E$ in which p phonon, each of energy

$$\hbar\omega = \Delta E/p \quad , \tag{1.53}$$

---

$\hbar = h/2\pi$ (normalized Planck's constant).

is given by

$$
W_p = \frac{2\pi}{\hbar} \left| \left(\frac{\hbar}{2M\omega}\right)^{p/2} \sum_{\substack{i,j,\ldots p \\ =1,2,\ldots m}} (n_i+1)^{p/2} \right.
$$

$$
\left. \times \frac{<\psi_b|V_i|\psi_{a,b}><\psi_{a,b}|V_j|\psi_{a,b}>\ldots<\psi_{a,b}|V_p|\psi_a>}{[E_a-E_{a,b}-\hbar\omega][E_a-E_{a,b}-2\hbar\omega]\ldots[E_a-E_{a,b}-(p-1)\hbar\omega]} \right|^2 \quad , \tag{1.54}
$$

where M is the reduced mass. The number of modes m that contribute to multi-phonon relaxation is found by taking the number of nearest-neighbor glass-forming units times the number of high-energy modes of the glass-forming unit. It has been estimated that in an oxide glass $m \gtrsim 10$, hence in (1.54), the relaxation is dominated by terms involving $(n+1)^p$. In (1.54), the subscript a,b seen on the intermediate states indicates that all $2^{p-1}$ electronic sequences must be included in the sum. If the simplifying assumption is made that each matrix element in (1.54) may be replaced by an average matrix element $|<a|V'|b>|$, we obtain

$$
W_p = \frac{2\pi}{\hbar} \left(\frac{\hbar}{2M\omega}\right)^p (n+1)^p \, 2^{2(p-1)} \, m^{2p} \, \frac{|<a|V'|b>|^{2p}}{|\hbar\omega|^{2(p-1)}} \quad . \tag{1.55}
$$

This expression yields the temperature dependence of $W_p$ in the $p^{th}$-order process coupled to m phonon modes, if m is large. The occupation number n of a phonon mode is described by the Bose-Einstein relationship

$$
n(T) = (e^{\hbar\omega/kT}-1)^{-1} \quad . \tag{1.56}
$$

Thus in a $p^{th}$-order process the temperature dependence of $(n+1)^p$ permits identification of the order of the process and the energy $\hbar\omega$ of the phonons.

We can also use (1.55) to find the dependence of the multi-phonon rate on the energy gap by comparing the rate for a $p^{th}$ order process with that for p-1 phonon decay. Since in the same glass network $\hbar\omega$ and m are identical, we take the ratio $W_p/W_{p-1}$ and obtain

$$
W_p/W_{p-1} = 4\left(\frac{\hbar}{2M\omega}\right)(n+1)m^2 \, \frac{|<a|V'|b>|^2}{(\hbar\omega)^2} \quad . \tag{1.57}
$$

For a weak perturbation,

$$
W_p/W_{p-1} = \varepsilon \ll 1 \tag{1.58}
$$

and an expression results that shows the explicit experimental dependence of $W_p$ on the energy gap

$$W_p = W_0 \epsilon^p = W_0 \exp\{[\ln(\epsilon)/\hbar\omega]\Delta E\} \quad . \tag{1.59}$$

Using (1.56) we can also obtain

$$W_p = C[n(T)+1]^p \exp(-\alpha\Delta E) \quad , \tag{1.60}$$

the rate of decay for a $p^{th}$-order process. Here $\alpha = -\ln(\epsilon)/\hbar\omega$ and C is a host-dependent constant. Equation (1.60) is an important result of the multi-phonon theory because it shows the exponential dependence of $W_p$ on $\Delta E$ as well as the temperature dependence. As we shall see below, the exponential dependence has been verified experimentally in many rare-earth systems and investigations of the multiphonon rate dependence on temperature have yielded both the order of the process and the dominant phonon energy.

The lifetime $\tau_a$ of an excited J state is governed by the rate $A_{ab}$ of radiative and $\omega_{ab}$ nonradiative transitions according to

$$\frac{1}{\tau_a} = \sum_b (A_{ab}+\omega_{ab}) \quad , \tag{1.61}$$

where the summation includes all final states b. Note that $\omega_{ab}$, in addition to multiphonon processes, also includes relaxation by ion-ion transfer and energy migration to quenching centers. The radiative quantum efficiency $\eta_a$ is defined by

$$\eta_a = \frac{A_{ab}}{\sum_b (A_{ab}+\omega_{ab})} = \tau_a A_{ab} \quad . \tag{1.62}$$

From (1.61,62) it can be seen that if $\omega_{ab} \gg A_{ab}$, $\eta_a$ is very small and $\tau_a$ very short, hence no fluorescence will normally be observed, a condition that is met in rare-earth systems with small energy gaps. We see from (1.59) that, if only a single phonon is involved, the decay rate will be rapid (typically <1 ns in rare earths), whereas larger energy gaps have longer lifetimes and higher quantum efficiencies. If the gap is large, $A_{ab} \gg \omega_{ab}$, then strong fluorescence is observed. This condition is met, for instance, in the $^4F_{3/2}$ metastable level in $Nd^{3+}$, resulting in a radiative quantum efficiency $\eta_a \gtrsim 95\%$ [1.38] for ED-2 laser glass in the absence of concentration quenching.

Another important aspect of the multi-phonon theory is its assumed single-frequency dependence. Although all crystals and glasses show a wide range of phonon frequencies in their spectrum, an assumption of the theory (1.59,60) is that only a single frequency is important in the multiphonon process, and that the decay is determined primarily by the highest-energy vibrational mode. As we will see, it has been verified experimentally that, in most cases of interest, the single-frequency model is valid and that the dominant multiphonon process is the one of lowest order in which the contribution of the most-energetic phonons dominates.

We now turn to examination of the experimental data relevant to relaxation in Nd:glass. The general subject of radiative and nonradiative transitions of rare-earth ions in glass has been covered recently by RISEFELD [1.39] in an excellent review. Much useful information may also be found in the review by RISEBERG and WEBER [1.22]. The dependence of the multi-phonon rate on energy gap has been investigated by LAYNE [1.11] and LAYNE et al. [1.38]. The rare earths Nd, Er, Pr, and Tm were doped into silicate ED-4 (undoped ED-2) laser glass and the multi-phonon rates measured experimentally, with the results shown in Fig.1.12. There, an approximate exponential dependence upon energy gap is found, independent of the rare-earth ion or electronic level. Also shown in Fig.1.12 is the extrapolated nonradiative decay rate for the $^4F_{3/2}$ metastable level, assuming an energy gap of $\simeq 4800$ cm$^{-1}$. The radiative decay rate is $\simeq 3 \times 10^3$/s whereas the nonradiative rate is $\simeq 200$/s, thus the quantum efficiency is $\simeq 95\%$ with no concentration quenching in agreement with [1.38].

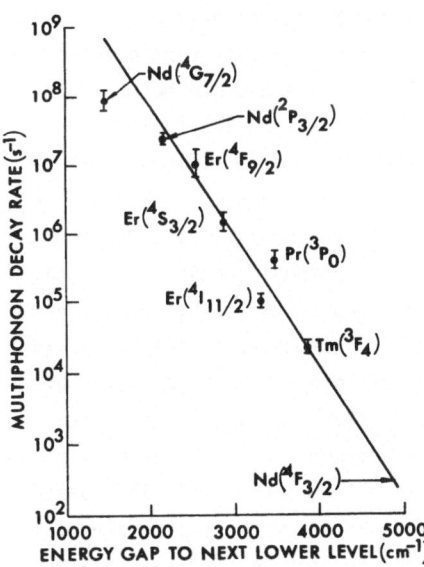

Fig. 1.12. Multiphonon decay rates as a function of energy gap to the next lower level for rare earths in ED-4 silicate glass [1.38]

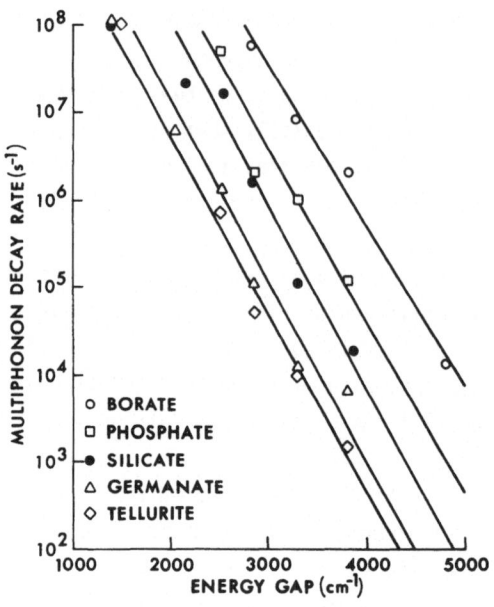

Fig. 1.13. Multiphonon decay rates for rare-earth ions in five oxide glasses plotted as a function of the energy gap to the next-lower level. The approximate frequencies of the high-energy phonons in each glass are: borate - 1350 cm$^{-1}$; phosphate - 1100 cm$^{-1}$; silicate - 1000 cm$^{-1}$; germanate - 900 cm$^{-1}$ and tellurite - 800 cm$^{-1}$ [1.38]

The $^4I_{11/2}$ terminal level, with a gap of $\simeq$1500 cm$^{-1}$ has a predicted lifetime of $\simeq$10 ns. In view of the recent experimental results presented in Sect.1.3, it would seem that this lifetime is inaccurate by a factor of almost 9. The cause of the discrepancy is not apparent at present, but probably involves the approximate nature of the multi-phonon theory.

Because the multi-phonon process is dominated by the highest-energy phonons, and the phonon spectrum is known to vary widely in different glass compositions, significant variations in the relaxation rate would be expected with glass type. This point has been investigated by RISEFELD [1.39], by LAYNE [1.11], and LAYNE et al. [1.38]. We show the result obtained by the latter in a study of multi-phonon rates in borate, phosphate, silicate, germanate, and tellurite glasses. Their results, shown in Fig.1.13, show the same approximate exponential dependence of multi-phonon rate on energy gap as in Fig.1.12, for each glass composition measured. The fastest decays are found in the borate glass, followed in order by phosphate, silicate, germanate, and tellurite. The tellurite glasses decay more slowly by a factor of $\simeq 10^{-3}$ for any level, but are ten times greater than the fastest rates measured in crystalline hosts. This observation accounts for the typically richer fluorescence spectra found in rare-earth crystalline systems. It is apparent from Fig.1.13 that the $^4I_{11/2}$ terminal-level lifetime will be strongly host dependent. As an example, $\tau_T$ should be shorter in phosphate than silicate

Table 1.8. Phonon energies of various oxide glasses [1.39]

| Glass | Bond | Phonon energy [cm$^{-1}$] |
|---|---|---|
| Borate | B-O | 1340 - 1480 |
| Phosphate | P-O | 1200 - 1350 |
| Germanate | Ge-O | 975 - 800 |
| Tellurite | Te-O | 750 - 600 |

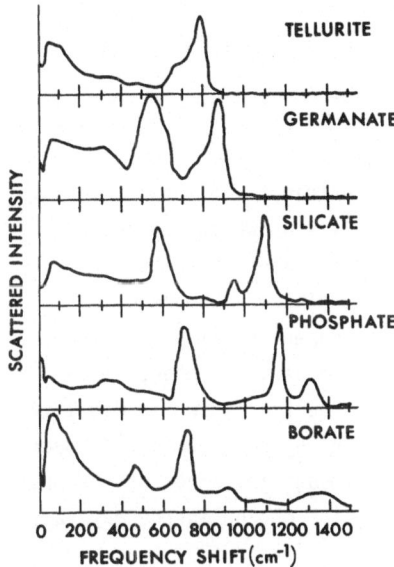

Fig. 1.14. Polarized Raman spectra of five oxide glasses [1.38]

compositions owing to the larger multi-phonon relaxation rate in phosphate glass, assuming the same energy gap.

The primary cause of the host-dependent multi-phonon rate is the phonon spectrum of the glass. There are two main groups of phonons in rare-earth-doped glasses. The first are high-energy phonons that arise from X-O stretching vibrations, where X = B, P, Ge, Te, Si, etc. Lower-energy phonons are associated with vibrations of the type Y-O, where Y = Na$^+$, K$^+$, Li$^+$, etc. or a rare-earth ion. Generally the highest-energy phonons associated with the X-O vibrations are responsible for multi-phonon decay. RISEFELD [1.39] has listed the phonon energies of importance in various glasses and they are reproduced in Table 1.8. The ranges are in agreement with the results presented by LAYNE et al. [1.38] which are shown in Fig.1.13. These authors have also obtained Raman spectra for the five glasses investigated and found the result shown in Fig. 1.14. The high-energy phonons are located in a range in agreement with Table 1.8. It is clear from Fig.1.14 that the high multi-phonon rates observed in borate glasses, for example, are consequences of the high-energy-phonon spectrum in the range of 1200-1500 cm$^{-1}$. Such a glass would have low efficiency for a Nd:laser glass, because the multi-phonon rate from the $^4F_{3/2}$ level is significant. An important result that emerges from this work is the strong correlation of decay rate with the highest-frequency vibrations of the glass network. This has also been observed in crystalline hosts and justifies the approximations made in the derivation of the multi-phonon theory.

We now investigate the temperature dependence of the multi-phonon rate. Recall, from (1.60) that the form of the derived rate with temperature is

$$W_p = W_0(n+1)^p \quad , \tag{1.63}$$

where the occupation number n is given by Bose-Einstein law (1.56). LAYNE [1.11] and LAYNE et al. [1.38] have published results in which the multi-phonon decay rate was measured as a function of temperature for the rare earth Tm$^{3+}$, and the transition ($^3F_4 \rightarrow {}^3H_5$) in oxide glasses. One result is shown in Fig.1.15 for Tm$^{3+}$ in silicate glass. The experimental points were modeled using (1.63) with p restricted to integer values, and phonon frequencies chosen to be near the peaks of the Raman spectra, shown in Fig.1.14. The best fit was found for four phonons of equal energy near 1000 cm$^{-1}$, near the Raman peak at 1100 cm$^{-1}$ in silicate glass. Although the order of the process seems well identified by use of this process, the phonon energy may vary

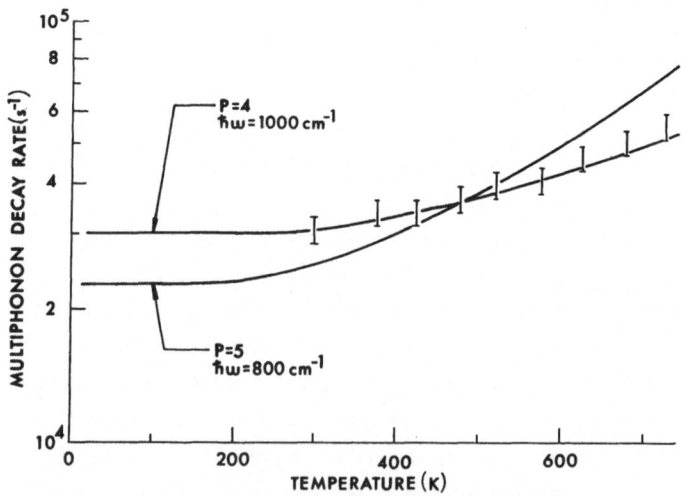

Fig. 1.15. Temperature dependence of the $^3F_4 \rightarrow {}^3H_5$ multi-phonon emission rate for Tm$^{3+}$ in silicate glass. Two fits of the data using (1.63) are shown [1.38]

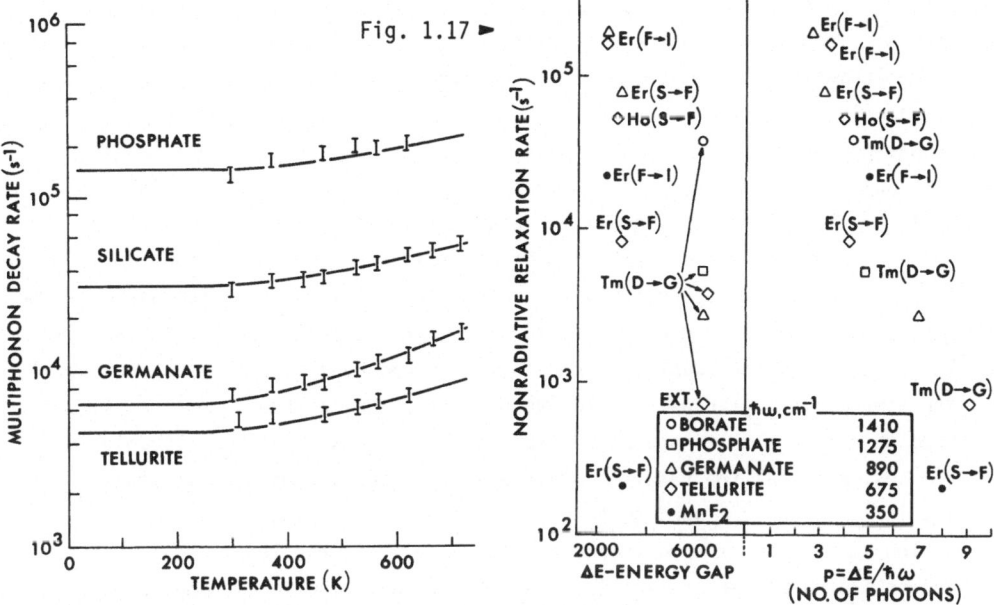

Fig. 1.16. Comparison of the temperature dependences of the $^3F_4 \rightarrow {}^3H_5$ multi-phonon emission rates for $Tm^{3+}$ in four different oxide glasses. The curves are temperature dependences predicted from (1.63) using the following numbers and energies of phonons: phosphate glass - 4 phonons, 1100 cm$^{-1}$; silicate glass - 4 phonons, 1000 cm$^{-1}$; germanate glass - 5 phonons, 825 cm$^{-1}$; tellurite glass - 5 phonons, 760 cm$^{-1}$ [1.38]

Fig. 1.17. Logarithmic plot of multi-phonon relaxation rates, W, for different rare earths in various host matrices versus energy gap, $\Delta E$, and versus number of phonons matching the energy gap [1.39]

in the range $950 \leq \hbar\omega \leq 1100$ cm$^{-1}$ and give the same result. In fact, from the Raman spectra there is evidence that three high-energy peaks take part in the multi-phonon process. Because the relaxation rate, ion-lattice coupling, and probably the order of the process change from site to site in the glass, it is impossible to assign a single frequency to the vibration. From the fit to the data in Fig.1.15, it is clear, however, that phonons of energy $\simeq 1000$ cm$^{-1}$ are responsible for the process and that $p = 4$. The multi-phonon decay rate for the same transition in $Tm^{3+}$ was measured for four oxide glasses (phosphate, silicate, germanate, tellurite) with the result shown in Fig.1.16. The best fit yielded the order of the process and the average phonon energy responsible for the decay, and agree well with the phonon spectra shown previously (Fig.1.14, Table 1.8). RISEFELD [1.39] has published results similar to these, using the same procedures, and obtained the data shown in Fig.1.17. There, the multi-phonon decay rate is shown both as a function of energy gap

($\Delta E$) and number of phonons $p = \Delta E / \hbar \omega$ for selected transitions in $Er^{3+}$, $Ho^{3+}$, and $Tm^{3+}$ in borate, phosphate, germanate and tellurite glass as well as $MnF_2$. The energies of the phonons are in good agreement with the results discussed previously.

Based upon the experimental investigation of multi-phonon decay rates, in particular their dependence upon energy gap and phonon energy, it is clear that the single-frequency model in which multiphonon relaxation is of the lowest order, and the contribution of the most energetic phonons dominate, seems to model the observed results rather well. Owing to the dominance of the energy gap and phonon energy in determining the relaxation rate, it can also be assumed that the coupling strength, described by $\varepsilon$ (1.58), is approximately the same for all glasses studied. We have seen that a theory that was first used to describe multi-phonon relaxation in crystals is also applicable to glasses, and that the results obtained are in good agreement with the Raman and infrared spectra used to find the phonon distribution.

The study of multi-phonon decay rates in glasses has led to an increased understanding of the physics involved. It has also led to a number of practical results, for instance, the realization that borate compositions are of little use in the design of efficient amplifiers for large systems.

## 1.6 The Nonlinear Index in Nd:Glass

In this book we are concerned primarily with high-peak-power Nd:glass laser systems. Such devices commonly propagate optical intensities in the range of 5-15 $GW/cm^2$ with corresponding pulse widths (FWHM) or 50-100 ps ($1\,ps = 10^{-12}\,s$). As will be seen below, such intensities, primarily through the third-order nonlinear-susceptibility tensor of glass, give rise to an intensity-dependent change of the local refractive index. The existence of a nonlinear index $n_2$ in glass, is responsible for the now well-known effects of small-scale self-focusing, whole-beam self-focusing, focal zooming, etc. (Chap.7), routinely observed in large laser systems; these effects impose severe constraints on the obtainable power output and on the fundamental design. To help alleviate nonlinear effects in glass lasers, researchers have concentrated much of their efforts on reducing the value ($n_2$) of the nonlinear index by the development of new laser glasses. This section is devoted to that subject. We shall see that the laser designer now has at his disposal a large variety of materials

that possess a wide range of nonlinear indices that allow the optimum staging of components for good beam propagation.

We begin by examining the origin of the nonlinear index $n_2$. In general,
the induced polarization density P of a material may be expanded in a power
series of the electric field $E(r,t)$

$$P = p^{(0)} + p^{(1)} + p^{(2)} + p^{(3)} + \ldots \quad . \tag{1.64}$$

The $n^{th}$ term $p^{(n)}$ is proportional to the $n^{th}$ power of the electric field

$$p^{(n)} = C(E)^n \quad . \tag{1.65}$$

For power densities at which irreversible damage to the material is not produced (typically $<10^{11}$ W/cm$^2$), this series converges rapidly and effects are
rarely seen beyond the third order. The first order $p^{(1)}$ term describes effects normally seen in classical optics, where

$$p_i^{(1)} = \sum_j x_{ij}^{(1)} E_j \quad , \tag{1.66}$$

which describes ordinary reflection, refraction, and other effects and is
linear in $E_j$. For most materials, which possess a center of inversion, $p_i^{(2)}$=0,
and the induced polarization density created by the fields $E_i$ expressed in
odd powers of E. The third-order-polarization density $p^{(3)}$ is responsible for
at least nine nonlinear effects, as described recently by HELLWARTH [1.40],
including the two phenomena of most interest here, self-focusing and self-
phase modulation. Thus, for most materials which possess a center of inversion, third-order nonlinear effects are the only ones of importance before
irreversible damage occurs. The third-order susceptibility term $x_{ijk\ell}$ is a
fourth-order tensor that has 81 components, but by arguments that involve
symmetry, the number of independent components is greatly reduced. The third-
order term in (1.64) is given by

$$p_i^{(3)} = \sum_{j,k,} x_{ijk\ell}^{(3)} E_j E_k E_\ell \quad , \tag{1.67}$$

where, in general,

$$x_{ijk\ell}^{(3)} = x_{ijk\ell}^{(3)}(-\omega_4, \omega_1, \omega_2, -\omega_3) \quad ; \tag{1.68}$$

the $\omega_i$ refer to the frequencies of the applied electric fields $(\omega_1, \omega_2, \omega_3)$ and $\omega_4$ the frequency of $p_i^{(3)}$. Here, we will be interested in the case $(\omega_4 = \omega_1 = \omega_2 = \omega_3 = \omega)$ of a monochromatic beam, thus

$$p_i^{(3)} = \chi_{1111}^{(3)}(-\omega, \omega, \omega, -\omega)|E|^2 E_i \quad , \tag{1.69}$$

which demonstrates the intensity-dependent $(|E|^2)$ third-order polarization density $p^{(3)}$. The nonlinear refractive index for a linearly polarized monochromatic beam of frequency $\omega$ in an isotropic medium, such as glass, can be shown to be

$$n_2 = \frac{12\pi}{n_0} \text{Re}\left\{\chi_{1111}^{(3)}(-\omega, \omega, \omega, -\omega)\right\} \quad , \tag{1.70}$$

where Re denotes the real part of $\chi_{1111}^{(3)}$. The imaginary part of $\chi_{1111}^{(3)}$ is related to the two-phonon absorption coefficients for a medium, and is assumed to be negligible in all that follows. The total index n for a medium can then be written

$$n = n_0 + n_2 <E^2> = n_0 + \gamma I \quad , \tag{1.71a}$$

where I is the intensity, related to the time average $<E^2>$ in electrostatic units by

$$I = \frac{n_0}{4\pi c} <E^2> \quad . \tag{1.71b}$$

Because the value of $n_2$ is usually given in electrostatic units (esu), $\gamma$ is given by

$$\gamma = \frac{4\pi n_2}{n_0 c} (\times 10^7) = \frac{(4.19 \times 10^{-3}) n_2}{n_0} \left[\frac{cm^2}{W}\right] \quad . \tag{1.72}$$

In general, a number of separate effects may contribute to the total value of $n_2$. The physical mechanism and their typical time response have been reviewed recently by SVELTO [1.41]. Solid-state laser systems designed for short-pulse propagation (50-100 ps) suffer from an induced $n_2$ due to two effects. First, the so-called "electronic" contribution is due to a nonlinear distortion of the electron clouds around the nucleus. The relaxation time for this process is typically the duration of a few optical cycles. The second nonlinearity is due to the nuclei as they attempt to minimize the total matter-field interaction energy; the relaxation time here is the time involved for a nucleus to execute a vibrational or rotational cycle. In glasses, the primary medium of interest here, the electronic contribution to $n_2$ ($n_2^e$) is typically in the range of 70-85%. For liquids, the nuclear contribution ($n_2^n$) is usually much larger. It has been experimentally verified that in glass, the total nonlinear response time is short compared to picosecond pulses normally propagated in glass-laser systems. The total nonlinear index in glasses or liquids ($n_2$) can thus be written

$$n_2 = n_2^e + n_2^n \quad ; \tag{1.73}$$

$n_2$ has also been found to depend upon the polarization state of the optical wave. Linear and circularly polarized beams propagate unchanged in an isotropic material such as glass. The value of the nonlinear index for circular polarized $n_{2,c}$ has been found to be less than the value for the linear index $n_{2,1}$ in glass. This fact has resulted in the design of a rod-amplifier laser system whose performance in propagating a circularly polarized beam resulted in higher achievable peak power than that which would have resulted from the propagation of a linearly polarized beam [1.42]. HELLWARTH [1.40] has shown that it is possible to write the total nonlinear indices for linear and circular polarization in the Born-Oppenheimer (B-O) approximation, as

$$n_{21} = \frac{12\pi}{n_0} \text{Re}\left\{\chi_{1111}^{(3)}\right\} = \frac{2\pi}{n_0} \left(\frac{3}{4}\sigma + A_0 + B_0\right) \quad , \tag{1.74}$$

and

$$n_{2c} = \frac{24\pi}{n_0} \text{Re}\left\{\chi_{1122}^{(3)}\right\} = \frac{\pi}{n_0} \left(\sigma + B_0 + 2A_0\right) \quad . \tag{1.75}$$

In (1.74,75), $\sigma$, $A_0$ and $B_0$ are (B-O) coefficients, which describe the electronic ($\sigma$) and nuclear ($A_0$,$B_0$) nonlinearities. They are related to the electronic ($\sigma_{ijk\ell}$) and nuclear ($d_{ijk\ell}$) response tensors in isotropic media through the relations,

$$\sigma_{ijk\ell} = \frac{1}{6}\sigma(\delta_{ij}\delta_{k\ell} + \delta_{ik}\delta_{j\ell} + \delta_{i\ell}\delta_{jk}) \tag{1.76}$$

$$d_{ijk\ell} = a(t)\delta_{ij}\delta_{k\ell} + \frac{1}{2}b(t)(\delta_{i\ell}\delta_{jk} + \delta_{ik}\delta_{j\ell}) \quad , \tag{1.77}$$

where $\delta_{ij}$ is the Kronecker delta function, and $A_0$ and $B_0$ are the Fourier transforms of $a(t)$ and $b(t)$, respectively, at zero frequency. Note that from (1.74,75) for no nuclear contribution ($A_0 = B_0 = 0$),

$$\frac{n_{2,c}}{n_{2,1}} = \frac{2}{3} \quad , \tag{1.78}$$

or for electronic contribution only, the circular nonlinear index is lower by a factor of 2/3 than the linear value, in agreement with observations made earlier [1.40].

Using (1.73-75) we can also write

$$n_{21}^e = \frac{3}{2} \frac{\pi}{n_0} \sigma \tag{1.79}$$

$$n_{2c}^e = \frac{\pi}{n_0} \sigma \tag{1.80}$$

for the electronic nonlinear indices and

$$n_{21}^n = \frac{2\pi}{n_0} (A_0 + B_0) \tag{1.81}$$

$$n_{2c}^n = \frac{\pi}{n_0} (2A_0 + B_0) \tag{1.82}$$

for the nuclear nonlinear indices. It is also useful to consider electronic and nuclear separation of the third-order susceptibility $\chi_{1111}^{(3)}$ of interest here:

$$\chi_{1111}^{(3)} = \chi_{1111}^{e(3)} + \chi_{1111}^{n(3)} \tag{1.83}$$

because often only the electronic or nuclear parts are measured separately. Then the nonlinear indices in (1.79-82) may also be written, using (1.74,75), as

$$n_{21}^e = \frac{12\pi}{n_0} \operatorname{Re}\left\{\chi_{1111}^{e(3)}\right\} \tag{1.84}$$

$$n_{2c}^e = \frac{24\pi}{n_0} \operatorname{Re}\left\{\chi_{1122}^{e(3)}\right\} \tag{1.85}$$

$$n_{21}^n = \frac{12\pi}{n_0} \operatorname{Re}\left\{\chi_{1111}^{n(3)}\right\} \tag{1.86}$$

$$n_{2c}^n = \frac{24\pi}{n_0} \operatorname{Re}\left\{\chi_{1122}^{n(3)}\right\} \quad . \tag{1.87}$$

It is also obvious that the various $\chi_{ijk\ell}^{(3)}$ values may be written in terms of $\sigma$, $B_0$, and $A_0$, but for brevity we will not do so here.

We now examine the effect that $n_2$ has on the propagation of a beam in a first-order sense. Of interest here is the phase of the wave that propagates in an intensity-dependent medium. The consequences for the propagation of beams in high-peak-power laser systems will be reserved for Chap.7. The phase difference between a wave travelling in a vacuum and in a medium of index n is known to be

$$\Delta\phi = \frac{2\pi}{\lambda} \int_0^L n(z)dz \quad , \tag{1.88}$$

where $\lambda$ is the vacuum wavelength and the beam propagates a distance L in the z direction. The total index n is in general a function of Z. From (1.71), we see that

$$\Delta\phi = \frac{2\pi}{\lambda} n_0 L + \frac{2\pi}{\lambda} \int \gamma I(z)dz \quad . \tag{1.89}$$

Thus, we have a constant phase term $\phi_0$ as in linear optics where $\phi_0$ is $2\pi n_0 L/\lambda$, and an intensity-dependent phase B given by

$$B = \frac{2\pi}{\lambda} \int \gamma I(z)dz \quad , \tag{1.90}$$

which represents the phase difference between a zero intensity and high intensity [I(z)] beam in a nonlinear medium. Using (1.72) we have

$$B = \frac{8\pi^2 \times 10^7}{\lambda c} \int_0^L \frac{n_2}{n_0} I(z)dz \quad . \tag{1.91}$$

This equation, widely known in high-peak-power laser circles as the "B integral" or break-up integral, is a fundamental parameter used in their design, and will be discussed in detail in Chap.7. Physically, it is easy to see that, if the propagating beam is a function of radius r as well as z, $I = I(r,z)$ and the phase is radially dependent. Since I is usually maximum in the center (r=0), the phase delay is greatest there; the overall result is that the beam collapses, which causes the phenomenon of whole-beam self-focusing. The effect presently limits the achievable power obtainable from any glass-laser system that operates in the picosecond regime; if not controlled, the effect can lead to severe permanent damage to the optical components.

We have seen in previous sections the compositional dependence of such laser parameters as stimulated-emission cross-section, quantum efficiency, and branching ratio. $n_2$ has also been found to be similarly dependent. The first tractable approach to calculating the electronic contribution $n_2^e$ to $n_2$ is due to BOLING et al. [1.43] and we will discuss that theory here. The results are particularly useful because of the explicit dependencies obtained on glass composition.

A molecular solid (transparent dielectric) is assumed to consist of a number of constituent types, each possessing a unique linear polarizability $\alpha^i$ and mean second hyperpolarizability $\gamma^i$. For a local harmonic field of frequency $\omega$ interacting with the solid, given by $E_1 \cos(\omega t)$, the microscopic polarization induced for each $i^{th}$ constituent is given by

$$p^i = \alpha^i E_1 \cos(\omega t) + \frac{1}{6} \gamma^i (E_1 \cos\omega t)^3 \quad , \tag{1.92}$$

where we have obviously ignored terms second order in $E_1$. The average polarizability $\bar{\alpha}$ and hyperpolarizability $\bar{\gamma}$ are given by

$$\bar{\alpha} = \sum_i \alpha^i N^i / \sum_i N^i \tag{1.93}$$

and

$$\bar{\gamma} = \sum_i \gamma^i N^i / \sum_i N^i \quad . \tag{1.94}$$

Here $N^i$ is the concentration of the $i^{th}$ constituent, and $N = \sum_i N^i$ the total concentration. The macroscopic dipole moment induced in the solid is

$$P = \sum_i N^i p^i \quad , \tag{1.95}$$

which, after applying the local-field correction factor $f = (n^2+2)/3$ can be written

$$P = fN\bar{\alpha}E_1 \cos(\omega t) + N \frac{\bar{\gamma}}{24} f^4 E_1^3 [3\cos(\omega t) + \cos(3\omega t)] \quad . \tag{1.96}$$

In terms of the electronic linear $\left[\chi_{11}^{e(1)}(-\omega,\omega)\right]$ and nonlinear susceptibility $\left[\chi_{1111}^{e(3)}(-\omega,\omega,\omega,-\omega)\right]$, we have, from (1.64-67),

$$P(\omega) = \chi_{11}^{e(1)}(-\omega,\omega)E(\omega) + 3\chi_{1111}^{e(3)}(-\omega,\omega,\omega,-\omega)E^3(\omega) \quad . \tag{1.97}$$

It can be seen, comparing (1.96,97) that

$$\chi_{1111}^{e(3)}(-\omega,\omega,\omega,-\omega) = \frac{f^4}{24} N\bar{\gamma} \quad . \tag{1.98}$$

The quantities $\bar{\alpha}$ and $\bar{\gamma}$ are known to be related through

$$\bar{\gamma} = q\bar{\alpha}^2 \quad , \tag{1.99}$$

where q is a parameter. Equation (1.98) is then written

$$\chi_{1111}^{e(3)} = \frac{f^4}{24} (N\bar{\alpha})^2 \left(\frac{q}{N}\right) = \frac{1}{24} \left(\frac{n^2+2}{3}\right)^2 \left(\frac{n^2-1}{4\pi}\right)^2 \frac{q}{N} \quad , \tag{1.100}$$

which shows the explicit dependence of the third-order nonlinear susceptibility on the refractive index of the material. To calculate q, BOLING et al.
[1.43] have employed a semiclassical derivation that models a classical nonlinear oscillator described by the usual equation

$$\frac{d^2x}{dt^2} + \omega_0^2(x-\lambda x^3) = S \frac{eE}{m} \cos(\omega t) \quad , \tag{1.101}$$

where S is the effective oscillator strength and $\lambda$ a measure of the nonlinear
coupling. It is shown that the free oscillation of the system occurs at a frequency $\omega$ given by

$$\omega = \omega_0 \left[ 1 - \frac{3}{4}\lambda \frac{S^2 e^2 E^2}{m^2 (\omega_0^2 - \omega^2)^2} \right] \quad , \tag{1.102}$$

which shows the dependence of the shifted frequency on the optical Stark effect. It is also shown that the parameter q defined in (1.99) above may be
written

$$q = \frac{6\lambda S \omega_0^2}{m(\omega_0^2 - \omega^2)^2} \quad , \tag{1.103}$$

which reduces in the limit $\omega \ll \omega_0$ to

$$q = \frac{6\lambda S}{m\omega_0^2} \quad . \tag{1.104}$$

The nonlinear parameter $\lambda$ must be determined and if a Gaussian potential well
of the form

$$V = V_0 \left[ 1 - \exp\left(\frac{m\omega_0^2}{2V_0}\right) x^2 \right] \tag{1.105}$$

is used, where $V_0$ is the well depth or ionization potential and $m\omega_0^2/2V_0 = \lambda$,
the Stark shift is

$$\frac{\Delta\omega}{\omega_0} = -\frac{3}{4}g\ \frac{\alpha E^2}{V_0} \quad , \tag{1.106}$$

and q is

$$q = \frac{6g}{\hbar}\ \frac{\omega_0^3}{(\omega_0^2-\omega^2)^2} \quad , \tag{1.107}$$

where $q = S\hbar\omega_0/2V_0$ for the Gaussian well. Leaving q as an adjustable parameter to be determined by experiment, (1.100) can be written as

$$\chi_{1111}^{e(3)} = \frac{g}{4}\left(\frac{n^2+2}{3}\right)^2\left(\frac{n^2-1}{4\pi}\right)^2\ \frac{1}{N\hbar\omega_0} \quad . \tag{1.108}$$

This work has demonstrated that the Stark shift and the nonlinear suscept-ibility $\chi_{1111}^{(3)}$ both arise from the nonlinearity $\lambda$ in the potential well. The effect of the field is to distort the charge distribution of the atom or ion so that it is extended further into the region of nonlinearity. Equation (1.108) is the principal result for this theory, although further embellish-ments to be discussed later make the calculation of $n_2^e$ more straightforward. It has been found that the best value of g is ($g \simeq 3$) for glasses, as deter-mined from the ellipse rotation measurements of OWYOUNG [1.44] and leads to excellent agreement with experimental data. The quantities N and $\omega_0$ in (1.108) are determined by fitting the refractive-index data for a particular glass to the Lorentz-Lorenz equation given by

$$\frac{n^2-1}{n^2+2} = \frac{4\pi}{3}\ \frac{Ne^2}{m(\omega_0^2-\omega^2)} \quad . \tag{1.109}$$

Here the oscillator strength is assumed to be =1. From (1.74,108) the elec-tronic nonlinear index for linear polarization is

$$n_{21}^e = \frac{(n^2+2)^2(n^2-1)^2}{16\pi N\hbar\omega_0 n} \quad . \tag{1.110}$$

GLASS [1.10] has cast this equation in a more useful form. Glass optical properties are generally specified by the Abbe value and the index $n = n_d$ at the Na "D" lines. The Abbe value is defined as

$$\nu_d = \frac{n_d-1}{n_F-n_e} \quad , \tag{1.111}$$

where $n_d$, $n_F$, $n_e$ are measured at the corresponding wavelengths $\lambda_d = 587.6$ nm, $\lambda_F = 486.1$ nm and $\lambda_e = 656.3$ nm. The quantity $1/\nu_d$ is a measure of the dispersive power of a glass. A derivation [1.10] that we shall not repeat here shows that

$$N = \frac{\pi}{r_0} \frac{(n_d^2-1)(n_d+1)(X_F-X_e)\nu_d}{2n_d} \quad , \tag{1.112}$$

and

$$X_0 = \omega_0^2 = X_d + \frac{(n_d^2+2)(n_d+1)(X_F-X_c)\nu_d}{6n_d} \quad ; \tag{1.113}$$

$r_0$ is the Bohr radius ($r_0=0.528 \times 10^{-8}$ cm) and $X = 1/\lambda^2$. Substitution of (1.112, 113) into (1.110) and calculation gives an expression for the electronic non-linear index for linear polarization defined at the Na "D" lines

$$n_{2,1,d}^e(10^{-13} \text{ esu}) = \frac{K(n_d-1)(n_d^2+2)^2}{\nu_d\left\{1.517+\left[(n_d^2+2)(n_d+1)\nu_d\right]/6n_d\right\}^{\frac{1}{2}}} \tag{1.114}$$

where K is a constant. Because glass manufacturers normally measure $\nu_d$ and $n_d$, comparison of the calculated results (1.114) with experimental results is fairly easy. GLASS [1.10] has made such a comparison using the value K = 68 in (1.114) which represents a best fit to the data. His results are shown in Fig.1.18 for seven glasses; the agreement is best for low-dispersion glasses as would be expected from the analysis that led to (1.112,113). It should

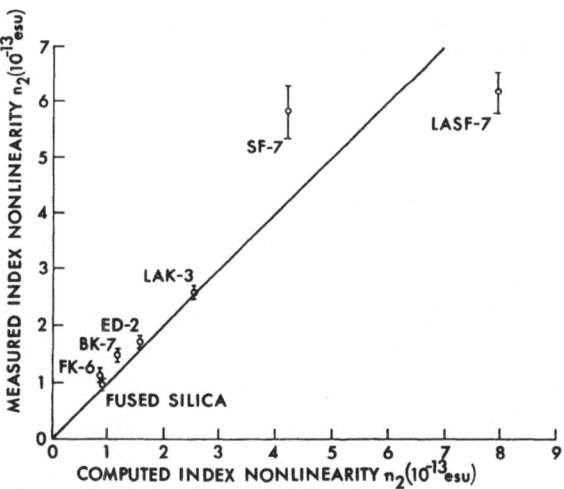

Fig. 1.18. Plot of measured $n_2$ values against computed $n_2$ values for several optical glasses [1.10]

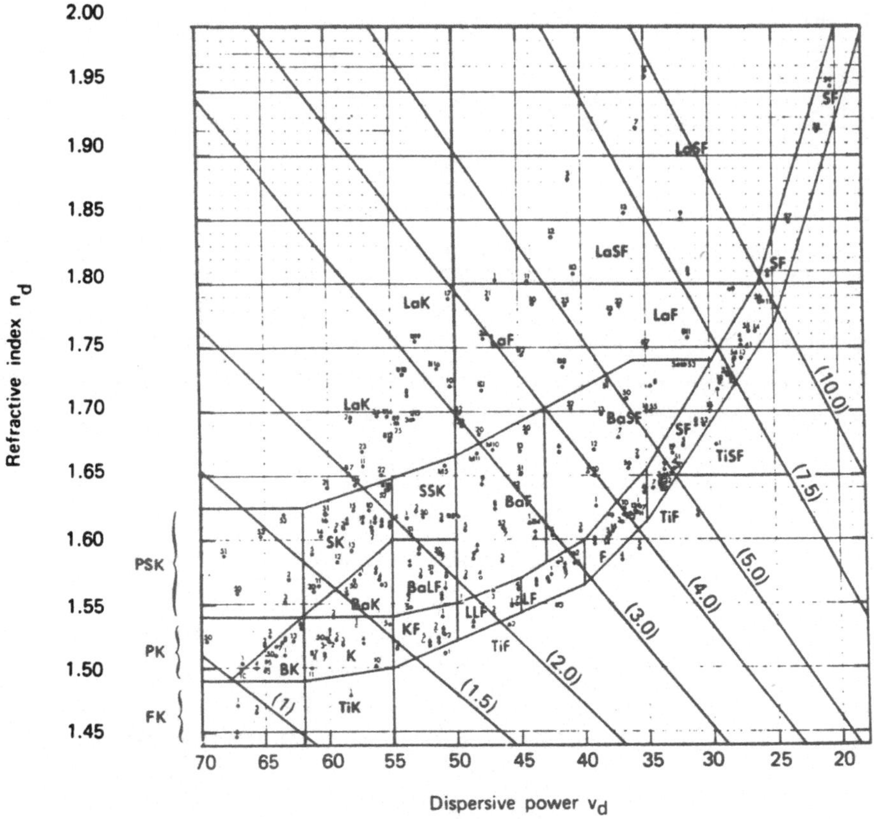

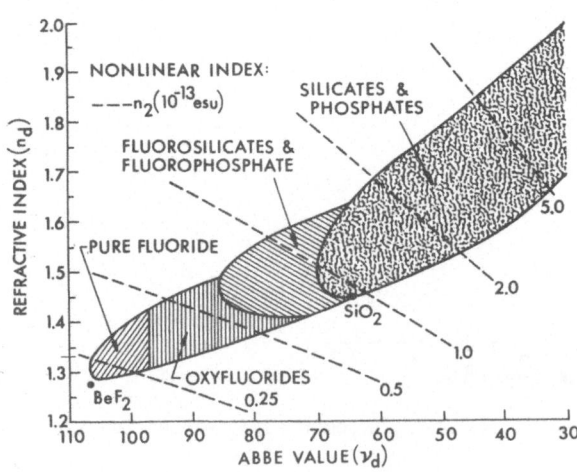

Fig. 1.19. Refractive index versus dispersive power for commercial optical glasses. Numbers in parentheses indicate nonlinear index $n_2$ in units of $10^{-13}$ esu [1.10]

Fig. 1.20. Refractive indices for various fluoride and ordinary laser glasses [1.45]

also be noted that the calculated electronic nonlinear index is really compared with the *total* nonlinear index in Fig.1.18 because the measurements do
not discriminate between the electronic and nuclear parts. Nevertheless, the
agreement is excellent, which is probably due either to the fact that the
nuclear contribution is small in glass (typically 15-30%) or that the nuclear
contribution increases with the electronic. From the above analysis of $n_{2,1,d}^e$
we conclude that the most promising materials for obtaining low nonlinear indices are those that display large Abbe values (low dispersion) and low linear
indices. An example of a compositional chart in $\nu_d$, $n_d$ space is shown in Fig.
1.19, where lines of constant $n_2$ have been included [1.10]. It may be concluded from Fig.1.19 and similar results that the nonlinear index of fluoride
glasses should be very low. Phosphate and fluorophosphate compositions are
also low compared to silicate glasses. This may be shown as in Fig.1.20 [1.45]
where the refractive index ($n_d$) and Abbe values have been used to generate
compositional contours. Also plotted are lines of constant $n_2$ determined from
(1.114). The pure fluorides should display the lowest $n_2$ values, followed in
increasing order by the fluorophosphates, phosphates, and silicates. We will
tabulate known values of $n_{2,1,d}^e$ in Chap.2, but mention here that these predictions have been verified by actual measurements on laser glasses. MILAM
and WEBER [1.46] have measured the total $n_{2,1,d}$ in a silicate, 3 phosphates,
and a fluorophosphate composition with the results shown in Table 1.9.

The fluorophosphate displayed the lowest nonlinearity and the silicate
the largest. The nonlinear indices of fluoride crystals have been measured
by MILAM et al. [1.47]; those crystals possess nonlinear indices among the
smallest reported in solids. A comparison of the measured total $n_{2,1,d}$ in
various glasses was published recently by WEBER et al. [1.48], whose results
are shown in Table 1.10. There, the nonlinear coefficient $\gamma$, related to
$n_{2,1,d}$ through (1.72) is shown for various glasses, Nd:glass compositions,
and three Faraday rotator glasses. The lowest nonlinear coefficient is displayed by the $BeF_2$ glass, the largest by the silicate SF-6. The silicate,

Table 1.9. Nonlinear refractive-index coefficients for Nd phosphate laser
glasses measured interferometrically by the use of linearly polarized 1064-nm
laser pulses of 125 ps duration [1.46]

| Glass | Type | $n_d$ | $\nu_d$ | $n_2[10^{-13}$ esu] | $\gamma[10^{-16}$ cm$^2$/W] |
|---|---|---|---|---|---|
| Silicate | ED-2 | 1.567 | 54 | 1.41 ± 0.14 | 3.77 ± 0.38 |
| Phosphate | LHG-5 | 1.541 | 63 | 1.16 ± 0.12 | 3.15 ± 0.32 |
| Phosphate | LHG-6 | 1.532 | 66 | 1.01 ± 0.10 | 2.76 ± 0.28 |
| Phosphate | EV-1 | 1.507 | 69 | 0.91 ± 0.09 | 2.53 ± 0.25 |
| Fluorophosphate | | 1.492 | 81 | 0.71 ± 0.14 | 2.0 ± 0.4 |

Table 1.10. Nonlinear refractive-index coefficients of glasses measured interferometrically by the use of linearly polarized 1064-nm laser pulses [1.48]

| Glass | Type | $n_d$ | $\gamma[10^{-20}\ m^2/W]$ |
|---|---|---|---|
| Fused Silica | 4000 | 1.458 | 2.73 ± 0.27 |
| Beryllium Fluoride | BeF$_2$ | 1.275 | 0.75 ± 0.10 |
| Silicate | SF-6 | 1.805 | 21.2 ± 2.1[a] |
| Borosilicate | BK-7 | 1.517 | 3.43 ± 0.34 |
| Fluorosilicate | FK-5 | 1.487 | 3.01 ± 0.30 |
| Fluorophosphate | FK-51 | 1.487 | 1.94 ± 0.19 |
| *Nd Laser Glasses* | | | |
| Silicate | C-835 | 1.50 | 5.10 ± 0.51[a] |
| Silicate | C-2828 | 1.53 | 5.69 ± 0.57 |
| Silicate | ED-2 | 1.572 | 4.05 ± 0.80 |
| Phosphate | EV-1 | 1.507 | 2.53 ± 0.25 |
| Phosphate | LHG-5 | 1.541 | 3.15 ± 0.32 |
| Phosphate | LHG-6 | 1.532 | 2.76 ± 0.28 |
| Phosphate | Q-88 | 1.536 | 3.11 ± 0.40 |
| Fluorophosphate | A86-82 | 1.492 | 2.0 ± 0.4 |
| Fluorophosphate | LG-812 | 1.426 | 1.44 ± 0.23 |
| Fluoroberyllate | B-101 | 1.340 | 1.03 ± 0.31 |
| *Faraday Rotator Glass* | | | |
| Phosphate (Ce) | FR-4 | 1.572 | 5.23 ± 0.80 |
| Silicate (Tb) | EY-1 | 1.626 | 4.57 ± 0.77 |
| Silicate (Tb) | FR-5 | 1.686 | 5.2 ± 0.8 |

[a] Measured relative to ED-2 with $\gamma = 4.05 \times 10^{-20}\ m^2/W$.

phosphate, fluorophosphate, and fluoroberyllate Nd:laser glasses are seen to display, in that order, lower nonlinear coefficients in agreement with the results mentioned previously concerning the Abbe value and index $n_d$.

Some measurements have appeared recently concerning the relative contributions of electronic and nuclear nonlinearity in some glass compositions. WEBER et al. [1.49] obtained the data shown in Table 1.11 for BeF$_2$ glasses. There the nuclear contribution is shown in terms of the coefficients $A_0$, $B_0$ defined previously along with the total nonlinear index $n_{2,1,d}$. It is a simple matter to determine that in these glasses the nuclear contribution is only 20-30% of the total nonlinearity. WEBER et al. [1.48] have also compiled the nuclear and total nonlinear contribution in 10 glasses with results shown in Table 1.12. The fraction of electronic nonlinearity $f_e$ is also listed and ranges from 0.81 in the SF-6 and ED-4 silicate glasses to 0.57 in a phosphate one (LHG-6). The reader interested in compilations of calculated electronic nonlinear index may consult the LLL glass-data sheets mentioned previously [1.31], an example of which was shown in Table 1.6.

Table 1.11. Linear and nonlinear refractive indices of beryllium fluoride glasses. The nuclear contributions to $n_2$ are derived from Raman-scattering data; the total $n_2$ was measured by use of linearly polarized 1064-nm laser pulses of 150 ps duration. All quantities are in esu [1.49]

| Glass | $n_d$ | $10^{16} B_0$ | $10^{16} (A_0+B_0)$ | $10^{15} n_2[\text{nucl.}]$ | $10^{15} n_2$ |
|---|---|---|---|---|---|
| $BeF_2$ | 1.275 | $4 {}^{+2}_{-1}$ | $10 {}^{+5}_{-2}$ | $5 {}^{+2}_{-1}$ | $23 \pm 3$ |
| B102 | 1.340 | $14 {}^{+4}_{-2}$ | $20 {}^{+6}_{-3}$ | $9 {}^{+3}_{-1}$ | $33 \pm 10$ |
| B402 | 1.384 | $9 {}^{+10}_{-5}$ | $23 {}^{+25}_{-12}$ | $10 {}^{+12}_{-5}$ | $38(\text{est})^{a}$ |
| $SiO_2$ | 1.459 | $14 \pm 2$ | $57 \pm 6$ | $31 \pm 2$ | $95 \pm 10$ |

[a] No experimental data for the total $n_2$ were obtained for this sample because the optical quality was not adequate to obtain a usable fringe pattern

Table 1.12. Nuclear contributions to the nonlinear refractive index of glasses. Parameters $A_0$ and $B_0$ were determined from Raman-scattering spectra by Heiman and Hellwarth; the total $n_2$ values are from Table I. $f_e$ is the fraction of $n_2$ which is electronic in origin. All quantities are in esu [1.48]

| Glass | Type | $10^{16} A_0$ | $10^{16} B_0$ | $10^{14} n_2[\text{nucl.}]$ | $10^{14} n_2(\text{total})$ | $f_e$ |
|---|---|---|---|---|---|---|
| Silicate | SF-6 | 203 | 277 | 16.7 | 90.0 | 0.81 |
| Silicate | ED-4 | 42 | 26 | 2.74 | 14.1 | 0.81 |
| Silica | $SiO_2$ | 43 | 15 | 2.49 | 9.5 | 0.74 |
| Borosilicate | BK-7 | 32 | 35 | 2.78 | 12.4 | 0.78 |
| Phosphate | LHG-5 | 51 | 59 | 4.5 | 11.6 | 0.61 |
| Phosphate | LHG-6 | 37 | 67 | 4.3 | 10.1 | 0.57 |
| Fluorosilicate | FK-5 | 32 | 22 | 2.28 | 10.7 | 0.79 |
| Fluorophosphate | FK-51 | 15 | 24 | 1.64 | 6.9 | 0.76 |
| Fluoroberyllate | B-101 | 6 | 14 | 0.94 | 3.3 | 0.72 |
| Beryllium Fluoride | $BeF_2$ | 6 | 4 | 0.49 | 2.3 | 0.79 |

It is often of interest to estimate the nonlinear index for circular polarization for glass compositions for which the nuclear parameters $A_0$ and $B_0$ are not available. JACOBS [1.50] has obtained a plot of $n_{2,1}$ as a function of $n_{2,c}$, reproduced in Fig.1.21, which shows an almost linear relationship between the two values. The data points are taken from [1.49]; $n_{2c}$ was calculated according to

$$n_{2c} = \frac{2}{3}\left[n_{21} + \frac{\pi}{n_0}\left(A_0 - \frac{B_0}{2}\right)\right] , \qquad (1.115)$$

which may be derived from (1.74,75). The straight line is a least-squares fit to the nine data points and has a slope of 1.44. We can accurately relate $n_{2c}$ to $n_{21}$ by the use of

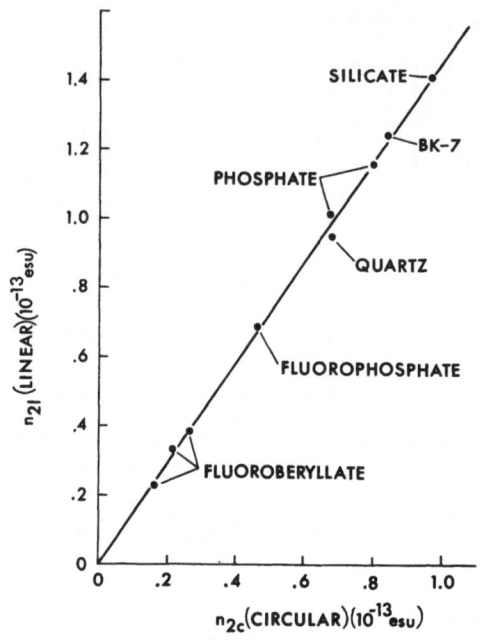

Fig. 1.21. Nonlinear index of refraction for linear ($n_{2l}$) and circular ($n_{2c}$) polarization [1.50]

$$n_{2c} = n_{2l} / 1.44 \quad . \tag{1.116}$$

For glasses for which $n_{2l}$ has been measured or calculated as indicated above, (1.116) will yield $n_{2c}$, which is of interest in laser systems that may propagate circular polarization.

We have not discussed here the interferometric techniques employed by investigators to measure the total $n_2$ of various glasses and crystals. Instead, we refer the reader to the definitive articles by MORAN et al. [1.51], BLISS et al. [1.52] and WEBER et al. [1.48].

Equation (1.114) is also expected to be applicable to gases and liquids. BROWN [1.53] calculated the electronic nonlinear index for air to assess possible significant B integral contributions in a high-peak-power laser system. For linear polarization, the value

$$n_{2,l,d}^{e} \simeq 1.68 \times 10^{-17} \text{ esu} \tag{1.117}$$

was obtained. This is approximately four orders of magnitude less than for typical glasses; it would probably be greater if the nuclear contribution were included. The B-integral contribution from (1.117) is small, however, if a high-peak-power beam is propagated from the output of a laser system over a rather long path length to the target chamber used in laser-fusion research, significant B integral contributions may occur [1.53]. Other gases, such as He or $N_2$ that display low $n_2^e$ values may be substituted for air to reduce B-integral effects. The value in (1.117) was obtained by use of $K-68$ in (1.114), a value that is applicable only to glass. Experimental values of $n_{2,1,d}^e$ would have to be obtained experimentally for many gases and the appropriate value of K determined by a least-squares fit of (1.114) to the data. Nevertheless, the order of magnitude indicated in (1.117) is probably correct.

Nonlinear indices have also been obtained for liquids. In particular, many values for $\sigma$, $A_0$, and $B_0$ are presently known. The reader is referred to the review outline by HELLWARTH [1.40] for a tabulation. It is also possible to apply the theory developed by BOLING et al. [1.43] and GLASS [1.10] to the calculation of $n_2^e$ in liquids. Such a program has been undertaken by RINEFIERD et al. [1.54] who measured $\nu_d$, $n_d$, $n_F$ and $n_e$ for twelve liquids and using (1.114) generated values for $n_2^e$. A value of $K = 68$ was assumed in that work. More recently, BROWN et al. [1.55] have extended these results to find the total nonlinear indices $n_2^T$ of a number of liquids. Based upon the measurement of linear indices and calculation of the Abbe value, the electronic contribution $n_2^e$ to $n_2^T$ was calculated by normalizing (1.114) to the well-known value for benzene, yielding $K = 101.5$. Using the measurements tabulated by HELLWARTH [1.40], both the nuclear contribution $n_2^n$ and $n_2^T$ were obtained. The results

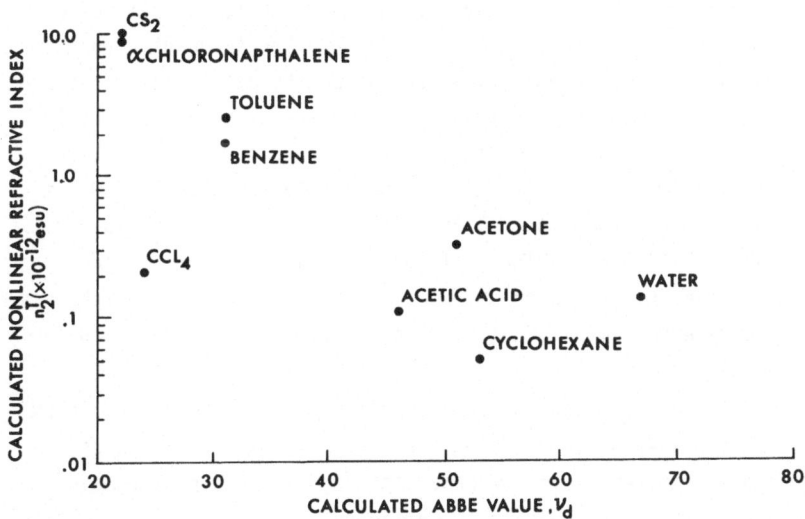

Fig. 1.22. Calculated total nonlinear index as a function of Abbe value for selected liquids

are shown in Fig.1.22 where the total nonlinear index $n_2^T$ is plotted as a function of linear index $n_d$. The same general trends are observed as with glasses; $n_2^T$ is lowest for large $\nu_d$ and small $n_d$, although the order of magnitude of $n_2^T$ is about a factor of 10 larger in liquids. These data are important because they allow us to assess the nonlinear properties of liquids used in laser systems for cooling or index-matching applications.

It should be noted that all of the values of $n_2$ reported to date in glasses and liquids are positive. It is well known, however, that in the region of an electronic transition the nonlinear susceptibility $\chi_{1111}$ changes value and becomes negative. Negative values of $\chi_{1111}$ or equivalently $n_2$ lead to the phenomena of self-defocusing. LEHMBERG et al. [1.56] have investigated negative $n_2$ values in Cs vapor. It is hoped that a device such as a Cs cell inserted in the proper location in a laser system will be able to compensate partially for the effects of whole-beam self-focusing and to a lesser degree small-scale self-focusing. Practical problems, such as maintaining a uniform Cs vapor density in a large cell, have so far prevented the deployment of such cells in large laser systems.

In the past year, much attention in the laser-fusion community has been focused on frequency doubling or tripling the output of Nd:glass systems, since those processes may be achieved with high efficiency and the laser-fusion process is apparently more favorable at shorter wavelengths. A problem associated with this scheme is that the nonlinear problems become much worse. If we frequency triple, for example, the B-integral, formally given by (1.92) becomes at least a factor of three larger. The frequency dependence of the nonlinear index, neglected in the work of BOLING et al. [1.43], has not been investigated experimentally, although such measurements are currently being pursued at LLL. It has been estimated that for fused silica the nonlinear index at 0.35 μm is at least a factor of 1.5 larger than at 1.06 μm [1.57], leading to an enhancement in B by a factor of 4.5. Little has been done theoretically; SPARKS and DUTHLER [1.58] have calculated the frequency behavior of $n_2$ in LiF. Clearly what is needed in this area is a comprehensive model that includes enhancement of $n_2$ from normal dispersion and two-photon and Raman resonances.

We have, in this chapter, considered properties of the rare-earth ion $Nd^{3+}$ interacting with a glassy host. Although the treatment has in all cases, by necessity, not been comprehensive, we have attempted to present the main results obtained by researchers to date, accompanied by numerous examples from the literature. All of the most important parameters and effects relevant to the design of a modern high-peak-power Nd:glass system have been presented.

One area of great current interest, not essential to this work, is energy transfer in rare-earth systems. The field is in its infancy and good reviews of work to date have been given by RISEBERG and WEBER [1.22] and by LAYNE [1.11].

# 2. Optical and Physical Properties of Laser Glasses

In Chap.1 we considered the basic properties of laser glass and introduced many of the fundamental parameters used in specifying such glasses. Many other optical properties are, of course, important in the design of components for laser systems and we treat those in this chapter. To understand completely the operation of laser amplifiers in a real system and to employ a glass composition that is optimal for a given application, however, requires that consideration be given to thermal, elasto-optic, and mechanical constants of laser glasses. Here we first define all of the fundamental parameters of interest (Sects.2.1,2) and then introduce figures of merit for the evaluation of laser glasses (Sect.2.3). In Sect.2.4, we show correlations between the optical and physical properties of laser glasses. Finally, Sect.2.5 will investigate a number of effects which have been found to be important in the consideration of laser glasses for practical applications. We refer particularly to impurity problems, thermal-shock resistance, durability, and intrinsic stress, all of which have been found to be important for consideration, in addition to the more fundamental properties previously discussed.

## 2.1 Fundamental Optical Constants

It is our intention here to review optical parameters that we have previously discussed, to introduce others necessary for a complete description of a laser-amplifier or system, and finally to give a tabulation of the data presently known for a representative variety of commercial and experimental laser glasses in the silicate, phosphate, fluorophosphate, and fluoroberyllate compositions. We begin by listing the parameters defined in Chap.1:

1) $\Delta\lambda$ Fluorescence bandwidth FWHM at 1.06 $\mu$m, $({}^4F_{3/2} \to {}^4I_{11/2})$ [nm]
2) $\Delta\lambda_e$ Effective fluorescence bandwidth at 1.06 $\mu$m, $({}^4F_{3/2} \to {}^4I_{11/2})$ [nm]

3) $\lambda_p$     Peak fluorescence wavelength for the transition $(^4F_{3/2} \rightarrow {}^4I_{11/2})$ [nm]

4) $\sigma_p$     Peak stimulated-emission coefficient at $\lambda_p$   [cm$^2$]

5) $\alpha_0$     Specific-gain coefficient, defined as $\alpha_0 = \sigma_p/h\nu_p = \sigma_p\lambda_p/hc$ [cm$^2$/J]

6) $\alpha_L$     Passive (unpumped)-loss coefficient at 1.06 $\mu$m   [cm$^{-1}$]

7) $E_{SAT,3}$     Saturation flux in a three-level system, given by $E_{SAT,3} = 1/2\alpha_0 = hc/2\lambda_p\sigma_p$   [J/cm$^2$]

8) $E_{SAT,4}$     Saturation flux in a four-level system, given by $E_{SAT,4} = 1/\alpha_0 = hc/\lambda_p\sigma_p$   [J/cm$^2$]

9) $\tau_T$     Terminal-level lifetime $(^4I_{11/2} \rightarrow {}^4I_{9/2})$   [s]

10) $\tau_0$     Fluorescence lifetime $(^4F_{3/2} \rightarrow {}^4I_{11/2})$ for zero or low Nd doping [s]

11) $\Omega_2, \Omega_4, \Omega_6$ Judd-Ofelt parameters for Nd$^{3+}$ in a given composition   [cm$^2$]

12) $\beta$     Branching ratio for $(^4F_{3/2} \rightarrow {}^4I_{11/2})$ transition, dimensionless

13) $n_d$     Linear refractive index at the Na"D" lines ($\lambda$=589 nm), dimensionless

14) $n_0$     Linear refractive index at 1.06 $\mu$m, dimensionless

15) $\nu_0$     Abbe value at the Na"D" lines, $\lambda = 589$ nm, dimensionless

16) $n_{2,1}$     Nonlinear index of refraction at the Na"D" lines for linear polarization   [esu]

17) $n_{2,c}$     Nonlinear index of refraction at the Na"D" lines for circular polarization   [esu]

18) $\gamma_1$     Nonlinear coefficient for linear polarization, related to $n_{2,1}$ through $\gamma_1 = cn_{2,1}/n_0$   [cm$^2$/W]

19) $\gamma_c$     Nonlinear coefficient for circular polarization, related to $n_{2,c}$ through $\gamma_c = cn_{2,c}/n_0$   [cm$^2$/W]

It is of interest in evaluating glasses for laser amplifiers to consider a number of properties that are relevant to thermal effects, amplified spontaneous emission or parasitic oscillations (Chap.4), and the relative absorption efficiency for Xe flashlamp radiation. Xe flashlamps are the primary pump source now used in glass-laser systems. A typical output spectrum, which may be found in Chap.3, when convolved with the absorption spectrum appropriate to a given laser glass, yields the absorption, stored-energy density, and heat density as a function of penetration or optical depth. The derivative (slope) of the stored energy density and heat density may also be obtained. The width, location, and strength of the absorption bands change with glass composition, hence convolution of that spectrum with a standard flashlamp output spectrum yields a way of comparing the relative merits of different compositions when certain important effects occur. We begin by noting that monochromatic radiation of intensity $I_0$ is absorbed in an isotropic homogeneous medium according to Beer's law

$$I = I_0 \ e^{-\alpha x} \quad , \tag{2.1}$$

where $\alpha$ is the absorption coefficient [cm$^{-1}$] and I the reduced intensity at a distance x into the medium. In laser amplifiers, the pump source is a Xe flashlamp whose output consists of the superposition of pressure-broadened discrete lines as well as a broad continuous background, which may be described by a "greybody" distribution. For such a source the incident flux in a laser amplifier must be described by a spectral distribution $I_{0\lambda}$ [W/cm$^2$-nm] [spectral emittance]. Then the absorption $A_E$ in the form of stored energy density and that ($A_H$) in the form of heat density are given by

$$A_E(x,t) = c_1 \int_{\lambda_1}^{\lambda_2} n_\lambda I_{0\lambda}\left(\frac{\lambda}{\lambda_p}\right)(1 - e^{-\alpha_\lambda x})d\lambda \tag{2.2}$$

$$A_H(x,t) = c_2 \int_{\lambda_1}^{\lambda_2} n_\lambda I_{0\lambda}\left(1 - \frac{\lambda}{\lambda_p}\right)(1 - e^{-\alpha_\lambda x})d\lambda \quad , \tag{2.3}$$

where $c_1$ and $c_2$ are constants, $n_\lambda$ the wavelength-dependent quantum efficiency, $\lambda_1$ and $\lambda_2$ the wavelength limits, $\lambda_p$ the peak laser wavelength, and $\alpha_\lambda$ the wavelength-dependent absorption coefficient. Note that (2.3) gives the heat deposited due to the quantum defect, or difference of energy between the laser photon at $\lambda_L$ and the absorbed photon at $\lambda$, but does not consider the heating due to the ultraviolet edge, which is usually significant. The separate background absorption due to the edge may be calculated from

$$A = \int_{\lambda_1}^{\lambda_2} I_{0\lambda}(1 - e^{-\alpha_\lambda x})d\lambda \quad . \tag{2.4}$$

Here $\lambda_1$ is the wavelength to which the edge absorbs strongly, and $\lambda_2$ is the shortest wavelength of significance emitted by the Xe flashlamp. The absorption per Nd$^{3+}$ ion may be assumed to be independent of Nd concentration [2.1] and (2.2,3) may be parameterized to obtain $A_E$ and $A_H$ as a function of optical thickness $\nu = Px$ where P is the wt-% Nd doping

$$A_E(\nu,t) = c_1 \int_{\lambda_1}^{\lambda_2} n_\lambda I_{0\lambda}\left(\frac{\lambda}{\lambda_p}\right)(1 - e^{-\alpha_\lambda' \nu})d\lambda \quad , \tag{2.5}$$

$$A_H(\nu,t) = c_2 \int_{\lambda_1}^{\lambda_2} \eta_\lambda I_{0\lambda}\left(1 - \frac{\lambda}{\lambda_p}\right)(1 - e^{-\alpha_\lambda' \nu})d\lambda \quad . \tag{2.6}$$

As will be seen in Chap.3, $I_{0\lambda}$ is a function both of the current density J and the diameter D of the flashlamp. To first order, however, it has been shown [2.1] that the distributions $A_E$ and $A_H$ are insensitive to J; hence it is sufficient to obtain those quantities for a single current density J. Physically, the stored-energy density $E(\nu)$ and heat density $H(\nu)$ are proportional to the derivatives of (2.5,6), respectively. Then we have

$$E(\nu) = c_1 \int_{\lambda_1}^{\lambda_2} \alpha_\lambda' \eta_\lambda I_{0\lambda}\left(\frac{\lambda}{\lambda_p}\right)e^{-\alpha_\lambda' \nu}d\lambda \tag{2.7}$$

$$H(\nu) = c_2 \int_{\lambda_1}^{\lambda_2} \alpha_\lambda' \eta_\lambda I_{0\lambda}\left(1 - \frac{\lambda}{\lambda_p}\right)e^{-\alpha_\lambda' \nu}d\lambda \quad . \tag{2.8}$$

DE SHAZER and KOMAI [2.2] have obtained measurements for ED-2 laser glass which indicate that $\eta_\lambda \simeq$ constant, independent of $\lambda$. Although we have seen that the quantum efficiency can vary widely across an absorption band, as demonstrated by FLN experiments (Sect.1.2), in a macroscopic sense we take $\eta_\lambda$ as an average over the ensemble of available sites. Therefore we may take $\eta_\lambda$ outside the integrals in (2.5-8). For a given flashlamp spectrum characterized by $I_{0\lambda}$, in order to maximize $E(\nu)$ it is necessary to maximize the quantum efficiency $\eta_\lambda$, which physically represents the ratio of the number of ions in an excited state that decay to the $^4F_{3/2}$ metastable level, divided by the total number of absorbed photons.

Equation (2.7) is generally evaluated by convoluting a computer-generated flashlamp spectrum that approximates well the instantaneous output of a Xe flashlamp, with $\alpha_\lambda' \exp(-\alpha_\lambda' \nu)$, where the $\alpha_\lambda'$ are obtained from passive absorption spectra generated for each glass. The details have been discussed elsewhere [2.1,3]. In Fig.2.1, we show the absorption into inversion ($A_E$), heat ($A_H$), and total absorption $A_T$ as a function of $\nu$ for EV-2 phosphate glass; this result is typical and was generated by evaluating (2.5,6). In Fig.2.2, the stored-energy density $E(\nu)$, as determined by (2.7) is shown for a silicate (ED-2) and two phosphate (EV-2, LHG-5) compositions. The fact that the phosphate glasses display a steeper fall-off with $\nu$ (slope) than the silicate is a consequence of the narrowness and intensity of the absorption bands in phosphate vs silicate glass. Figure 2.2 represents, to first order, the one-sided (uncoupled) stored-energy density distributed in an active-mirror amplifier (Chap.5), and is useful for performing a variety of analyses. The

Fig. 2.2

OPTICAL THICKNESS $\nu$(%-cm)

RELATIVE STORED ENERGY DENSITY

ED-2

EV-2

LHG-5

Fig. 2.1

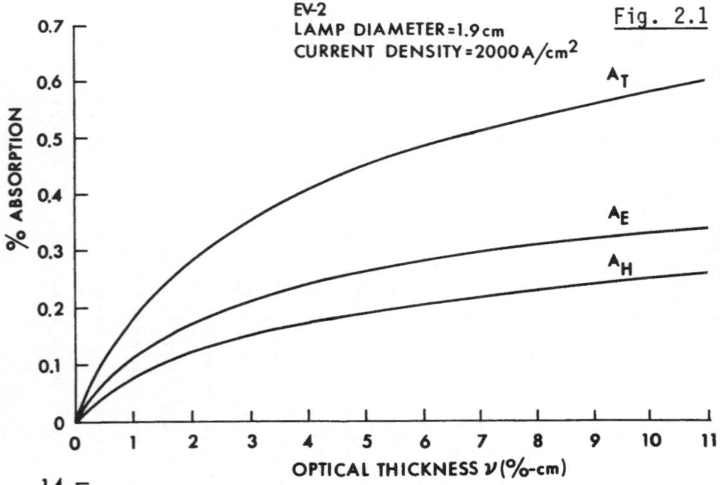

EV-2
LAMP DIAMETER=1.9 cm
CURRENT DENSITY=2000 A/cm²

$A_T$

$A_E$

$A_H$

OPTICAL THICKNESS $\nu$(%-cm)

% ABSORPTION

ED-2
LAMP DIAMETER=1.9 cm
CURRENT DENSITY=2000 A/cm²

RELATIVE STORED ENERGY DENSITY

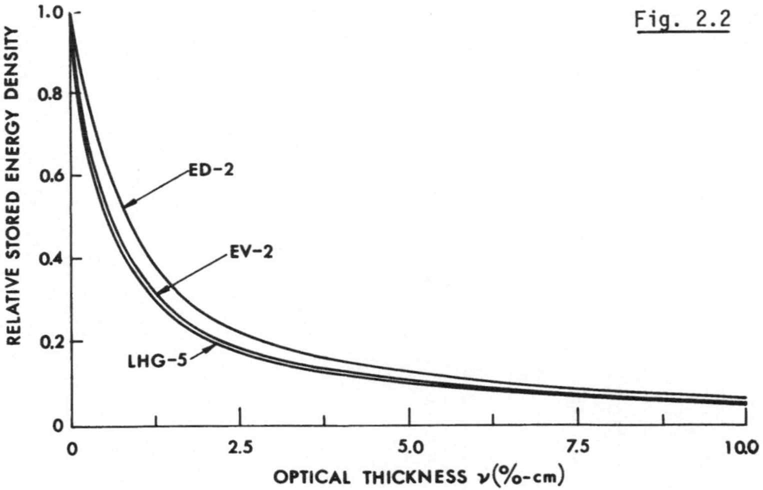

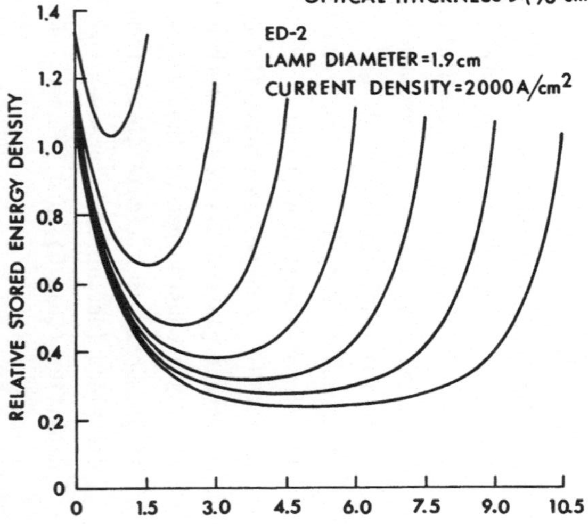

Fig. 2.1. Absorption of Xe flashlamp radiation in Nd:glass into stored-
energy density ($A_E$) and heat density ($A_H$), and total absorption ($A_T$) as a
function of optical thickness, for EV-2 phosphate glass

Fig. 2.2. Relative stored-energy density as a function of optical thickness for ED-2, EV-2, and LHG-5 laser glasses

Fig. 2.3. Relative stored-energy density in a disk as a function of optical thickness for ED-2 glass

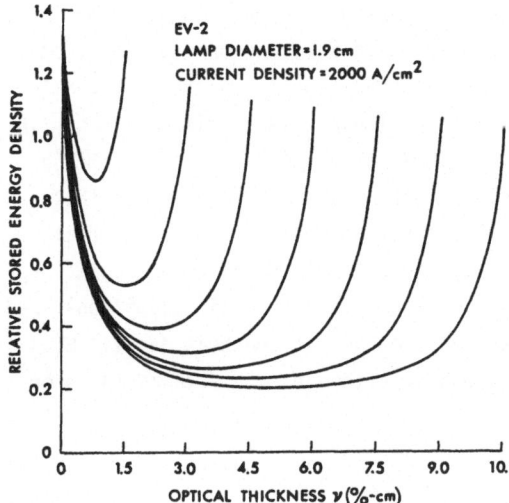

Fig. 2.4. Relative stored-energy density in a disk as a function of optical thickness for EV-2 glass

form of the $E(\nu)$ curve has been found to be important in the threshold for parasitic oscillations and in the ASE problem [2.3,4] (see Chap.4). In disk amplifiers, which are pumped from two sides, similar curves may be generated by evaluating the sum

$$E_0(\nu) = E(\nu) + E(\nu_M-\nu) \quad , \tag{2.9}$$

where $\nu_M$ is the maximum optical thickness. We show in Figs.2.3,4 the curves obtained for a silicate (ED-2) and a phosphate (EV-2) composition, respectively, using (2.9). The curves in Figs.2.2-4 have been modeled in a variety of analytical formulations [2.1]; good accuracy may be obtained by using a polynomial approximation [2.4] of the form

$$E(\nu) = K_1 \sum_{i=0}^{n} a_i \nu^i \tag{2.10}$$

for $E(\nu)$, and

$$H(\nu) = K_2 \sum_{i=0}^{n} b_i \nu^i \tag{2.11}$$

for $H(\nu)$. $K_1$ and $K_2$ are normalization constants and the various coefficients $a_i$ and $b_i$ were found by a least-squares fit to the data obtained by direct evaluation of (2.7,8). By use of a sufficient number of terms in (2.10,11), maximum errors of typically 0.05% resulted [2.4]. The first attempt to verify the predictions of (2.7) was made by HOLZRICHTER and DONICH [2.5] who mea-

64

sured the total absorption in a variety of doping-thickness combinations of ED-2 laser glass. The only direct measurement of $E(\nu)$ was made by ABATE et al. [2.6] in ED-2, EV-2, Q-88, and LHG-7 glasses. The latter three are all phosphates whereas the former is a silicate. The results, which are shown in Fig.2.5, confirm the prediction made by BROWN et al. [2.4], and by McMAHON [2.7] that $|dE/d\nu|$ is greater in phosphate than in silicate glasses; that fact is important in the evaluation of parasitics and ASE as well as heat deposition in active-mirror and rod amplifiers. Furthermore, the results presented in [2.6] completely confirm all of the aforementioned assumptions used in the derivation of (2.5-8). Similar measurements of the heat density $H(\nu)$ have not, so far as we know, been performed, although such data could be obtained from Raman experiments.

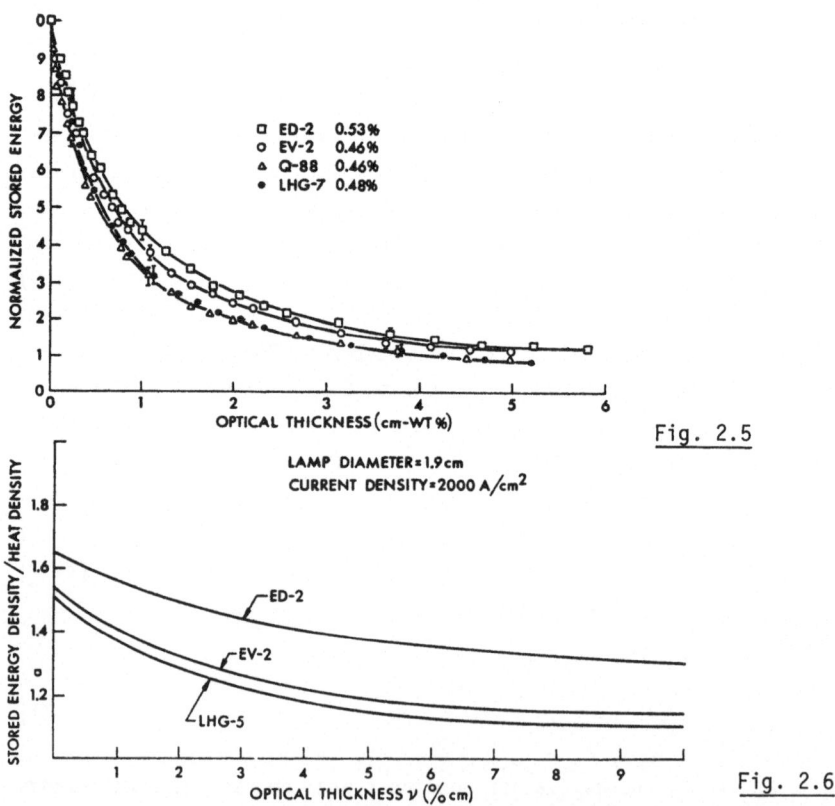

Fig. 2.5

Fig. 2.6

Fig. 2.5. Normalized stored-energy density profiles in ED-2, EV-2, Q-88, and LHG-7 laser glasses as a function of optical thickness. Solid curves are calculated whereas the data were obtained from experiment [2.6]

Fig. 2.6. Ratio of stored-energy density to heat density in ED-2, EV-2, and LHG-5 laser glasses as a function of optical thickness [2.4]

Another quantity of interest in the evaluation of laser glasses is the ratio E/H of stored-energy density to heat density of a glass. BROWN et al. [2.4] have obtained E/H as a function of $\nu$ by evaluating (2.7,8) for a silicate (ED-2) and two phosphate glasses (EV-2, LHG-5), with the results shown in Fig.2.6. It can be seen that all three curves fall off weakly with $\nu$, in approximately the same manner. The silicate displays the maximum E/H while the phosphates generate more heat per unit inversion density than the silicate. That is a consequence of the narrowness of the absorption bands as well as a consequence of a uniform shift of the band peaks towards the blue in the phosphate glasses; it is an important consideration in evaluating thermal properties of laser glasses.

The use of polynomials to describe $E(\nu)$ and $H(\nu)$ (2.5) is cumbersome; a more tractable approach in the evaluation of large numbers of laser glasses has been taken by workers at LLL [2.8] where the absorption efficiency of a glass normalized to a perfect absorber is obtained by the computer program VODAC. This program simulates the transfer of energy from the capacitor bank through the flashlamp and finally into the laser transition of a glass sample. The absorption obtained as a function of optical depth (Nd concentration [ions/cm$^3$] × glass thickness) represents the *total* absorption of the sample from a standard Xe flashlamp output, and is equivalent to the sum of (2.5,6) with the appropriate $I_{0\lambda}$. The Nd concentration $\rho$ can be related to the doping P[wt-%] through the relationship

$$\rho \, [\text{ions/cm}^3] = \frac{2PNd}{M} \quad , \tag{2.12}$$

where N is Avogadro's number ($6.023 \times 10^{23}$), M the molecular weight of $Nd_2O_3$ (336.48) and the factor of two accounts for the two $Nd^{3+}$ ions per molecule of $Nd_2O_3$. In general, the density d = d(P) varies in a linear fashion with p, hence

$$d = d_0(1+K) \quad , \tag{2.13}$$

where $d_0$ is the glass density at zero concentration and K is a constant. Values of $d_0$ and K have been obtained by KRUPKE [2.9] and ABATE et al. [2.6]. The resulting curves of absorption efficiency $A_T$ versus optical depth for a large number of laser glasses [2.8], an example of which is shown in Fig.2.7 for LHG-5 phosphate laser glass, may be accurately represented by the so-called A, B, C, D function

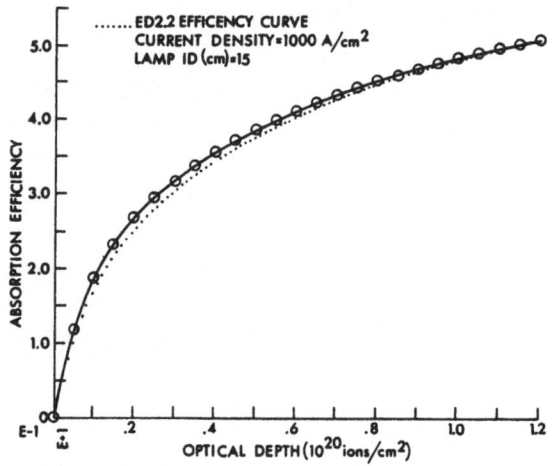

Fig. 2.7. Efficiency of absorption of LHG-5 glass relative to ED-2 glass as a function of optical depth [2.8]

$$A_T = A\left[B(1 - e^{-Cx}) + (1-B)(1 - e^{-Dx})\right] \quad , \tag{2.14}$$

where A, B, C, and D are constants and X is the optical depth. In all of the data summaries, including Fig.2.7, a standard absorption-efficiency curve for ED-2 laser glass is included to compare the absorption of the glass under study to the arbitrary standard. Values of the absorption efficiency $\eta_A$ of a glass relative to ED-2 are also included, where

$$\eta_A = \frac{A_T}{A_{ED-2}} \quad . \tag{2.15}$$

For phosphate compositions, $\eta_A > 1$, whereas for fluorophosphates and fluoroberyllates $\eta_A < 1$. Therefore, in evaluating the relative merits of a laser glass for an application it is possible to know over a wide range of compositions, how a particular glass will absorb relative to a "standard". It is also important to realize, however, that in accordance with the foregoing results, different compositions also display varying E/H values. In the phosphates, for instance, a larger $A_T$ is obtained in phosphates than in silicates, but a larger fraction of the absorbed energy is converted into heat. The separation of $A_T$ into stored-energy density or heat is not presently accounted for in the data sheets [2.8], thus it is difficult to compare widely varying compositions on a first-order basis.

Equation (2.14) is useful for comparing glass compositions when problems such as parasitic oscillations or ASE are present, or thermal effects are important. For parasitic oscillations and ASE the quantity $dE(\nu)/d\nu$ is important (Chap.4), hence

$$\frac{dE(\nu)}{d\nu} \sim \frac{d^2A_T}{d\nu^2} \sim \frac{d^2A_T}{dx^2} \quad , \tag{2.16}$$

or

$$\left|\frac{dE(\nu)}{d\nu}\right| = KA\left[BC^2 e^{-Cx} + D^2(1-B)e^{-Dx}\right] \quad . \tag{2.17}$$

In considerations of thermal effects, $dH/d\nu$ is important. By making the assumption that $E(\nu) \sim H(\nu)$ which is justified only to first order, relative values of $dH/d\nu$ may also be obtained from (2.17).

It is clear from the preceding that the comparison of laser glasses for a particular application is complicated, and in fact should be done by a computer program that takes into account all of the effects mentioned. The number of variables that must be evaluated is somewhat bewildering when many laser glasses are considered. Nevertheless, laser designers have found it profitable to consider figure-of-merit (FOM) calculations to at least reduce the number of glasses considered before a final analysis is done; we will treat such calculations in Sect.2.4.

We now turn to the tabulation of parameters relevant to various laser glasses. Only silicate, phosphate, fluorophosphate, and a few fluoroberyllate compositions are listed because they represent the main laser glasses presently being developed for application in large fusion systems. In Table 2.1, we list the optical parameters for 27 laser glasses. The silicate and phosphate compositions are generally commercially available. The fluorophosphates, which are currently undergoing intensive development, are really experimental compositions chosen at random from the LLL Nd Laser Glass Data Sheets [2.8], as have the data for all other glasses, unless otherwise indicated. The $BeFl_2$ glasses are also experimental and presently unavailable in optical quality in large sizes. In general, all of the glasses included in Table 2.1 were chosen to demonstrate the wide range of available values of the parameters of interest and to show how they change from composition to composition. The passive absorption coefficient $\alpha_L$ has not been included in Table 2.1 because of its wide variation from melt to melt within a single composition space, due to the presence of various impurities. Typical values of $\alpha_L$ are in the range $0.001 \le \alpha_L \le 0.005$ cm$^{-1}$ for commercially available laser glasses. The most glaring lack of data is for the terminal-level lifetime $\tau_T$.

Table 2.2 is concerned with linear and nonlinear refractive indices for a variety of glasses. Many compositions have been chosen not because they are currently commercially important, but to illustrate again the range of

Table 2.1.

| Glass | $\Delta\lambda_e$ [nm] | $\Delta\lambda$ [nm] | $\lambda_p$ [nm] | $\sigma_p$ [$10^{-20}$ cm²] | $\alpha_0$ [cm²/J] | $E_{SAT,3}$ [J/cm²] | $E_{SAT,4}$ [J/cm²] | $\tau_T$ [$10^{-9}$ s] | $\tau_0$ [$10^{-9}$ s] | $\Omega_2$ | $\Omega_4$ [$10^{-20}$ cm²] | $\Omega_6$ | $\beta$ | Commercial source |
|---|---|---|---|---|---|---|---|---|---|---|---|---|---|---|
| *Silicates* | | | | | | | | | | | | | | |
| ED-2 | 34.43 | 27.79 | 1061.0 | 2.698 | 0.130 | 3.85 | 7.70 | 1.25[a] | 359 | 3.24 | 4.59 | 4.80 | 0.484 | Owens-Illinois |
| ED-3 | 34.71 | 28.18 | 1061.0 | 2.872 | 0.154 | 3.25 | 6.50 | | 333 | 3.37 | 4.99 | 5.13 | 0.482 | |
| ED-8 | 34.00 | 28.05 | 1061.0 | 3.024 | 0.162 | 3.09 | 6.17 | | 316 | 2.98 | 4.90 | 5.23 | 0.486 | |
| LSG 91H | 34.40 | 27.41 | 1061.5 | 2.422 | 0.130 | 3.85 | 7.70 | | 412 | 3.49 | 3.99 | 4.40 | 0.489 | Hoya |
| LG 650 | 34.30 | 23.53 | 1057.0 | 1.052 | 0.056 | 8.90 | 17.81 | | 926 | 3.69 | 2.24 | 1.81 | 0.461 | Schott |
| *Phosphates* | | | | | | | | | | | | | | |
| LHG-5 | 26.07 | 21.96 | 1053.0 | 4.148 | 0.221 | 2.27 | 4.53 | | 322 | 4.64 | 5.10 | 5.91 | 0.500 | Hoya |
| LHG-6 | 26.40 | 22.21 | 1053.0 | 4.125 | 0.220 | 2.28 | 4.56 | | 318 | 4.10 | 5.46 | 5.27 | 0.492 | |
| LHG-7 | 26.41 | 22.19 | 1053.0 | 3.924 | 0.209 | 2.40 | 4.79 | | 345 | 5.37 | 5.07 | 5.74 | 0.497 | |
| LHG-8 | 25.91 | 21.76 | 1053.0 | 4.000 | 0.213 | 2.35 | 4.79 | | 338 | 4.40 | 5.10 | 5.60 | 0.493 | |
| Q-88 | 26.27 | 21.89 | 1054.0 | 3.998 | 0.213 | 2.35 | 4.38 | | 326 | 3.33 | 5.12 | 5.63 | 0.494 | Kigre |
| EV-2 | 22.93 | 19.02 | 1054.0 | 4.294 | 0.229 | 2.19 | 4.27 | | 372 | 3.98 | 4.57 | 5.56 | 0.504 | |
| EV-3 | 23.70 | 19.48 | 1054.0 | 4.395 | 0.234 | 2.14 | 5.12 | | 340 | 4.02 | 5.25 | 5.69 | 0.504 | Owens-Illinois |
| LC 700 | 26.62 | 22.33 | 1053.5 | 3.673 | 0.195 | 2.56 | 4.95 | | 363 | 4.22 | 5.63 | 5.40 | 0.500 | |
| LG 710 | 24.92 | 20.87 | 1053.4 | 3.794 | 0.202 | 2.48 | 4.70 | | 367 | 3.84 | 4.52 | 5.14 | 0.497 | Schott |
| *Fluorophosphates* | | | | | | | | | | | | | | |
| LHG 104A | 30.69 | 25.92 | 1051.0 | 2.772 | 0.147 | 3.40 | 6.80 | | 465 | 1.77 | 4.18 | 5.03 | 0.505 | Hoya |
| LHG 104B | 31.25 | 26.56 | 1051.5 | 2.657 | 0.141 | 3.54 | 7.09 | | 464 | 1.71 | 4.03 | 4.83 | 0.504 | |
| LHG 105A | 30.19 | 25.28 | 1050.5 | 2.674 | 0.142 | 3.52 | 7.05 | | 493 | 1.60 | 3.96 | 4.80 | 0.506 | |
| LHG-10 | 31.25 | 26.47 | 1051.0 | 2.670 | 0.142 | 3.53 | 7.06 | | 461 | 1.67 | 4.10 | 4.86 | 0.504 | |
| LHG-11 | 31.2 | 26.44 | 1051.0 | 2.651 | 0.141 | 3.55 | 7.11 | | 468 | 1.51 | 4.14 | 4.81 | 0.502 | |
| FK 50 | 30.90 | 25.98 | 1054.5 | 2.703 | 0.144 | 3.47 | 6.95 | | 444 | 2.06 | 4.20 | 4.62 | 0.495 | Schott |
| LG 813 | 30.95 | 26.05 | 1051.0 | 2.668 | 0.142 | 3.53 | 7.06 | | 488 | 1.86 | 4.05 | 4.94 | 0.506 | |
| E111-1 | 27.69 | 23.38 | 1054.0 | 3.579 | 0.191 | 2.62 | 5.25 | | 368 | 2.70 | 4.83 | 5.54 | 0.498 | |
| E123-1 | 30.41 | 25.58 | 1052.5 | 2.946 | 0.157 | 3.19 | 6.39 | | 417 | 1.91 | 4.34 | 5.10 | 0.502 | Owens-Illinois |
| E133-5 | 30.75 | 26.11 | 1050.5 | 2.511 | 0.133 | 3.75 | 7.51 | | 507 | 1.59 | 3.73 | 4.55 | 0.507 | |
| E181-2 | 30.36 | 26.02 | 1050.0 | 2.563 | 0.136 | 3.68 | 7.36 | | 504 | 1.54 | 3.91 | 4.57 | 0.503 | |
| *Fluoroberyllates* | | | | | | | | | | | | | | |
| B402 | 28.56 | 23.84 | 1047.5 | 2.871 | 0.152 | 3.29 | 6.58 | | 516 | 0.726 | 4.33 | 5.02 | 0.504 | Experimental LLL(UCLA) |
| B101 | 23.19 | 19.35 | 1047.5 | 3.148 | 0.167 | 3.00 | 6.00 | | 613 | 0.226 | 3.92 | 4.60 | 0.505 | Experimental |

*Note:* Unless otherwise indicated, data is taken from 2.8
[a] W.E. Martin, D. Milam: Appl. Phys. Lett. *32*, 816 (1978)

Table 2.2.

| Glass | $n_0^a$ | $n_d^b$ | $\nu_d^b$ | $n_{2,1}^{TM,c}$ [$10^{-13}$ esu] | $\gamma_1^{M,c}$ [$10^{-20}$ m²/W] | $n_{2,c}^{M,C,c}$ [$10^{-13}$ esu] | $\gamma_c^{M,C,c}$ [$10^{-20}$ m²/W] | $n_{2,1}^{ec}$ [$10^{-13}$ esu] | $A_0$ [$10^{16}$]$^c$ | $B_0$ [$10^{16}$]$^c$ | Source |
|---|---|---|---|---|---|---|---|---|---|---|---|
| *Silicates* | | | | | | | | | | | |
| ED-2 | 1.5554 | 1.5716 | 57.5 | 1.52 | 4.05 | 1.056 e | 2.81 e | 1.596 | | | Owens-Illinois |
| LSG-91H | 1.5487 | 1.559 | 57.3 | | | 1.074 d | 2.89 d | 1.546 | | | Hoya |
| *Phosphates* | | | | | | | | | | | |
| EV-1 | 1.5037 | 1.507 | 72.2 | 0.91 | 2.53 | 0.632 e | 2.53 e | 0.934 | | | } Owens-Illinois |
| EV-2 | | 1.507 | 65.8 | | | 0.649 d | 1.804 d | 1.182 | | | |
| LHG-5 | 1.5291 | 1.5382 | 69.4 | 1.16 | 3.15 | 0.806 e | 2.196 e | 1.068 | 51 | 59 | |
| LHG-6 | | 1.531 | 67.6 | 1.01 | 2.76 | 0.701 d | 1.918 e | 1.050 | 37 | 67 | Hoya |
| LHG-7 | 1.5050 | 1.5132 | 70.2 | | | 0.729 d | 2.019 d | 1.234 | | | |
| LHG-8 | | 1.5234 | 64.8 | | | | | | | | |
| Q-88 | 1.5310 | 1.5449 | | 1.15 | 3.11 | 0.799 e | 2.167 e | | | | Kigre |
| *Fluorophosphates* | | | | | | | | | | | |
| LG-812 | | 1.426 | | 0.49 | 1.44 | 0.340 e | 0.999 e | 0.55 | | | Schott |
| A86-82 | | 1.492 | | 0.71 | 2.00 | 0.493 e | 1.384 e | 0.59 | | | Experimental |
| LHG-10VA | | 1.4428 | 88.9 | | | 0.382 d | 1.109 d | 0.50 | | | |
| LHG-104B | | 1.4621 | 89.0 | | | 0.410 d | 1.175 d | 0.50 | | | Hoya |
| LHG-105A | | 1.4376 | 94.5 | | | 0.347 d | 1.011 d | 0.93 | | | |
| E111-1 | | 1.5021 | 71.7 | | | 0.646 d | 1.802 d | 0.70 | | | Owens-Illinois |
| E123-1 | | 1.4845 | 83.5 | | | 0.486 d | 1.372 d | 0.55 | | | |
| E133-5 | | 1.4511 | 91.5 | | | 0.382 d | 1.103 d | 0.53 | | | Owens-Illinois |
| E181-2 | | 1.4427 | 92.2 | | | 0.368 d | 1.069 d | 0.38 | | | |
| B402 | | 1.3838 | 100.0 | | | 0.264 d | 0.799 d | | 6.0 | 14.0 | Experimental (UCLA) |
| B101 | | 1.3459 | 96.0 | 0.43 | 1.34 | 0.299 e | 0.931 e | 0.34 | | | |
| B111 | | 1.3439 | 105.0 | | | 0.208 d | 0.649 d | 0.30 | | | |
| *Optical Glass* | | | | | | | | | | | |
| BK-7 | 1.5032 | 1.517 | 63.4 | 1.24 | 3.43 | 0.861 e | 2.378 e | 1.18 | 32.0 | 35.0 | Schott |
| FK-5 | | 1.487 | | 1.07 | 3.01 | 0.743 e | 2.094 e | | 32.0 | 22.0 | Schott |
| SF-6 | | 1.805 | | 9.13 | 21.2 | 6.34 e | 14.717 e | | 203.0 | 277.0 | |
| BeF2 | | 1.275 | | 0.23 | 0.75 | 0.16 e | 0.526 e | | 6.0 | 4.0 | Corning |
| *Fused Silica* (Type 4000) | | 1.458 | 84.0 | 0.95 | 2.73 | 0.66 e | 1.897 e | | 43.0 | 15.0 | Multiple sources |
| *Faraday Rotator* | | | | | | | | | | | |
| FR-4 | | 1.572 | | 1.96 | 5.23 | 1.361 e | 3.528 e | | | | } Hoya |
| FR-5 | | 1.686 | | 2.09 | 5.20 | 1.451 e | 3.506 e | | | | |
| EY-1 | | 1.626 | | 1.77 | 4.57 | 1.229 e | 3.167 e | | | | Owens-Illinois |
| *Pockels Cell* | | | | | | | | | | | |
| KH2PO4 | | 1.468 | | 1.02 | 2.90 | 0.708 e | 2.021 e | | | | Multiple sources |

a Data taken from [2.6] and S. Jacobs: private communication
b Data taken from [2.8]
c M.J. Weber, D. Milam, W.L. Smith: UCRL-80779 (1978), and Optical Engineering 17, 463 (1978)
d Calculated according to (1.115)
e Calculated using (1.116)

Table 2.3.

| Glass | $\eta_A$ [a] | $\left(\dfrac{dE}{d\nu}\right)_{\nu=0}$ or $\left(\dfrac{dH}{d\nu}\right)_{\nu=0}$ [b] | $\left(\dfrac{E}{H}\right)_{\nu=0}$ [c] | Commercial source |
|---|---|---|---|---|
| *Silicates* | | | | |
| ED-2 | 1.00 | 1.00 | 1.64 | Owens-Illinois |
| LSG-91H | 0.91 | 0.85 | | Hoya |
| *Phosphates* | | | | |
| LHG-5 | 1.18 | 1.31 | 1.50 | }Hoya |
| LHG-6 | 0.89 | 1.56 | | |
| EV-2 | 1.05 | 1.12 | 1.53 | Owens-Illinois |
| Q-88 | 1.10 | 1.88 | | Kigre |
| *Fluorophosphates* | | | | |
| LHG-104A | 0.81 | | | |
| LHG-104B | 0.79 | 0.58 | | |
| LHG-105A | 0.77 | 0.57 | | }Hoya |
| LHG-10 | 0.80 | 0.57 | | |
| LHG-11 | 0.79 | 0.56 | | |
| FK-50 | 0.82 | 0.74 | | Schott |
| *Fluoroberyllates* | | | | |
| B 402 | 0.63 | 0.60 | | }Experimental |
| B 101 | 0.64 | 0.56 | | (UCLA) |

[a] Data taken from [2.8]; [b] Calculated from data in [2.8];
[c] Data taken from [2.4]

parameters available with present compositions. The value of the refractive index is given both at 1.06 µm and the Na "D" lines. Also shown are the Abbe value, the calculated electronic contribution ($n_{2,1}^{ec}$) to the total nonlinear index, the measured values of the total nonlinear index ($n_{2,1}^{TM}$) and the corresponding nonlinear coefficient ($\gamma_1^M$) for linear polarization. The measured values of the nuclear coefficients $A_0$ and $B_0$ are also listed where available. The total nonlinear index for circular polarization $n_{2,c}^{M,c}$ has been calculated for all glasses, as has the nonlinear coefficient $\gamma_c^{M,c}$. Where $n_{2,1}^{TM}$ and $A_0$ and $B_0$ were available, $n_{2,c}^{M,c}$ was calculated according to (1.115). In all other cases, the estimate was made that $n_{2,c}^{M,c} \simeq n_{2,1}^{ec}/1.44$, according to [1.50] and (1.116). In addition to laser glasses, Table 2.2 was expanded to include a number of optical glasses, some of which are commonly used in laser systems for lenses, windows, etc. Three Faraday rotator glasses and one Pockels cell crystal material have also been included for completeness.

Finally, in Table 2.3, a number of relative optical properties have been tabulated. The first is the total absorption efficiency $\eta_A$ relative to ED-2 laser glasses. It can be seen that silicates and phosphates, on the average, are equivalent, but fluorophosphates and fluoroberyllates are comparatively

less efficient in converting Xe flashlamp energy to absorbed energy. The slope
of the inversion density (dE/dν) and heat density (dH/dν) was calculated relative to ED-2 by use of data from [2.8] and (2.17), for $ν = 0$. The phosphates
generally fall off faster with ν than silicates, in agreement with [2.4,7].
The fluorophosphates and fluoroberyllates fall off less rapidly than silicates and phosphates, a fact of consequence for parasitics, ASE, and thermal
effects in amplifiers. We have also listed three values of $(E/H)_{ν=0}$, which
indicate that phosphates produce more heat per unit inversion density than
do silicates.

This concludes our discussion and tabulation of the major glass optical
properties of interest in laser-fusion system design. Most data have been included for the conventional laser glasses most frequently used in current
systems. For glasses not tabulated, the reader is encouraged to consult [2.8]
as well as vendor data sheets.


## 2.2 Fundamental Physical Constants


Section 2.1 and indeed all of Chap.1 were concerned with the optical properties of laser glasses. In most modern laser systems, however, intelligent
design demands consideration of other properties of interest. These include
thermal, photo-elastic, and mechanical parameters, all of which are included
here under the general category of physical constants. The relationship of
these parameters to physical effects of interest in various amplifiers will
be discussed in Sect.2.3 below. It is important to realize, however, that
which parameters the designer considers important or chooses to ignore depends
upon the end use of the laser system being designed. For systems such as
ARGUS or SHIVA at LLL which make extensive use of disk amplifiers and have
an associated thermal cycling time of 3-4 hours, only optical effects have
been considered to achieve the largest peak power/beamline. In the LLE at
the University of Rochester, however, a number of laser systems have been
constructed (GDL, ZETA, OMEGA) that have been designed to be user oriented.
Repetition rate was considered to be of importance; a firing was required
every half hour, hence consideration of thermal effects was included in the
design process and selection of laser glass.

We begin by listing the parameters to be considered here, along with a
brief explanation of why they are considered important. Details may be found
in Sect.2.3.

*Thermal Properties*

K        Thermal conductivity, important in considering the ability of a glass to dissipate heat, [cal/cm s °C].

c        Specific heat, partially determines the ability of a glass to dissipate heat and induced temperature change for unit heat input, [cal/g °C].

$\alpha$        Thermal-expansion coefficient, partially determines the change of linear index with temperature change, important in determining the curvature in an active-mirror amplifier induced by Xe flashlamp pumping, [$°C^{-1}$].

$K/\rho c$        Thermal diffusivity, $\rho$ is glass density [$g/cm^3$], determines the thermal decay time constant or heat removal time, [$cm^2/s$].

*Mechanical Properties*

Y        Young's modulus, determines the stiffness of a laser disk, related to glass hardness and photoelastic constants, [$kg/cm^2$].

p        Poisson's ratio, needed to calculate photoelastic constants, dimensionless.

$\rho$        Density, needed to evaluate $K/\rho c$, partially determines the induced temperature change for a given heat input, [$g/cm^3$].

H        Knoop Hardness, determines the abrasion resistance of a glass, directly related to rupture strength [see (2.63) below], [$kg/cm^2$].

*Thermo-Optic and Elasto-Optic Properties*

$\beta = \dfrac{dn}{dT}$        Refractive-index temperature coefficient, partially determines the change of linear index with induced temperature change, [$°C^{-1}$].

$\beta + \alpha(n-1)$        Thermo-optic coefficient, in the absence of stress in a glass gives the total change of linear index for a given $\Delta T$, [$°C^{-1}$].

$B_\perp, B_\parallel$        Photoelastic constants, partially determine the change of index or optical pathlength with applied stress in the perpendicular ($B_\perp$) or parallel ($B_\parallel$) directions, [$(nm/cm)/(kg/cm^2)$].

B        Stress-optic coefficient, given by $B = B_\perp - B_\parallel$, [$(nm/cm)/(kg/cm^2)$].

These constants represent the data needed to describe completely all of the known physical effects that occur in a typical laser system. Unfortunately, until recently the only data available were for silicate compositions, and were provided by commercial manufacturers who measured them under widely varying conditions and used different experimental techniques. Much as LLL has developed a single spectroscopic and laser facility to characterize the optical and nonlinear properties of laser glasses, the LLE has a similar program, GLAMP (Glass, Lamp and Materials Program) which has recently provided the first consistent data on the physical properties of silicate and phosphate laser glasses. That work is currently being continued to measure the physical properties of fluorophosphate and fluoroberyllate compositions. In particular, BROWN et al. [2.10] have presented physical data for two silicate and three phosphate glasses. A more comprehensive listing of physical properties will be presented here, with much data taken from unpublished results by JACOBS and RINEFIERD [2.11]. Table 2.4 lists values of the parameters for eleven laser glasses as well as for the commonly used commercial glass BK-7 and clear fused quartz. Some features of this table should be noted. Thermal conductivity K is greatest for silicate glasses; on the average, it is considerably less for phosphates, which results in a reduced capacity for heat dissipated in the latter. The specific heat c, can be seen to be virtually independent of composition. Thermal expansion, specified by $\alpha$, varies widely and is least in CFQ. It is also generally greater in phosphates than in silicates, which affects such problems as thermal lensing in active-mirror amplifiers and the application of dielectric coatings to such devices (see Chap.5). Young's modulus Y is greatest in silicates and considerably less in phosphate glasses; the value of Y directly affects the stiffness of a disk and determines the amount of possible mounting distortion of a disk. Poisson's ratio p is approximately independent of composition. Density $\rho$ is also approximately independent of composition, except for the fluorophosphates. The Knoop Hardness H is maximum for CFQ and least for the fluoroberyllate 816KY. The phosphates are generally much softer than the silicates, which considerably increases the difficulty in optically polishing phosphate components. The change of linear index with temperature $\beta$ was measured in some cases at two wavelengths, YAG (1.06 $\mu$m) and He-Ne (633 nm) to illustrate the lack of significant wavelength dispersion. It should be noted that $\beta$ may be positive or negative; it is positive for the two silicates measured (ED-2, LSG-91H), negative for the phosphates (EV-2, Q-88, LHG-5, LHG-7, LG-700, LG-710). As a consequence of this the thermo-optic coefficient $\beta+\alpha(n-1)$ may also be positive or negative, depending mostly upon the magnitude of $\alpha$. Because $\beta+\alpha(n-1)$ represents the change of optical pathlength per unit temperature change, it is easy to see that in the absence of stress, a temperature profile in, say, a rod amplifier, which is minimum in the center and maximum at the edge, causes radial focusing or defocusing of a plane parallel wavefront propagating through such a device. It has been shown [2.12] that the use of alternate phosphate compositions in a rod laser system can partially compensate the effects of thermally induced wavefront distortion. Finally, Table 2.4 also shows that the stress-birefringence coefficients $B_\perp$ and $B_\parallel$ are greater in phosphate than in the silicate compositions, but the stress-optic coefficient B is generally less. A consequence of this is that, although phosphate glasses dissipate heat

Table 2.4. Glass physical data

| Parameter | ED-2 | LSG-91H | EV-2 | Q-88 | LHG-5 | LHG-7 | LG-700 | LG-710 | LHG-104B | LG-800 | 816 KY | BK-7 | CFQ | Data Source |
|---|---|---|---|---|---|---|---|---|---|---|---|---|---|---|
| *Thermal* | | | | | | | | | | | | | | |
| $K[10^{-3}$ cal/cm s °C] | 3.0 | 2.5 | 1.08 | 1.76 | 1.75 | 1.46 | 1.65 | 1.12 | 2.00 | 1.95 | 1.83 | 2.66 | 2.35 | [2.11,12] and Vendor data |
| $C$[cal/g °C] | 0.22 | 0.15 | 0.15 | 0.21 | 0.17 | 0.17 | 0.20 | 0.17 | | | | 0.21 | 0.17 | Vendor data and [2.12] |
| $\alpha[10^{-7}$/°C] | 77 | 84 | 141 | 95 | 80 | 95 | 101 | 126 | 135 | 130 | 158 | 67 | 5 | [2.11,12] |
| $K/\rho c[10^{-3}$ cm²/s] | 5.4 | 6.0 | 2.7 | 3.2 | 3.9 | 3.3 | 3.17 | 2.3 | | | | 511 | 8.9 | |
| *Mechanical* | | | | | | | | | | | | | | |
| $Y[10^5$ kg/cm²] | 9.62 | 8.61 | 4.26 | 7.12 | 7.26 | 6.05 | 6.43 | 4.59 | 7.39 | 7.96 | | 8.30 | 7.41 | [2.11,12] |
| $p$ | 0.24 | 0.24 | 0.24 | 0.24 | 0.24 | 0.19 | 0.24 | 0.24 | 0.27 | | 0.29 | 0.21 | 0.17 | Vendor data |
| $\rho$[g/cm³] | 2.51 | 2.77 | 2.72 | 2.67 | 2.64 | 2.60 | 2.60 | 2.91 | 3.64 | 3.74 | 3.44 | 2.51 | 2.21 | [2.11,12] |
| $H[10^4$ kg/cm²] | 6.12 | 4.78 | 2.65 | 4.18 | 4.23 | 3.67 | 3.72 | 2.61 | 3.55 | 3.60 | 1.66 | 5.98 | 6.56 | [2.11,12] and Vendor data |
| *Thermo– and Elasto-Optic* | | | | | | | | | | | | | | |
| $\beta = \dfrac{dn}{dT}$ $[10^{-7}$/°C] (1.06μ) | +38.0 | +50.0 | -108.0 | -5.0 | -1.0 | -29.0 | -18.0 | | | | | +27.0 | +89.0 | [2.11,12] |
| $[10^{-7}$/°C] (633nm) | | | -111.0 | -6.0 | -2.0 | | -31.0 | -72.0 | | -80.0 | | +34.0 | | [2.11,12] |
| $\beta+\alpha(n-1)$ (1.06μ) $[10^{-7}$/°C] | +74.0 | +73.0 | -29.0 | +34.0 | +40.0 | +22.0 | +20.0 | +28.0 | | +6.0 | | +67.0 | +91.0 | |
| $B_\perp$ [(nm/cm)/(kg·cm²)] | 2.25 | 3.40 | 8.38 | 5.00 | 3.91 | 4.90 | | | | | | | | [2.11,12] |
| $B_\parallel$ (633nm) [(nm/cm)/(kg·cm²)] | 0.24 | 1.05 | 6.68 | 2.64 | 1.71 | 2.44 | | | | | | | | [2.11,12] |
| $B$ (1.06μ) [(nm/cm)/(kg·cm²)] | 1.92 | 1.99 | 1.45 | 2.07 | 2.17 | 2.14 | 2.06 | 3.09 | 0.54 | 0.59 | 0.55 | 2.59 | 3.22 | [2.11,12] |

slower (smaller K) and produce more heat per unit inversion density (E/H), they are also less sensitive to the effects of stress-induced birefringence [2.10]. Many of the effects mentioned will be treated in more detail in the following section.


## 2.3  Figures of Merit for Laser Glasses

We have, thus far in this chapter, listed the important optical and physical properties of laser and optical glasses and tabulated data for a number of compositions, including most of those that are at present commerically important. We now turn to the standard method used for first-order analysis of the suitability of a laser or optical glass for a particular system. The procedure involves an assessment of the important physical phenomena involved in a given situation and leads to the definition of figures of merit for each optical component considered for use in a high-peak-power laser system. This includes amplifiers (we include three types: rods, disks, and active mirrors), lenses, windows, Faraday rotators, and Pockels cells. The special function of each of these devices will be discussed briefly here but in more detail in Chaps.5 and 8. In what follows, we first discuss the simplest case of amplifiers that operate in a pump or parasitic-limited mode and the corresponding figures of merit (FOM) for rod, disk, and active-mirror amplifiers (Sect.2.3.1), followed by the FOM for passive (nonamplifying) devices such as lenses, windows, Faraday rotators, and Pockels cells (Sect.2.3.2). In Sect.2.3.3 we show a general FOM for laser glasses, which includes a consideration of physical as well as optical parameters. An evaluation of the general FOM in selected cases is shown in Sect.2.3.4.

### 2.3.1  Parasitic and Pump-Limited Figures of Merit

We first consider the calculation of a FOM for the three major types of amplifiers now in common use in most Nd:glass laser systems. They are the rod amplifier, disk amplifier, and active-mirror amplifier, all of which will be described in detail in Chap.5. A restriction here is that only cross-section ($\sigma$) [or specific gain coefficient ($\alpha_0$)] and nonlinear effects are considered important ($n_2$). The resulting FOM were first described by colleagues at LLL [2.13] and are useful in evaluating laser glasses for a system such as ARGUS or SHIVA where other effects, such as repetition rate, are not considered.

The first factor $F_1$ of obvious importance in any amplifier is

$$F_1 = \sigma \qquad (2.18)$$

because maximization of the gain (small-signal or saturated) is of interest. There are, however, two important cases. The first is that in which the gain or stored-energy density is limited by the available pump source. That is, if it is possible to increase the optical pumping, a corresponding increase of stored-energy density can result. Such a condition is said to be *pump limited*, and (2.18) applies to that case. The second is said to be *parasitic limited*; it is possible, if the grain across the width of a disk becomes great enough, to sustain an oscillation that results from the finite reflectance at the edge of the disk. The resulting parasitic oscillation will then lead to a depletion of gain above a threshold that may be predetermined (see Chap.4). Consequently, any increase of optical pumping will result in little or no net increase of gain. In this case, the cross section $\sigma$ becomes unimportant and we set $F_1 = 1$.

The second factor in the FOM results from the analysis of the power-handling capability of an amplifier. It has been shown [2.14] that the ultimate power $P_u$ obtainable from a given component in a system whose power output is limited by the existence of the cubic nonlinearity $n_2$ is given by

$$P_u = X \Delta B F \quad , \qquad (2.19)$$

where $\Delta B$ is the B integral accumulated in a given component (or stage), F the fill factor, and X the so-called X factor. The derivation of the X factor will be given in Sect.2.3.3. We mention, however, that the second factor $F_2$ of importance in the FOM is $F_2 \propto X$ which may be shown to be, for each amplifier

$$F_2 = \frac{n_d}{n_2^T} \quad \text{(rod amplifier, active-mirror amplifier)} \qquad (2.20)$$

$$F_2 = \frac{n_d^2}{n_2^T} \quad \text{(disk amplifier)} \quad . \qquad (2.21)$$

The difference between (2.20,21) is the linear index $n_d$ which results from the larger area of a disk placed at Brewster's angle (the area is larger by the factor $\sqrt{n_d^2 + 1} \simeq n_d$). The total FOM $\eta$ is defined as

$$\eta = F_1 F_2 \quad , \qquad (2.22)$$

Table 2.5.

| Component | Figure of merit $\eta$ |
|---|---|
| Rod amplifier<br>Active-mirror amplifier<br>(pump limited) | $\sigma n_d / n_2^T$ |
| Rod amplifier<br>Active-mirror amplifier<br>(parasitic limited) | $n_d / n_2^T$ |
| Disk amplifier<br>(pump limited) | $\sigma n_d^2 / n_2^T$ |
| Disk amplifier<br>(parasitic limited) | $n_d^2 / n_2^T$ |

which, when evaluated by use of the foregoing results, gives Table 2.5. These
FOM are in fact only a subset of a more general FOM, which will be presented
in Sect.2.3.3. It is fairly obvious that in selecting glass for a pump-limited
amplifier, a glass with a maximum $\sigma$ and a minimum $n_2$ should be selected, while
on the parasitic-limited case $\sigma$ may be compromised, consistent with the value
of that limit, to obtain a low $n_2$ glass.

## 2.3.2 Figures of Merit for Optical Components

In any real laser system, a variety of optical components are used in addition to amplifiers. They include lenses used in spatial filters, plates which
may be windows on Pockels cells, Faraday rotators, and Pockels cells. Since
all of these devices are nonamplifying, only the factor $F_2$, derivable from
the X factor, is important and $\eta = F_2$. Once again, anticipating results which
are derived in Chap.7, we list here the FOM or $F_2$ for four optical components.
   A Faraday rotator is a device used for isolation in a laser system. In addition to providing isolation between amplification stages, where necessary
to prevent oscillation, they are commonly used to provide protection from
pulses back-reflected from a target. The devices operate by the Faraday effect,
in which certain materials subjected to a magnetic field H rotate the plane
of polarization of linearly polarized light through an angle $\theta$ according to

$$\theta = VHl \quad , \tag{2.23}$$

where V is the Verdet constant [arc min/Gauss-cm], H the component of the
magnetic field in the direction of propagation [Gauss] and l the pathlength
of the light in the material [cm]. For a Faraday rotator, it is of importance

Table 2.6.

| Component | Figure of merit $\eta$ |
|-----------|------------------------|
| Lens | $n_d(n_d-1)/n_2^T$ |
| Window | $n_d/n_2^T$ |
| Faraday rotator | $V\, n_d/n_2^T$ |
| Pockels cell | $R\, n_d/n_2^T$ |

to reduce the path length in the material 1 to minimize nonlinear effects. That is accomplished by maximizing the Verdet constant V, which is therefore included in the FOM in Table 2.6. Pockels cells are devices that utilize the Pockels effect, in which an electric field $E_Z$ [V/cm] applied along the direction of light propagation (Z) causes a difference of indices of refraction for light polarized in the x and y directions, according to

$$\Delta n = n^3 R E_Z \quad , \tag{2.24}$$

where R is the electro-optic coefficient [cm/V] and n the linear index of the material. The phase difference $\delta$ in a crystal of length 1 is given by

$$\delta = \frac{2\pi}{\lambda} n^3 R V_Z \quad , \tag{2.25}$$

where $V_Z = E_Z 1$. With the axes of the analyzer oriented orthogonal to the input polarization direction, the phase and transmitted light intensity I are related according to

$$I = I_0 \sin^2\!\left(\frac{\delta}{2}\right) \quad , \tag{2.26}$$

where $I_0$ is the incident intensity. Pockels cells have found wide use in laser systems to suppress ASE, which might otherwise arrive at the target for hundreds of microseconds before the main laser pulse and either destroy it or reduce its efficiency. As in all other optical components in a laser system it is necessary to minimize the length 1 of a crystal and nonlinear effects and that is accomplished by maximizing R. We present in Table 2.7 an evaluation of the mentioned figures of merit, excluding Pockels cells and Faraday rotators, as computed by workers at LLL [2.13]. The values are nor-

Table 2.7. Figures of merit of optical materials [2.12]

| Component | Figure of merit | Silicate glass (ED2-2) | Phosphate glass (EV-2) | Fluoro-phosphate glass (LG-812) | Fluoro-beryllate glass (B-101) |
|---|---|---|---|---|---|
| Laser rod (pump limited) | $n/n_2$ | 1.0 | 2.3 | 2.7 | 4.1 |
| Laser disk (pump limited | $\sigma n^2/n_2$ | 1.0 | 2.2 | 2.5 | 3.6 |
| Laser disk (parasitic limited) | $n^2/n_2$ | 1.0 | 1.4 | 2.3 | 3.1 |
| Window | $n/n_2$ | 1.0 | 1.4 | 2.3 | 3.6 |
| Lens | $n(n-1)/n_2$ | 1.0 | 1.3 | 2.0 | 2.2 |

malized to ED-2 laser glass and represent parameters of advanced laser glasses
in the four major composition spaces of interest at the present time. Although
the phosphate, fluorophosphate, and fluoroberyllate compositions would seem
to offer a substantial improvement in short-pulse laser performance according
to the FOM given in Table 2.7, in practice, other important factors must be
evaluated as well. As an example, it has been demonstrated that the use of
phosphate laser glass in a system leads to a substantial increase in performance as predicted by Table 2.7; the same may not be true, however, for the
fluoroberyllates. The latter typically absorb only about 64% as much of the
flashlamp radiation as silicates and phosphates according to Table 2.3, and
have low cross sections. To find out as accurately as possible what the real
system-performance improvement will be, it is necessary to do a computer analysis that includes all of the relevant parameters. Such an analysis, as
applied to disk and active-mirror amplifiers, will be presented in Chap.5.

## 2.3.3 General Figures of Merit for Rod, Disk, and Active-Mirror Amplifiers

We have seen in the previous section how it is possible to define a FOM for
a given optical component in a laser system and evaluate it for selected
laser glasses. In general, however, other physical effects are important in
the selection of laser glasses for a system application. The FOM derived previously are too restrictive, and consider only gain and nonlinear properties.
Those results have been extended to include other effects in rod amplifiers,
and were given recently by BROWN et al. [2.10]. A FOM $\eta$ may be simply expressed
in its most general form by an equation of the form

$$\eta = \prod_{i=1}^{n} (J_i)^{W_i} \quad , \tag{2.27}$$

where n separate physical effects are described by the $J_i$ factors. $W_i$ is a weighting exponent which may be used to describe the relative importance of a $J_i$ factor in the general FOM ($0 \le W_i \le 1$). In this section we present the $J_i$ factors important to rod, disk, and active-mirror amplifiers that operate in the short pulse or high-peak-power regime. Finally, in Sect.2.3.4 those factors will be evaluated in selected cases of interest and the FOM introduced previously in Sect.2.3.2, will be shown to be special cases of (2.27).

### $J_1$

The first factor to be considered here is $J_1$, which involves the nonlinear effects known to plague high-peak-power glass lasers; the quantity of interest is the B integral, defined previously by (1.90). We evaluate the B integral in the case of a rod amplifier whose entire length L is uniformly inverted with a gain $\alpha$ [cm$^{-1}$]. The initial intensity [W/cm$^2$] entering the amplifier is $I_0$, and in the small-signal limit, (1.90) becomes

$$B = \frac{2\pi}{\lambda_p} C \frac{n_2^T}{n_d} \int_0^L I_0 \, e^{\alpha z} \, dz \tag{2.28}$$

or

$$B = 2\pi C \left( \frac{n_2^T}{n_d \lambda_p} \right) \frac{I_0}{\alpha} (e^{\alpha L} - 1) \quad . \tag{2.29}$$

Defining the small-signal gain $G_0 = \exp(\alpha L)$, and noting that the power $P_0$ is given by

$$P_0 = I_0 A = I_0 \left( \frac{\pi D^2}{4} \right) \quad , \tag{2.30}$$

where D is the rod diameter and A the rod face area, and the output power P is,

$$P = G_0 P_0 \quad , \tag{2.31}$$

we can rewrite (2.29) as

$$B = \frac{K n_2^T P}{n_d D^2 \alpha \lambda_p} \left( 1 - \frac{1}{G_0} \right) \quad . \tag{2.32}$$

If we define

$$X = \frac{P}{B} \quad ,$$ (2.33)

we have

$$X = K \frac{\alpha n_d D^2}{n_2^T \lambda_p} \left(1 - \frac{1}{G_0}\right)^{-1} \quad .$$ (2.34)

As has been noted previously, the X factor is a measure of the ultimate power output of an amplifier in terms of the accumulated B integral, and will be investigated further in Chap.7. It is clear, however, that to maximize the peak-power output of a rod amplifier, it is necessary to maximize the ratio $n_d/n_2^T \lambda_p$, here $J_1^R$ for a rod amplifier is

$$J_1^R = \frac{n_d}{n_2^T \lambda_p} \sim \frac{n_d}{n_2^T} \quad ,$$ (2.35)

where, because $\lambda_p$ varies little from composition to composition compared to $n_d/n_2^T$, it is normally neglected. For an active-mirror amplifier, it can be shown that the correct FOM factor $J_1$ is the same as (2.35) for rods, hence $J_1^A = J_1^R$. For a disk amplifier, it is easily shown that the area $A_0$ of a disk at Brewster's angle is $\approx n_d$ times the area orthogonal to beam propagation, hence from (2.29-31) we have

$$X = K \frac{\alpha n_d^2 D^2}{n_2^T \lambda_p} \left(1 - \frac{1}{G_0}\right)^{-1} \quad ,$$ (2.36)

so that $J_1^D$ for disks is

$$J_1^D = \frac{n_d^2}{n_2^T \lambda_p} \sim \frac{n_d^2}{n_2^T} \quad .$$ (2.37)

In comparing $J_1$ factors for rods and active-mirrors and disks, it should be noted that the extra factor $n_d$ in (2.37) [typically $n_d \sim 1.50$] gives disks an advantage in terms of ultimate power-handling capability and is due to the larger area, by a factor of $n_d$, than, say, a rod amplifier. That fact is somewhat mitigated, however, by the observation that a disk amplifier cannot propagate circular polarization, whereas rods and active mirrors operated at small angles of incidence may. Consequently, because $n_{2c} \sim n_{2l}/1.44$ (1.116), the nonlinear index may be reduced for circular polarization in (2.35).

$J_2$

The second factor, .common to all types of amplifiers, is obtained from (2.34, 36) and numerically equal to the specific gain coefficient $\alpha_0$. Hence

$$J_2^R = J_2^D = J_2^A = \alpha_0 \quad . \tag{2.38}$$

To maximize $\alpha$ in an amplifier means that for a given desired gain one may minimize the path length L in glass, or from (2.28) the B integral. By (1.24), this is equivalent to maximizing $\alpha_0$, and we ignore the passive loss coefficient.

$J_3$

Another factor common to all three amplifiers is the quantum efficiency $\eta$ of the $^4F_{3/2}$ level, which we wish to maximize for a given wt-% Nd doping. The principal reason for this is that to maximize the stored-energy density, $\eta$ should be maximized. When $\eta$ is maximized, so is the radiative lifetime $\tau$, thus the maximum energy storage is obtained. Then

$$J_3^R = J_3^D = J_3^A = \eta \quad . \tag{2.39}$$

$J_4$

In Chap.1 we considered the branching ratio $\beta$ for the $(^4F_{3/2} \rightarrow {}^4I_{11/2})$ transition located at 1.06 $\mu$m in $Nd^{3+}$. To obtain the maximum inversion which decays via that transition, it is necessary to maximize $\beta$ in all amplifiers, and thus to minimize losses that occur principally through the $^4F_{3/2} \rightarrow {}^4I_{9/2}$ transition at 880 nm. The factor $J_4$ is then

$$J_4^R = J_4^D = J_4^A = \beta \quad . \tag{2.40}$$

$J_5$

It has been seen previously that the relative absorption efficiency $\eta_A$ of a glass may change significantly from composition to composition (Table 2.3). Although different fractions of the absorbed energy go into stored-energy density for different glass compositions, to first order it is desirable to maximize $\eta_A$, thus

$$J_5^R = J_5^D = J_5^A = \eta_A \quad . \tag{2.41}$$

*J₆*

In Sect.2.1 we discussed the relevance of calculating the slope $(dE/d\nu)_0$ of the stored-energy density at zero optical thickness, which may change from composition to composition. We have not yet treated the subject of parasitic oscillations and ASE for which we refer the reader to Chap.4. To reduce the likelihood of a surface parasitic or significant ASE losses in disk or active-mirror amplifiers, it is desirable to minimize $(dE/d\nu)_0$. In those devices, beam propagation is in general in the same direction as the stored-energy density gradient. In rod amplifiers, however, the propagation direction is orthogonal to the gradient, leading to a radially dependent gain profile. In Chap.8, we will discuss the consequences of a radial profile $(dE/d\nu)$ in terms of beam propagation. It will be shown that in that case it is also of interest to minimize $(dE/d\nu)_0$. The second consideration here is that $(dH/d\nu) \sim (dE/d\nu)$; hence, minimizing $(dE/d\nu)$ means that the heat gradient is also minimized. This is important in disk amplifiers in which thermal lensing is caused by $dH/d\nu$, in active-mirror amplifiers in which induced curvature is minimized by minimizing $dH/d\nu$, and in rod amplifiers in which to minimize $dH/d\nu$ is to minimize induced birefringence losses. Then for all three amplifiers

$$J_6^R = J_6^D = J_6^A = \left| \frac{dE}{d\nu_0} \right|^{-1} . \tag{2.42}$$

*J₇*

In Sect.2.1, we examined the ratio $(E/H)_0$ of stored-energy density to heat density in a glass; the ratio was shown to be a weak function of $\nu$, and also that the quantity is composition dependent. It is also desirable to maximize that ratio in any amplifier; doing so means that H is minimized, so that at the same time the magnitude of $dH/d\nu$ is minimized. For all amplifiers, we have

$$J_7^R = J_7^D = J_7^A = \left( \frac{E}{H} \right)_0 . \tag{2.43}$$

*J₈*

The heating that occurs in all laser glasses is due to two sources. The first is the quantum defect between the absorbed and emitted photons. Another major contribution, however, is the background absorption due to the glass host itself. All glasses display a so-called ultraviolet edge, generally located in the range of (250-400 nm), where the optical density of the glass is high. Because most Xe flashlamps emit radiation well into the ultraviolet region

and such radiation is normally only partially filtered by lamp envelopes,
$H_2O$ jackets, etc., considerable heating may occur from the direct absorption
of such radiation by the ultraviolet edge. Owing to the high optical density
of glass in the ultraviolet region, heat is deposited in a short distance
(typically a few nm), which can result in severe birefringence losses in rod
amplifiers. Thermal lensing in disk amplifiers and active-mirror amplifiers
may also be enhanced by such heat, hence in all cases it is of interest to
reduce the absorption. The ultraviolet edge changes significantly from composition to composition, and is shifted towards the blue in phosphates, as
compared to silicates. The same is true of fluorophosphates. The fluoroberyllate glasses are optically clear down to below 250 nm. In an attempt to
quantify this effect, we define the wavelength $\lambda_{UV}$ of the ultraviolet edge
as the spectral position where the glass background absorption has become
more than 0.5 optical density unit per cm. The appropriate FOM for all three
amplifiers is then

$$J_8^R = J_8^D = J_8^A = |\lambda_{UV}|^{-1} \quad .$$
(2.44)

$J_9$

In many applications, it is necessary to consider the repetition-rate performance of a laser system that is limited primarily by thermal effects. The
thermal conductivity of glass is low compared to crystals or metals. To remove heat rapidly requires that the characteristic thermal-removal length l
for an amplifier be minimized. A thermal analysis of rod, disk, and active-
mirror amplifiers shows [2.12] that to first order, the initial temperature
$T_0$ decays according to

$$T \sim T_0 \, e^{-t/\tau} \quad ,$$
(2.45)

where the thermal decay time constant $\tau$ is given by

$$\tau \sim \left(\frac{\rho c}{K}\right) l^2 \quad .$$
(2.46)

The characteristic thermal-removal length l may be taken to be the thickness
of a disk or active mirror and the radius of a rod amplifier. Because we wish
to minimize the thermal recovery time $\tau$, we should maximize the thermal diffusivity $K/\rho c$, regardless of what cooling process is used in a diffusion limited amplifier. Therefore, for all three amplifiers

$$J_9^R = J_9^D = J_9^A = \frac{K}{\rho c} \quad . \tag{2.47}$$

$J_{10}$

The induced temperature change $\Delta T$ in an amplifier volume V due to heat input $\Delta H$ is

$$\Delta T = \Delta H / \rho c \, V \quad . \tag{2.48}$$

To reduce the effects of thermal distortions in any amplifier it is clearly desirable to minimize $\Delta T$ for a given $\Delta H$. Then $\rho c$ should be maximized and

$$J_{10}^R = J_{10}^D = J_{10}^A = \rho c \quad . \tag{2.49}$$

$J_{11}, J_{12}, J_{13}$

The three factors to be presented here are all involved with various thermo-optic effects that take place in rod, disk, or active-mirror amplifiers. Although we will not give the detailed results here (the reader is referred to Chap.5), the relevant results will be presented. To be considered first is the rod amplifier, followed by the disk and active-mirror amplifiers which are treated as one because disk and active-mirror amplifiers have the same geometry.

We derive a FOM for rod amplifiers by noting three effects that cause optical distortion. One is a finite change in the index of refraction due to $\beta(dn/dT)$; another is elongation of the rod due to the thermal expansion coefficient $\alpha$; the third, due to the thermal gradient, is stress, which changes the index and leads to birefringence. In general, the optical path length $P_r(r)$ for radial polarization is different from that for tangential polarization $P_\theta(r)$. The analysis will be presented in Chap.5, where it will be shown that the average change of optical path length $\bar{P}$ for a rod amplifier (per unit length — per unit temperature change) is given by

$$\bar{P} = \frac{\Delta P_r'' + \Delta P_\theta''}{2} = 2\beta + \frac{\alpha Y}{(1-p)} (3B_\perp + B_{\parallel}) \quad , \tag{2.50}$$

where $\beta$, $\alpha$, $Y$, $P$, $B_\perp$, and $B_{\parallel}$ have all been defined previously. The average change of optical path length $\bar{P}$ may be positive, negative, or zero and, depending upon the material, parameters may lead to focusing or defocusing of a beam traveling through a rod amplifier. We wish to minimize this effect and take the FOM $J_{11}^R$ to be

$$J_{11}^R = |\Delta P_r'' + \Delta P_\theta''|^{-1} \quad , \tag{2.51}$$

where we neglect the factor of two. This effect does not occur, to first order, in disk or active-mirror amplifiers; hence we set $J_{11}^D = J_{11}^A = 1$.

Another important phenomenon in rod amplifiers is thermally induced birefringence. This is investigated by calculating the optical path difference $P_D$ between r and θ polarized light. Derivation shows that

$$P_D = P_r'' - P_\theta'' = \frac{\alpha Y}{(1-p)} (B_\perp - B_{||}) \tag{2.52}$$

or, because the stress optic coefficient B is

$$B = B_\perp - B_{||} \quad , \tag{2.53}$$

we have

$$P_D = \frac{\alpha Y B}{(1-\nu)} \quad . \tag{2.54}$$

It is desirable to minimize birefringence losses, so we take

$$J_{12}^R = |\Delta P_r'' - \Delta P_\theta''|^{-1} \sim B^{-1} \tag{2.55}$$

as the FOM for a rod amplifier. Birefringence is also important in disk amplifiers, hence $J_{12}^D = J_{12}^R$, but not in active-mirror amplifiers for any angle of incidence; therefore we have $J_{12}^A = 1$. The final FOM, $J_{13}$, is related to disk warpage due to thermal gradients. If z is the thickness dimension, there is a nonuniform temperature profile T(z) in both active-mirror and disk amplifiers. The temperature distribution causes thermally induced focusing, which is caused by material distortions in those devices. The radius of curvature R of the induced distortion of an active mirror is given by

$$R = \left[ \frac{6\alpha}{l^2} \int_0^1 T(z)dz - \frac{12\alpha}{l^3} \int_0^1 zT(z)dz \right]^{-1} \quad , \tag{2.56}$$

where l is the active-mirror or disk thickness. We wish to maximize R so that, regardless of the temperature profile, α should be minimized. Then for active mirrors and, to first-order, disk amplifiers, the appropriate FOM is

$$J_{13}^D = J_{13}^A = \alpha^{-1} \quad . \tag{2.57}$$

Table 2.8. $J_i$ factors — short pulse

| Factor | Rod | Disk | Active mirror |
|---|---|---|---|
| $J_1$ | $\dfrac{n_d}{\dfrac{T}{n_{2\lambda p}}}$ | $\dfrac{n_d^2}{\dfrac{T}{n_{2\lambda p}}}$ | $\dfrac{n_d}{\dfrac{T}{n_{2\lambda p}}}$ |
| $J_2$ | $\alpha_0$ | $\alpha_0$ | $\alpha_0$ |
| $J_3$ | $\eta$ | $\eta$ | $\eta$ |
| $J_4$ | $\beta$ | $\beta$ | $\beta$ |
| $J_5$ | $\eta_A$ | $\eta_A$ | $\eta_A$ |
| $J_6$ | $\dfrac{dE}{d\nu_0}^{-1}$ | $\dfrac{dE}{d\nu_0}^{-1}$ | $\dfrac{dE}{d\nu_0}^{-1}$ |
| $J_7$ | $\left(\dfrac{E}{H}\right)_0$ | $\left(\dfrac{E}{H}\right)_0$ | $\left(\dfrac{E}{H}\right)_0$ |
| $J_8$ | $\lambda_{UV}^{-1}$ | $\lambda_{UV}^{-1}$ | $\lambda_{UV}^{-1}$ |
| $J_9$ | $\dfrac{K}{\rho c}$ | $\dfrac{K}{\rho c}$ | $\dfrac{K}{\rho c}$ |
| $J_{10}$ | $\rho c$ | $\rho c$ | $\rho c$ |
| $J_{11}$ | $|\Delta P_r + \Delta P_\theta|^{-1}$ | 1 | 1 |
| $J_{12}$ | $|\Delta P_r - \Delta P_\theta|^{-1}$ | $|\Delta P_r - \Delta P_\theta|^{-1}$ | 1 |
| $J_{13}$ | 1 | $\dfrac{1}{\alpha}$ | $\dfrac{1}{\alpha}$ |

This effect does not occur in rod amplifiers, so $J_{13}^R = 1$.

We have shown the derivations of the $J_i$ factors that have been found to be important in the evaluation of laser glasses for short-pulse applications using rod, disk, or active-mirror amplifiers. Other factors may be included where required; those mentioned in the foregoing are summarized in Table 2.8. The evaluation of these FOM for selected cases will be shown in the following Sect.2.3.4.

## 2.3.4 Evaluation of Figures of Merit — Selected Cases

We have seen in Sect.2.3.3 how it is possible to formulate a general FOM for rod, disk, and active-mirror amplifiers and define the individual $J_i$ factors for each. Although the list of $J_i$ factors could be expanded to include other lesser effects, Table 2.8 contains the factors that describe the most important phenomena so far encountered in glass-laser systems. What is not clear at present is how to evaluate (2.27) for application to any particular system. The weight $W_i$ that should be assigned to each separate factor is completely unspecified. The choice of $W_i$ is determined most easily by the designer's evaluation of the laser-system application which includes a good deal of intuition, insight, and experience. Table 2.5 gives the FOM defined by designers at LLL for rod, active-mirror, and disk amplifiers in pump- and parasitic-limited operation. If in (2.27) we set $W_i = 0$ for all $i \neq 1$, and $W_1 = 1$, we obtain

$$\eta = \frac{n_d}{n_2 \lambda_p^T} \quad , \quad \frac{n_d^2}{n_2 \lambda_p^T} \quad , \tag{2.58}$$

or the previously defined FOM for rod, active-mirror, and disk amplifiers in the parasitic-limited case. Similarly, if we take $W_i = 0$ for all $i \neq 1,2$, and $W_1 = W_2 = 1$, we have

$$\eta = \frac{n_d \alpha_0}{n_2 \lambda_p^T} \quad , \quad \frac{n_d^2 \alpha_0}{n_2 \lambda_p^T} \quad , \tag{2.59}$$

which is equivalent to the pump-limited FOM for rod, active-mirror, and disk amplifiers. Note that in Table 2.5, $\lambda_p$ has been ignored because of its small variation in the FOM. Thus, the familiar FOM for pump- and parasitic-limited operations are really only limiting cases of the more general FOM (2.27) and represent a very narrow parameter space. Thermal, thermo-optic, and many optical properties of laser glasses are totally ignored. The SHIVA and ARGUS systems at LLL have very long thermal cycling times. The decision to ignore thermal properties in those systems was based, in part, on the realization that the thermal-removal speed in disk amplifiers is geometry-limited; that is, because of the required design for other reasons (see Chap.5) such as cleanliness and damage, heat may be removed only by convection and radiation, both of which are relatively slow processes. As a consequence, the thermal removal speed is not limited by glass properties, hence those quantities ($J_7$, $J_8, J_9, J_{10}$) may be ignored ($W_7 = W_8 = W_9 = W_{10} = 0$). We may refer to this situation as convection-limited thermal behavior. In rod and active-mirror amplifiers, active cooling is accomplished by the use of liquids that circulate around the diameter or along one face, respectively (see Chap.5), which is a different situation. In rod and active-mirror amplifiers, the thermal-removal time is determined by glass properties; in other words, the heat is not trapped in a boundary between the glass and cooling medium, but is limited by the glass thermal-diffusion behavior, described by the thermal diffusivity $K/\rho c$. In the GDL and OMEGA laser systems at LLE which consist of rod amplifiers and will incorporate active-mirror amplifiers, some weight was assigned to thermal properties in the glass selection process. These systems are presently capable of being cycled in half an hour, unlike disk systems in use at LLL, Rutherford Laboratory, KMS Fusion, Inc., and Naval Research Laboratory, which generally have cycling times in the range of 2-4 hours.

Another problem in the evaluation of the $J_i$ factors is the almost complete lack of physical data for other than silicate glasses. This situation has been remedied recently for all of the phosphate laser-glass compositions of current importance [2.10,11], but still exists for fluorophosphate and fluoroberyllate compositions. A preliminary evaluation of $J_i$ factors in a comparison between silicate and phosphate glasses was presented recently by BROWN et al. [2.10]. Since that time, the number of $J_i$ factors has been expanded and the precision of the measurements improved. The values of the parameters used to evaluate each $J_i$ factor have been given previously in Tables 2.1-4. We have calculated the value of each $J_i$ factor in Table 2.8 for two silicate (ED-2, LSG-91H) and four phosphate glasses (EV-2, Q-88, LHG-5, LHG-7), normalized to ED-2 laser glass. The results are shown in Table 2.9 and show the consequences of changing glass composition from silicate to phosphate in a laser system. The power-handling capability ($J_1$) increases by an average factor of ≈1.39 for rods and active mirrors and ≈1.36 for disks. The specific gain coefficient ($J_2$) increases by ≈1.42; $J_3$ cannot be evaluated, owing to a lack of data at present. The branching ratio ($J_4$) is slightly greater in the phosphates (≈1.03) as is the absorption efficiency ($J_5$), by an average factor of ≈1.11. The slope of

Table 2.9. Component figures of merit — silicates and phosphates

| Factor | ED-2 | LSG-91H | EV-2 | Q-88 | LHG-5 | LHG-7 | Phosphate average |
|---|---|---|---|---|---|---|---|
| $J_1$ | 1.0 | 0.98 (0.97) | 1.57 (1.51) | 1.31 (1.29) | 1.29 (1.27) | 1.40 (1.35) | 1.39 (1.36) |
| $J_2$ | 1.0 | 0.84 | 1.49 | 1.38 | 1.44 | 1.36 | 1.42 |
| $J_3$ | 1.0 | - | - | - | - | - | - |
| $J_4$ | 1.0 | 1.01 | 1.04 | 1.02 | 1.03 | 1.03 | 1.03 |
| $J_5$ | 1.0 | 0.91 | 1.05 | 1.10 | 1.18 | - | 1.11 |
| $J_6$ | 1.0 | 1.08 | 0.73 | 0.49 | 0.69 | - | 0.64 |
| $J_7$ | 1.0 | - | - | - | 0.91 | 0.93 | 0.92 |
| $J_8$ | 1.0 | 1.0 | 1.09 | 1.09 | 1.09 | 1.09 | 1.09 |
| $J_9$ | 1.0 | 1.11 | 0.49 | 0.58 | 0.72 | 0.61 | 0.60 |
| $J_{10}$ | 1.0 | 0.75 | 0.74 | 1.02 | 0.81 | 0.80 | 0.84 |
| $J_{11}$ | 1.0 | 1.0 (0.81) | 1.0 (2.55) | 1.0 (1.04) | 1.0 (1.41) | 1.0 (1.67) | 1.0 (1.67) |
| $J_{12}$ | 1.0 | 1.0 (0.99) | 1.0 (1.63) | 1.0 (1.02) | 1.0 (1.20) | 1.0 (1.51) | 1.0 (1.25) |
| $J_{13}$ | 1.0 | 1.0 (0.95) | 1.0 (0.55) | 1.0 (0.86) | 1.0 (1.00) | 1.0 (0.85) | 1.0 (0.82) |

the stored-energy density or heat density ($J_6$) is substantially greater in phosphates, as is the heat generated per unit inversion density ($J_7$). The location of the UV edge is shifted towards the blue in phosphate glasses, resulting in a higher average FOM $J_8$, by a factor of $\simeq 1.09$. The thermal diffusivity ($J_9$) is substantially less in phosphates, by an average factor of $\simeq 0.60$, which increases the thermal-removal time. The induced temperature change ($J_{10}$) is also greater, resulting in a FOM that is on the average only $\simeq 0.84$ of that found in ED-2. The tendency for phosphate glasses to produce more heat and generate a larger temperature change than that found in silicates is somewhat mitigated by the fact that in phosphates the susceptibility to wavefront distortion ($J_{11}$) and birefringence ($J_{12}$) is substantially less by average factors of $\simeq 1.67$ and $\simeq 1.25$, respectively, for rod amplifiers. The phosphate glasses have been found to be perfectly acceptable in large laser systems (GDL and OMEGA) which must cycle every half hour. The value of $J_{13}$, which is a measure of the active thermal lensing present in disk and active-mirror amplifiers, is a factor of $\simeq 0.82$ less in phosphates than silicates. Such lensing is thus predicted to be worst with EV-2 glass and best with LHG-5.

The errors in the measurement of some quantities used in calculating Table 2.9 are large; and the $J_i$ values should therefore be used with caution, particularly those involving $B_\perp$ and $B_\parallel$ ($J_{11}$) (see [2.10,11]). Nevertheless, Table 2.9 provides a guide to the expected changes when silicate glass is replaced by phosphate. The highlights of this comparison are higher power output, gain, absorption efficiency, and stored-energy density slope, generation of more heat per unit volume due to the quantum defect but less due to the UV edge, and somewhat reduced thermal properties which are compensated to a certain degree by a decreased susceptibility to such distortions. Finally, thermal lensing is expected to be worse in phosphates than in silicates.

This concludes our treatment of the use of FOM in evaluating laser glasses. We now turn to the correlation of physical and optical properties.

## 2.4 Optical-Physical Properties Correlations in Laser Glasses

It has been the general experience in the high-peak power laser field that as glasses were optimized for high gain and low nonlinear index, other important properties were degraded. This may be seen in Fig.2.8 in which the

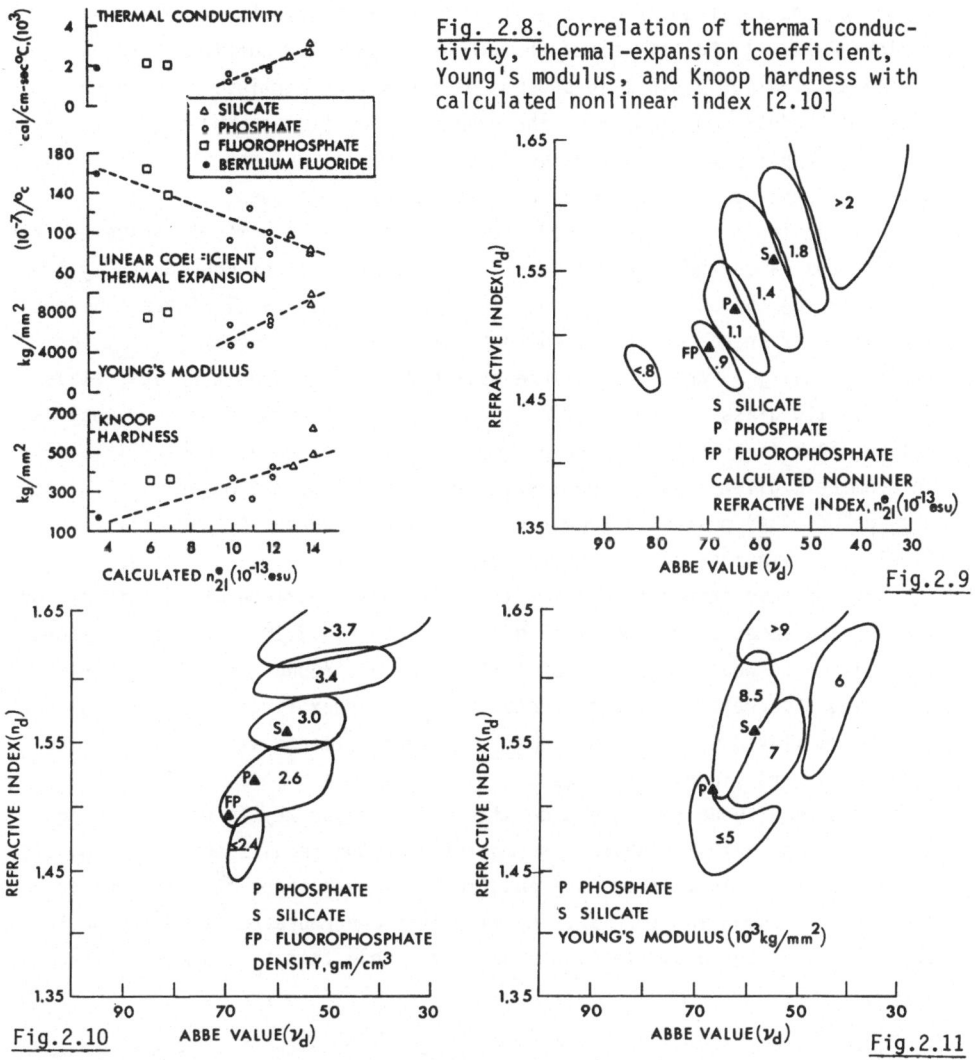

Fig. 2.8. Correlation of thermal conductivity, thermal-expansion coefficient, Young's modulus, and Knoop hardness with calculated nonlinear index [2.10]

Fig. 2.9. Regions of common nonlinear index as a function of linear index and Abbe value, for various optical glasses [2.10]

Fig. 2.10. Regions of common density as a function of linear index and Abbe value, for various optical glasses [2.10]

Fig. 2.11. Regions of common Young's modulus as a function of linear index and Abbe value, for various optical glasses [2.10]

the thermal conductivity (K), linear-expansion coefficient ($\alpha$), Young's modulus (Y) and Knoop hardness (H) have been plotted as functions of $n_2^e$ for silicate, phosphate, fluorophosphate and fluoroberyllate glasses [2.10]. There is a trend towards lower K, Y, and H whereas $\alpha$ can be seen to increase as $n_2^e$

is decreased. The fluorophosphates tend to display somewhat exceptional behavior; more will be said about that later. Figure 2.8 confirms what has been
presented previously, that in phosphates thermal properties, Young's modulus and hardness degrade while the expansion coefficient increases. We may
ask at this point whether there are any fundamental reasons why this degradation of physical properties occurs; to address this question we consider
the correlation of physical properties with more-basic optical parameters such
as the Abbe value and linear index. The problem was first considered by BROWN
et al. [2.10] who obtained the results shown in Figs.2.9-11. There, the phys-
ical-property trends are examined in a standard glass diagram $(n_d, \nu_d)$ for
$n_2^e$, the density $\rho$, and Y. Data were obtained from the Schott Optical Glass
Catalog (120 glasses were plotted which represent 16 composition groups) as
well as from the E series fluorophosphates investigated by Owens-Illinois.
To examine a given physical property, we circumscribe the majority of glass
data points that possess a common value of the property of interest. Figures
2.9-11 show a number of interesting results. First is that the range of
glasses is broad enough to display the well-known decrease of $n_2^e$ with $n_d$, or
when $\nu_d$ increases as may be seen in Fig.2.9. In Fig.2.10, there is a strong
correlation between $\rho$ and $n_d$, but little with $\nu_d$. Young's modulus, in Fig.2.11,
is strongly correlated with $n_d$ except for high-index large-dispersion glasses,
and can be linked via the high density of these glasses with the anomalous
behavior of the fluorophosphates.

It has been found that glass density is one major key to understanding
the glass-degradation problem. As suggested by the results shown in Fig.2.10,
glass density $\rho$ was plotted as a function of linear index $n_2^e$, with the dra-
matic result shown in Fig.2.12. The data that are representative of the 16
composition groups plus the fluorophosphates show that $n_d$ and $\rho$ are linked

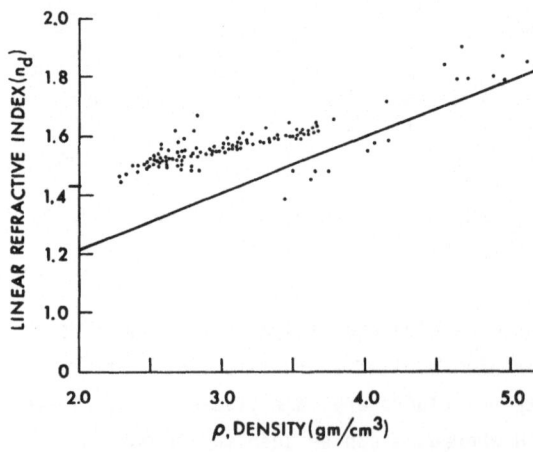

Fig. 2.12. Linear refractive
index as a function of density
for various optical glasses
[2.10]

with few exceptions via a linear relationship; a least-squares fit to the data is also shown in Fig.2.12. In fact, this would be expected from the well-known Lorenz-Lorentz equation relating $\rho$ to $n_d$

$$\frac{n_d^2-1}{n_d^2+2} = K\rho \quad , \tag{2.60}$$

where K is a constant. For small $n_d$, this becomes

$$n_d - 1 = K\rho \quad . \tag{2.61}$$

It is not difficult using (2.61) to show that the observed trends in Figs.2.8 and 2.11 for Young's modulus Y and the hardness H are related to changes of $\rho$. It can be shown that Y and $\rho$ are related by (2.11)

$$Y = \rho(V_E)^2 \quad , \tag{2.62}$$

where $V_E$ is the speed of an extensional acoustic wave in the material. Thus, if $V_E$ does not change dramatically from composition to composition, Y should be linearly related to $\rho$. This appears, from Fig.2.11, to be the case for glasses whose densities are $\leq 3$ g/cm$^3$. Similarly, there is a relationship between glass hardness H, Y, and the thermal rupture strength R, according to [2.10]

$$H = \frac{2}{3} R\left[1 + \ln \frac{Y}{3(1+p)R}\right] \quad . \tag{2.63}$$

Since $R \sim Y$ [2.10], H is given by

$$H = CY \quad , \tag{2.64}$$

where C is a constant. To understand Figs.2.8 and 2.11, we summarize the situation as follows. When glasses are optimized for low $n_2^e$, low $n_d$ values are sought. Because $n_d$ and $\rho$ are related according to (2.61), low $\rho$ values result. But when $\rho$ is decreased, (2.62,64) show that the values of Y and H are degraded. To date, no similar analysis has been presented to show why K decreases and $\alpha$ increases when glasses are optimized for low $n_2^e$.

As mentioned previously, the fluorophosphates do not seem to follow the observed trends very well. Experience has shown that the fluorophosphates are chemically more durable than the phosphates. This is consistent with the high-density — good-durability (hardness, stiffness, rupture strength) correlations just mentioned, since the fluorophosphates have rather high densities (typically 3.6 g/cm$^3$). High $\rho$, however, does not imply high index in fluorophosphates (typically $n_d \sim 1.45$). Therefore, the correlation ($n_d$-$\rho$) breaks down; we must consider the Lorenz-Lorentz law (2.61) to understand why. Rewritten, that becomes

$$[n] = \frac{(n_d^2-1)M}{(n_d^2+2)\rho} \quad , \tag{2.65}$$

where [n] is the total molar refractivity and M the molecular weight. For many glasses that consist of N constituents, the dependence of [n] on the refractivity is simply additive

$$[n] = \sum_{i=1}^{N} [n_i] \tag{2.66}$$

for constituents whose electrons do not strongly interact. This law may be broken in glasses in which electron-electron interactions are complicated by different chemical bonding. Such complications invalidate (2.66) and the ($n_d$-$\rho$) correlation that has been found for most optical glasses. It has been surmised that such is the case in fluorophosphates [2.10].


## 2.5 Properties Neglected in the FOM Formalism

We have discussed in the previous sections (2.1-4) the most important optical and physical parameters needed for a full description of a high-peak power Nd:glass laser system and its design. It is obvious, however, that other important effects must be considered in the choice of laser glass and its quality control before installation in a system. Here, in order, we present results on the following areas of major concern; thermal-shock resistance, passive birefringence, and impurity problems in phosphate compositions.

### 2.5.1 Thermal-Shock Resistance of Laser Glasses

One measure of the suitability of laser glasses for an application in which thermal cycling is involved is the thermal-shock resistance $T_{sr}$. This is of concern for systems that are actively cooled and may be subject to rapid temperature variations, which subject the glass to large induced stresses that result in fractures. Similar damage may result from rapid heat deposition

Table 2.10. Relative-thermal-shock resistance

| | Silicate | | Phosphate | | | |
|---|---|---|---|---|---|---|
| | ED-2 | LSG-91H | EV-2 | Q-88 | LHG-5 | LHG-7 |
| $T_{sr}$ | 1.0 | 1.0 | 0.4 | 0.4 | 0.8 | 0.6 |

during Xe flashlamp pumping. $T_{sr}$ may be calculated from the relationship [2.10]

$$T_{sr} = \frac{R}{Y\alpha} \left(\frac{K}{\rho c}\right)^{\frac{1}{2}} ,$$  (2.67)

where Y, $\alpha$, K, $\rho$, and c have been defined previously and R is the rupture strength previously discussed in connection with (2.63). Values of R have been presented [2.10] for silicate and phosphate laser glasses; more-recent data was presented by JACOBS [2.15]. The rupture strengths decrease, in general, in the phosphates relative to the silicate compositions. Using the values for R, and data presented previously for Y, $\alpha$, K, $\rho$, and c, we can calculate the thermal-shock resistance $T_{sr}$ relative to ED-2 silicate laser glass, with the result shown in Table 2.10 [2.10,16]. This table shows that increased care is wise when phosphate compositions are subjected to sudden temperature changes that may not affect silicates.

## 2.5.2  Passive Birefringence in Laser Glasses

Section 2.3.3 showed that thermally induced birefringence is a problem of concern in rod-amplifier geometries. Most rods also exhibit a certain amount of passive birefringence, which arises from permanent stresses within the material. The amount of residual stress is determined by the thermal cycling of the rod during manufacture; it may be removed, to a certain degree, by proper annealing. Passive birefringence has been measured as a function of rod radius by JACOBS [2.16] at the Laboratory for Laser Energetics. Results are shown in Fig.2.13 for phosphate (LHG-7) laser rods with diameters of 40, 64, 90 mm. The birefringence is proportional to $r^2$ for all sizes investigated [2.16]. The importance of this effect is that a linearly polarized beam, in propagating through a rod, suffers some retardation of the r and $\theta$ components relative to one another. In passing through a component which is polarization sensitive such as a $\lambda/4$ plate or polarizer, significant losses may occur if birefringence is not minimized. JACOBS [2.16] calculated the loss imparted to a linearly polarized beam in traveling through a 90 mm rod that is followed by one analyzer (polarizer). The result, shown in Fig.2.14, displays the sen-

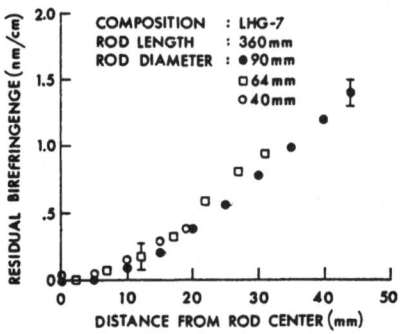
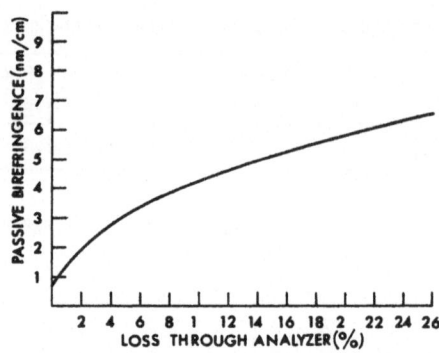

Fig. 2.13. Residual birefringence in LHG-7 laser rods as a function of distance from the rod center [2.16]

Fig. 2.14. Transmission loss through a 90 mm diameter rod due to passive birefringence [2.16]

sitivity of the loss to the value of the passive birefringence at the circumference (r=D/2); here D is the diameter of the rod. To limit the loss to less than 1%, residual birefringence must be $\lesssim$1.5 nm/cm at 1.06 $\mu$m; with birefringence of 5 nm/cm, 14% loss occurs. The calculation has assumed a 100% fill factor, which is never achieved in practice, and is thus somewhat pessimistic. In estimating the real birefringence loss in a rod amplifier, both the passive and actively induced birefringence due to optical pumping must be considered.

## 2.5.3 Impurities in Laser Glasses

In this last section, two impurities, which have been found to limit the performance of phosphate laser glasses will be discussed. The first is water, which has been found to cause serious fluorescence quenching in phosphates [2.10]. It is well known [2.17] that the O-H vibration located at $\approx$3600 cm$^{-1}$, relaxes rare-earth ions in glasses, crystals, and liquids. The result is that the multi-phonon emission from the $^4F_{3/2}$ band is enhanced, leading to a lower quantum efficiency or, equivalently, a shorter radiative lifetime. Phosphate compositions seem particularly susceptible to absorption of $H_2O$ during manufacture, as shown by recent measurements by JACOBS [2.16]; some results have been presented recently for LHG-8 glass with a low wt-% doping, for which concentration quenching is minimized. The severity of the effect can be as great as that due to concentration quenching at higher dopings. The presence of $H_2O$ is monitored by measuring the optical density (O.D.) of samples at 2.2 $\mu$m where $H_2O$ absorbs strongly. In Fig.2.15 the essential result is shown;

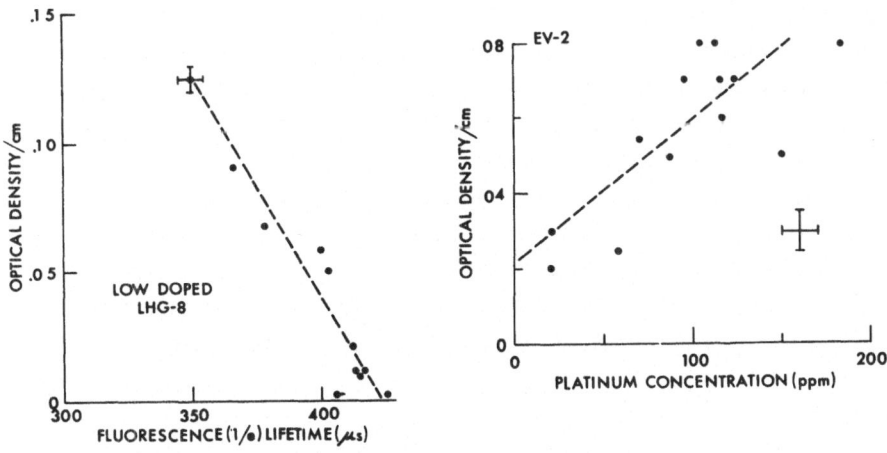

Fig. 2.15. Fluorescence lifetime ($^4F_{3/2}$) as a function of optical density at 2.2 μm in LHG-8 laser glass [2.16]

Fig. 2.16. Optical density at 370 nm as a function of dissolved Pt concentration [2.16]

the absorption at 2.2 μm, which is a measure of $H_2O$ content, is strongly correlated in an apparently linear fashion with a decrease of τ. Similar results on other glasses show the same trends [2.10].

A decrease of τ from a value that would be expected from a glass with little $H_2O$ content degrades amplifier performance significantly. Calculation has shown that a 10% reduction of peak stored-energy density results when τ is reduced from 350 to 240 μs [2.16] for one rod amplifier in current use at LLE.

Another impurity of particular concern in rod amplifiers is Pt which may be dissolved in laser glass from the Pt crucible during manufacture; it imparts an "orange" coloration to the glass. The effect is to shift the UV edge from that present in the base glass towards the visible. The absorption of laser glasses may be monitored at a single location, away from any $Nd^{3+}$ band, to measured Pt. JACOBS [2.16] plotted the absorption (O.D.) at 370 nm versus the Pt concentration (Fig.2.16). This effect has two serious consequences in rod amplifiers and to some extent in disk and active-mirror amplifiers. The first is that because Xe flashlamps produce a great deal of radiation in the ultraviolet, increased heating occurs. Because the absorption is strong, the heat is deposited in short distances from the circumference of the rod which leads to severe stress and consequently induced birefringence losses. The second effect is to mask the absorption at the $Nd^{3+}$ band located at ≈ 355 nm. Gain measurements [2.10] on 90 and 64 nm rod amplifiers with high and low Pt dopings obtained the result that if the measured stored-energy density is plotted as a function of normalized radius $(r/r_0)$, the effect of the Pt is to reduce considerably the stored-energy density in the center of the rod. That is exactly what would be expected if the 355 nm band were masked, because it penetrates well into the center of the rod. The other deleterious effect is to change the stored-energy density profile $E_s(r/r_0)$, increasing its slope. It is desirable, in order to achieve a large fill factor in a rod amplifier (Chap.8) and thus to maximize its output, to reduce $dE_s/dr$ to a minimum. With proper control over the glass-manufacturing process, acceptable levels of Pt are now routinely achieved in phosphate compositions.

# 3. Optical-Pump Sources for Nd : Glass Lasers

## 3.1 General Pumping Considerations

It is well known that Xe flashlamps are commonly used to pump Nd:glass lasers. The reasons are varied but include the following considerations:

1) The output of Xe flashlamps extends spectrally from the ultraviolet region to the infrared, thus overlapping the absorption spectra of $Nd^{3+}$.
2) Xe flashlamps are high-brightness sources and because of a high efficiency for converting electrical to spectral output they can provide the flux necessary to achieve reasonable stored-energy densities in Nd:glass lasers.
3) The lifetime limits of Xe flashlamps are fairly well understood; such lamps are durable and have, when driven properly, an acceptable lifetime for present Nd:glass laser systems.
4) The technology for producing high energy Xe flashlamps is advanced; commercial products are offered by at least four companies.

There are, however, a number of difficulties in applying Xe flashlamps to the optical pumping of Nd:glass. They are:

1) Because a Xe flashlamp has an output that extends continuously from the ultraviolet to the infrared, and the $Nd^{3+}$ absorption spectrum consists of a series of discrete bands, much of the incident flashlamp radiation is not absorbed in the glass and is eventually converted to heat. This is the major cause of the low efficiency of Nd:glass lasers.
2) Xe flashlamps cause a large amount of heat to be deposited in the glass. This arises from the quantum defect (energy difference between an absorbed and emitted photon) as well as the absorption of flashlamp ultraviolet output by the glass background absorption. The result is that the repetition rate of any system is seriously degraded.
3) As we will see, because of lifetime limits, flashlamps must be run with a current pulse width typically comparable to the fluorescence lifetime $\tau$ of the glass being pumped, thus, fluorescence decay losses can be severe.
4) The spectral output of a Xe flashlamp is emitted into $4\pi$ steradians; that is, the radiation is not directional. To obtain efficient coupling between a Xe flashlamp and an amplifier, the pump lamp must be located physically near the material to be pumped; reflectors are commonly used to direct a portion of the radiation; such use of mirrors is complicated by the fact that the Xe plasma absorbs some of the radiation that passes through it and reradiates it into $4\pi$ steradians.

In this chapter we will cover all of those aspects of Xe flashlamps which are important in their application to Nd:glass laser systems.

## 3.2  Xe Flashlamp Spectra

In designing a high-peak-power Nd:glass laser system, it is necessary to consider the spectral properties of Xe flashlamps. This is important for understanding heat deposition and thermal effects as well as describing the magnitude and temporal evolution of the stored-energy density. Various models have
been developed to describe the spectral distribution of Xe flashlamps; we
will investigate one in detail in Sect.3.3. Here we describe the experimentally
measured output of such lamps, show how we may describe the emissivity of the
Xe plasma, and give measured values of the lamp efficiency in converting electrical input to spectral output.

The spectral output of a Xe flashlamp is determined primarily by two parameters, the inside diameter D of the lamp, and the instantaneous current density J. To a lesser extent, the Xe fill pressure also affects the observed
spectrum. Due to the difficulty of the measurement, time-resolved spectra of
Xe flashlamps are not easily obtained. Recently obtained results of time-resolved spectra of 13-mm-bore Xe flashlamps will be described in Sect.3.7. Here,
we describe the essential features of the spectrum by referring the reader to
the work of GONCZ and NEWELL [3.1]. Time-integrated spectra at various aver-
age-on dc-current densities were obtained by using a spectral radiometer. An
example of this, reproduced in Fig.3.1, shows the characteristic feature of a
low-current-density Xe discharge; it is the preponderance of discrete (line)
radiation superimposed upon a broad background. The observed lines are considerably broadened (pressure and Stark broadening) and consist of a series
of lines known to correspond to bound-bound transitions in atomic Xe and Xe
in various states of ionization [3.2]. As the current density is increased,
however, the discrete lines gradually disappear, buried in the background radiation that gradually comes to predominate in a high-current-density discharge. An example of this is shown in Fig.3.2, where the absolute spectral
output is plotted for average current densities of 1700 and 5300 A/cm$^2$. Although line radiation is apparent at 1700 A/cm$^2$, at 5300 A/cm$^2$ it has almost
completely disappeared. The broad background radiation, characteristic of Xe
plasmas at high temperature, is due to bound-free and free-free transitions.
In fact, it may be well fitted by a blackbody distribution, which we will
describe below. Note that owing to its thermal nature, the amount of radiation in the ultraviolet increases with J because the blackbody peak shifts
toward the blue in accordance with Wein's law. Time-resolved spectroscopy on
13 mm and 19 mm bore lamps, to be described in Sect.3.7, shows that line
spectra are involved only on the positive-slope side of a current pulse. By

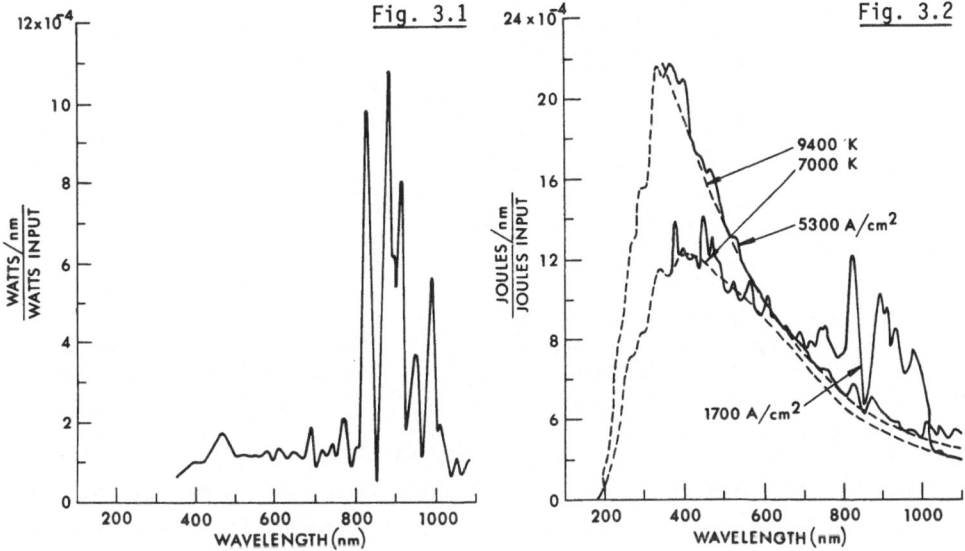

Fig. 3.1. Spectral emission, experimental dc lamp (1.7 atm Xe; 37 A/cm$^2$). Spectral bandwidth equal to 10 m$\mu$. After GONCZ and NEWELL [3.1]

Fig. 3.2. Spectral emission, FX-47A flashtube at two current densities (0.4 atm Xe; 1700 and 5300 A/cm$^2$). Spectral bandwidth equal to 10 m$\mu$. Coarse broken lines are relative spectral emission of blackbodies at 7000 and 9400 K. Fine broken lines represent measurements made in the ultraviolet and are not as accurate as those made in the visible and infrared [3.1]

the time the negative-slope side of the pulse is reached, the plasma has achieved its full ionization and pressure, broadening has removed the discrete lines and it may be described accurately by a blackbody distribution.

We may describe the emissivity $\varepsilon_\lambda$ of a plasma in general by the equation

$$\varepsilon_\lambda = 1 - \exp(-\alpha_\lambda \ell) \quad , \tag{3.1}$$

where $\alpha_\lambda$ is a wavelength-dependent absorption-emission coefficient and $\ell$ is the length of the plasma. Note that if $\alpha_\lambda \ell$ is large, $\varepsilon_\lambda \rightarrow 1$ and the emissivity of a blackbody is approached. The plasma is said to be optically "thick" at that wavelength. Similarly, if $\alpha_\lambda \ell \rightarrow 0$, $\varepsilon_\lambda \rightarrow 0$ and the plasma is said to be optically "thin". The absorption-emission coefficient $\alpha_\lambda$ varies with the current density J [$\alpha_\lambda = \alpha_\lambda(J)$] in a predictable manner for Xe-arc discharges. In particular, EMMETT et al. [3.3] measured the transmittance of a Xe flashlamp as a function of current density and wavelength. The wavelengths varied from 250-1000 nm and current densities up to 5000 A/cm$^2$. A summary of their results is shown in Fig.3.3, where the 30% and 60% transmittance points are

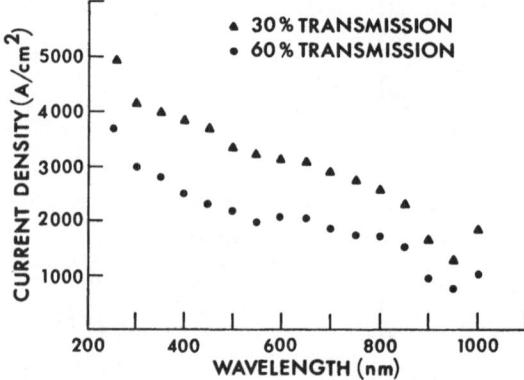

**Fig. 3.3.** Current density for constant transmission as a function of wavelength [3.3]

plotted as functions of current density and wavelength. Study of these data shows that, with the exception of points that overlap well-known discrete Xe lines, the background or continuum absorption is governed by the approximate equations

$$\alpha_\lambda(J_2) = \alpha_\lambda(J_1)\left(\frac{J_2}{J_1}\right)^2 \quad , \tag{3.2}$$

$$\alpha_\lambda(\lambda_2) = \alpha_\lambda(\lambda_1)\left(\frac{\lambda_2}{\lambda_1}\right)^\gamma \quad , \tag{3.3}$$

where $1 \leq \gamma \leq 1.5$. It should be emphasized that these relationships have been found to be valid not only for continuum absorption but also for line radiation. Programs such as GENEFF or FLASH, to be described in Sect.3.3, make extensive use of the scaling indicated by (3.2). More precise measurements of $\alpha_\lambda$ performed at LLL have been found to be at variance with those in [3.3], but more complete data is required [3.4].

As mentioned previously, at high current densities the spectral output of a Xe flashlamp approaches that of a blackbody. The conditions under which this happens have been discussed by FINKELNBURG [3.5], who mentions that every isothermal gas emits a continuous spectrum, the energy distribution of which is described by the Planck blackbody formula; as a consequence, the spectral distribution depends exclusively upon the temperature and not on the specific nature of the gas. The surface brightness or spectral radiance $M$ [W/cm$^2$-steradian-$\mu$m] is given by

$$M(\lambda,T) = \varepsilon_\lambda M_B(\lambda,T) \quad , \tag{3.4}$$

where $M_B(\lambda, T)$ is the spectral emittance of a blackbody, given by the Planck formula

$$M_B(\lambda, T) = \frac{K}{\lambda^3}\left(\frac{1}{e^{hc/\lambda kT}-1}\right) \tag{3.5}$$

and $\varepsilon_\lambda$ was defined by (3.1). Therefore,

$$M(\lambda, T) = M_B(\lambda, T)\left[1 - e^{-\alpha_\lambda \ell}\right] \quad . \tag{3.6}$$

In general, the absorption-emission coefficient $\alpha_\lambda$ is a function of wavelength, pressure, and absolute temperature T.

Xe flashlamps are efficient converters of electrical into spectral energy. The first measurements of this efficiency $\eta$ were made by GONCZ and NEWELL [3.1] on a wide variety of flashtubes. Using an FX-47A type linear flashlamp with a diameter of 1.3 cm, they found that the total efficiency for emission or radiation into the spectral region 0.35-1.1 $\mu$m was $\eta = 64.8\%$ at a current density of 5300 A/cm$^2$. For the same tube, but a current density of 1700 A/cm$^2$, $\eta = 64.6\%$ in the same spectral region. Another measurement, performed by HOLZ-RICHTER and DONICH [3.6] provided the value $\eta = 79.8\%$ for a 1 cm Xe flashlamp run at a peak current density of 4.5 kA/cm$^2$. They suggested that the output from a Xe flashlamp triggered in series rather than in parallel, as theirs was, has $\approx 10\%$ greater output. We can conclude from these measurements that Xe flashlamps are very efficient, in the range of $\approx 65-80\%$, in converting electrical energy to spectral output. The major cause of inefficiency in solid-state Nd:glass lasers is not the Xe flashlamp, but the fact that it is poorly matched spectrally to the material ($Nd^{3+}$) it is pumping. We could make the process more efficient by modifying the Xe spectrum in some way, increasing the breadth of the $Nd^{3+}$ absorption bands, or by utilizing an approximately monochromatic source, which is difficult to obtain.

## 3.3 A Xe Flashlamp Radiative Model

To completely design a Nd:glass laser amplifier for use in a high-peak-power Nd:glass laser system, it is necessary to have a computer model that simulates the spectral intensity of commonly used Xe flashlamps. Various attempts have been made to do this. CHURCH [3.7] obtained computer-generated spectra by use of a fundamental approach that calculated, for instance, transition probabilities in Xe for free-free and bound-free transitions. The method was cumbersome but resulted in spectra in qualitative agreement with those actually observed. The most straightforward approach was taken by TRENHOLME and EMMETT [3.8] and will be discussed in detail here. It is based largely upon experimental data and makes a number of simplifying assumptions. Among these are that the plasma is radially uniform, that the plasma is at a constant temperature with respect to its radius and length, and that the arc is wall stabilized. In fact, to entirely fill the bore the arc requires a time which is a substantial fraction of the total pulse duration typical of common Xe flashlamps. A radial variation is present even if the arc is wall stabilized, but it is small enough that it can be taken into account in a phenomenological way as we shall discuss below. Another important assumption is that the plasma is in local thermodynamic equilibrium, that the Planck blackbody distribution then applies, and that each point in the cylindrical volume may be characterized by an absorption-emission coefficient $\alpha_\lambda$. TRENHOLME and EMMETT mentioned [3.8] that the limits of validity of the model are a pulse duration $\tau$ in the range $(10\,\mu s \leq \tau \leq 10\,ms)$, current densities in the range $(300 \leq J \leq 10^4$ $A/cm^2)$ and a diameter D from 0.2-2 cm. We are interested in the spectral emittance W $[W(watts)/cm^2\text{-}\mu m]$ which may be computed by integrating the spectral radiance over all output angles from an arbitrary surface element of the plasma. It can be shown that W may be calculated from

$$W = \int M \cos\psi \, d\Omega \quad , \tag{3.7}$$

where M is given by (3.6), $\Omega$ is the solid angle, and $\psi$ the angle between the surface normal and an arbitrary ray. In general,

$$\cos\psi = \cos\phi \, \sin\theta \quad , \tag{3.8}$$

where $\phi$ and $\theta$ are the azimuthal and polar angles, respectively, and

$$d\Omega = \sin^2\theta \, d\theta \, d\phi \quad . \tag{3.9}$$

From (3.7),

$$W = \int_0^\pi d\theta \int_{-\pi/2}^{\pi/2} M \cos\phi \, \sin^2\theta \, d\phi \tag{3.10}$$

or

$$W = M_B \int\limits_{0}^{\pi} \int\limits_{-\pi/2}^{\pi/2} \cos\phi \, \sin^2\theta \left\{ 1-\exp\left[ -\left(\frac{\alpha D \, \cos\phi}{\sin\theta}\right)\right]\right\} d\phi \quad , \tag{3.11}$$

Evaluation of this integral yields

$$W = 2\pi M_B \int\limits_{0}^{\pi/2} \sin^2\theta \left[ I_1\left(\frac{\beta}{\sin\theta}\right) - L_1\left(\frac{\beta}{\sin\theta}\right)\right] d\theta \quad , \tag{3.12}$$

where $\beta = \alpha D$ and $I_1$ and $L_1$ are the first-order modified Bessel and Struve functions, respectively. Using the fact that the spectral emittance of a blackbody is $\pi$ times its spectral radiance,

$$W_B = \pi M_B \quad , \tag{3.13}$$

we have finally that

$$M = M_B F(\beta) \quad , \tag{3.14}$$

where

$$F(\beta) = 2 \int\limits_{0}^{\pi/2} \sin^2\theta \left[ I_1\left(\frac{\beta}{\sin\theta}\right) - L_1\left(\frac{\beta}{\sin\theta}\right)\right] d\theta \quad . \tag{3.15}$$

To find M then we must evaluate $F(\beta)$. In practice $F(\beta)$ is calculated by using numerical methods, for which a number of polynomial expansions are available. It has been plotted in Fig.3.4 as a function of $\beta$, which shows that when $\beta \approx 5$, $F(\beta) \rightarrow 1$. Thus for an optically thick plasma, $F(\beta) \rightarrow 1$, which of course should be the case because $F(\beta)$ is identical with the emissivity [see (3.4)]. After $F(\beta)$ has been calculated initially, its values may be stored in an array and recalled at will. To calculate M for a Xe flashlamp we must know the diameter D, the absorption-emission coefficients $\alpha_\lambda$ at the current density J of interest

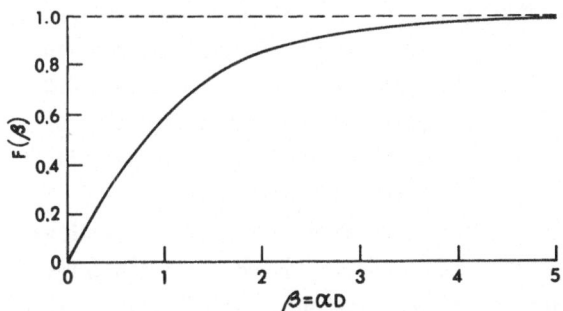

Fig. 3.4. Fraction of black-body output from a cylindrically distributed emitter as a function of the product of the diameter and the absorption-emission coefficient [3.8]

and the plasma temperature T. By fitting experimental data with the model, TRENHOLME and EMMETT [3.8] obtained an empirical expression that gives the absolute temperature T in terms of the current density and diameter,

$$T = [(9450\ D^{0.03}\ J^{0.01})^6 + (93\ D^{0.27}\ J^{0.34})^6]^{1/6}\ , \tag{3.16}$$

whence the units are D [m] and J [A/cm$^2$]. Radiative efficiency measurements performed by KELLY et al. [3.9] have confirmed the validity of (3.16), but with the replacement ($J^{0.34} \rightarrow J^{0.35}$). Although the change is small, for a current density of 4000 A/cm$^2$ it results in an 8% change in blackbody temperature which is a 35% change in the total blackbody output. They also found that the actual inside diameter D of the flashlamp should not be used, but rather the reduced diameter $D_R$ given by

$$D_R = \frac{D}{1 + \dfrac{0.08}{D}}\ , \tag{3.17}$$

when D is expressed in cm. This effective diameter is less than the actual diameter D of the radiating column owing to the cool nonradiating sheath surrounding the hot plasma. The remaining parameter to be determined is the absorption-emission coefficient $\alpha_\lambda$. Using results previously discussed [3.3] as well as previously published data, TRENHOLME and EMMETT [3.8] obtained the result shown in Fig.3.5 which shows $\alpha_\lambda$ in the spectral range of $\approx 0.2$-$1.8\ \mu$, for a current density of 1000 A/cm$^2$. For any other J, $\alpha_\lambda$ is simply scaled according to (3.2).

The prescription for finding M(D,J) is now clear. We obtain the $\alpha_\lambda$ from a table of stored values and scale them as $J^2$. Similarly, the appropriate F($\beta$) value is obtained from a stored table after each $\alpha_\lambda D$ has been calculated. The blackbody function $M_B$ is calculated by use of the Planck formula (3.5), the temperature T obtained from (3.16,17). This model has been used extensively at LLL in numerous laser design codes such as GENEFF and VODAC and at LLE where similar programs (FLASH, INVDEN) have been developed. An example of the output of FLASH is shown in Fig.3.6. There, M is plotted as a function of $\lambda$ for a one-cm inside-diameter Xe flashlamp at J = 1000, 2000, and 3000 A/cm$^2$. Some general features to be noted are that the peak shifts towards the blue as J is increased, that the integrated output increases significantly and that line radiation becomes less and less prominent.

This radiative model has been found to give excellent agreement with experiment where such data is available. Section 2.1 showed inversion curves that were obtained by convoluting FLASH generated spectra with glass absorp-

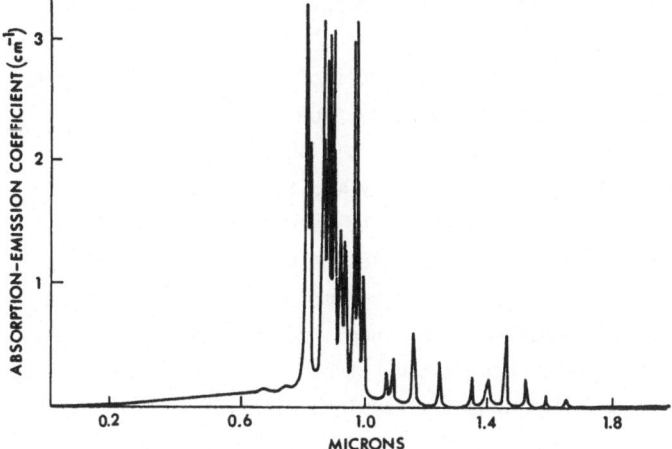

Fig. 3.5. Model absorption coefficient of xenon plasma as a function of wavelength at a current density of 1000 A/cm$^2$ [3.8]

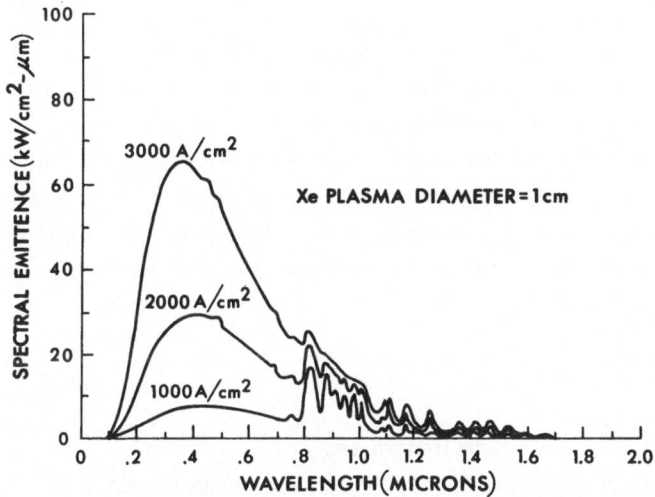

Fig. 3.6. Flashlamp spectral emittance as a function of wavelength for a 1 cm diameter Xe plasma and current densities of 1000, 2000, and 3000 A/cm$^2$

tion spectra; the results were presented in [2.5]. Direct measurement of the inversion density [2.7] in four laser glasses showed agreement with the predictions of the model, which indicates either accuracy of the model or the insensitivity of the inversion density to the detailed distribution obtained in the model. It will be seen in the next chapter that FLASH has also made it possible to assess the importance of glass composition in the thresholds for parasitic oscillations.

The spectral emittance M calculated in the foregoing is for a bare plasma with no wall effects considered. The thin quartz tube that surrounds the plasma does affect the output of a lamp in a predictable way. It has been shown [3.8] that the function F($\beta$) is reduced by $\approx 13.5\%$ and varies slightly with $\beta$ for an index of 1.46. The way in which the output of the flashlamp varies cannot be calculated analytically because the effect of the wall is not just to reduce the spectral output but to reflect a portion of it back into the plasma, where each ray may be reabsorbed. This process of absorption and subsequent reemission is very complicated; it can be approximated only by use of a large computer simulation code that employs Monte Carlo methods.

## 3.4  Xe Flashlamp Pulse-Forming Network Design

We now turn our attention to the pulse-forming network (PFN) design needed to drive Xe flashlamps in Nd:glass laser systems. To do this, we should take into account that a Xe flashlamp is an inherently nonlinear device, electrically. That is, during the evolution of the discharge, the resistance of the tube changes dramatically, from an almost infinite value at $t = 0$ to a value of typically 1 $\Omega$ at the peak of the current pulse. The present description of the impedance of Xe flashlamp is due to the work of GONCZ [3.10]. The voltage drop across the tube is given by

$$V_T = \frac{\gamma(J)i}{LA} \quad , \tag{3.18}$$

where i is the current, L and A the tube length and cross-sectional area, respectively, and $\gamma(J)$ the plasma resistivity, which was found by GONCZ to be a function of J. In particular, $\gamma$ was measured and found to obey

$$\gamma(J) = k/i^{\frac{1}{2}} \quad , \tag{3.19}$$

or to vary as the inverse $\frac{1}{2}$ power of the current i or current density J; k is a constant. By combining (3.18,19), it is usual to express the voltage drop across the tube as

$$V_T = K_0 |i|^{\frac{1}{2}} \quad , \tag{3.20}$$

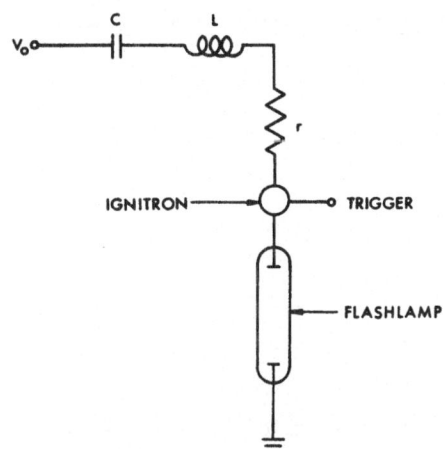

Fig. 3.7. Typical series mesh pulse-forming network for Xe flashlamps

where $K_0$ is a constant that is characteristic of a given lamp and is referred to as the impedance parameter. It is easily seen that $K_0$ scales according to

$$K_0 = k \frac{L}{D} \quad , \tag{3.21}$$

where k is a constant that depends only upon gas type and pressure. More recently, it has been shown that $K_0$ is accurately represented by [3.11]

$$K_0 = C\left(\frac{L}{D}\right)\left(\frac{P}{450}\right)^\gamma \quad , \tag{3.22}$$

where C is a constant, P the fill pressure in torr, and $\gamma$ an exponent. NOBLE and KRETSCHMER [3.12] have reported the values C = 1.27 and $\gamma$ = 0.2, whereas NEWELL [3.11] obtained C = 1.36 and $\gamma$ = 0.18.

Most Xe flashlamps in use today are driven by the use of single mesh L-R-C networks because of the economy of that design, which considerably simplifies our analysis. A typical PFN shown in Fig.3.7 consists of an energy-storage capacitor C which is charged to $V_0$ volts, an inductor L which is used to control the pulse duration, a resistor r which represents the dc resistance in the circuit from cables, electrodes, etc., a nonlinear Xe flashlamp and an ignitron device which is a high-current switch triggered by an external pulse. The voltage $V_0$ is distributed in the circuit according to the usual differential equation

$$L \frac{di}{dt} + ri \pm K_0 |i|^{\frac{1}{2}} + \frac{1}{C} \int_0^t i \; d\bar{\tau} = V_0 \quad . \tag{3.23}$$

Following the classic formulation of MARKIEWICZ and EMMETT [3.13], we make
the following substitutions in (3.23)

$$\begin{cases} Z_0 = \left(\dfrac{L}{C}\right)^{\frac{1}{2}} & T = (LC)^{\frac{1}{2}} & \alpha = \dfrac{K_0}{(V_0 Z_0)^{\frac{1}{2}}} \\ i = I\left(\dfrac{V_0}{Z_0}\right) & \tau = t/T & \beta = r/Z_0 \end{cases} \tag{3.24}$$

to obtain

$$\frac{dI}{d\tau} \pm (\alpha + \beta |I|^{\frac{1}{2}})|I|^{\frac{1}{2}} + \int_0^\tau I \, d\bar{\tau} = 1 \quad . \tag{3.25}$$

Here $Z_0$ is the characteristic impedance, T the time scale, $\alpha$ the so-called
damping parameter, $\beta$ the ratio of dc to characteristic circuit impedance, I
the normalized current and $\tau$ the normalized time. The normalized power $P_N$ is

$$P_N = \frac{P}{V_0^2/Z_0} = \alpha |I|^{3/2} \quad . \tag{3.26}$$

Similarly, the normalized energy is given by

$$E_N = \frac{E_L}{E_0} = 2\alpha \int_0^\tau |I|^{3/2} \, d\bar{\tau} \quad . \tag{3.27}$$

Here, $E_L$ is the energy dissipated by a lamp up until time $\tau$. Note that as $\tau$
is approached, both $P_N$ and $E_N \to 1$. Solutions of (3.25) have not been obtained
in analytical form because of its nonlinearity. Therefore, it is necessary
to use numerical techniques. Using a digital computer, MARKIEWICZ and EMMETT
[3.13] solved (3.25-27) for the particular case $\beta = 0$. Here, we follow closely
more recent work by BROWN and NEE [3.14] in considering solutions of these
equations including $\beta \neq 0$. The pulse of most interest to laser designers is
the so-called critically damped pulse. Underdamped current pulses are characterized by more than a single zero-current crossing (ringing). They are obviously less than optimum, because energy is dissipated in the flashlamp long
after the termination of the pulse that is amplified. For overdamped pulses,
although only a single zero-current crossing occurs, the peak current is always less than that for a critically damped pulse which results in less than
the maximum obtainable stored-energy density. A critically damped pulse is
defined as one for which all zero-current crossings vanish or no negative
values of $I(\tau)$ occur for $\tau > 0$. MARKIEWICZ and EMMETT [3.13] found that the
value of the damping parameter $\alpha$ for a critically damped pulse was $\alpha \simeq 0.75$.
In related work, HOLZRICHTER and EMMETT [3.15] found that $\alpha \simeq 0.78$. Using more

accurate numerical methods, BROWN and NEE [3.14] found $\alpha \approx 0.84$ to be the appropriate value for critically damped pulses. Solutions of (3.25-27) obtained by them are shown in Figs.3.8-3.10, respectively, for $\alpha = 0.84$ and $\beta$ varying in the range $0 \le \beta \le 0.5$. For the current I, Fig.3.8 shows that the effect of a finite resistance r in a circuit is to decrease the peak current, shift the peak towards shorter times, overdamp the pulse and increase the pulse width.

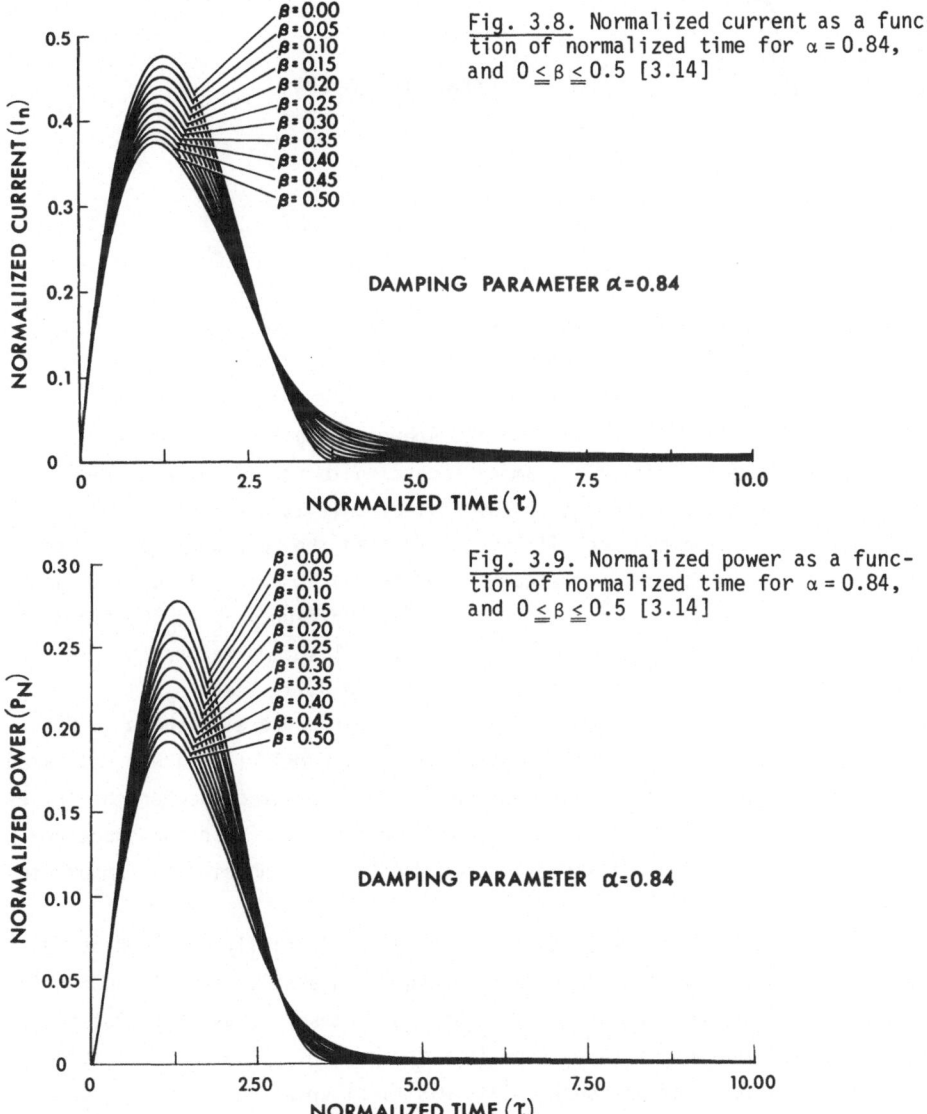

Fig. 3.8. Normalized current as a function of normalized time for $\alpha = 0.84$, and $0 \le \beta \le 0.5$ [3.14]

Fig. 3.9. Normalized power as a function of normalized time for $\alpha = 0.84$, and $0 \le \beta \le 0.5$ [3.14]

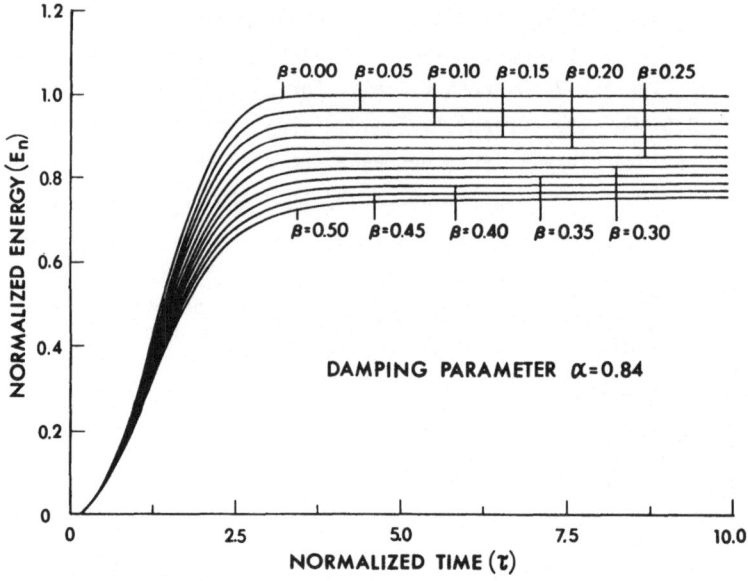

Fig. 3.10. Normalized energy as a function of normalized time for $\alpha = 0.84$, and $0 \le \beta \le 0.5$ [3.14]

Similar conclusions may be drawn for $P_N$, shown in Fig.3.9. Therefore, circuits designed to be critically damped according to ($\beta = 0$), but with neglect of the finite r that is present in any real circuit as was done in [3.13], will always be overdamped and will give less than the peak power or stored-energy density. In Fig.3.10, the normalized energy $E_N$ which $\rightarrow 1$ for $\beta = 0$ is always <0 for $\beta > 0$. We can define the transfer efficiency $\eta_E$ as the fraction of the initial energy that is dissipated in the flashlamp rather than in $i^2 r$ loss in the circuit, and from Fig.3.10 obtain Fig.3.11. There $\eta_E(\%)$ has been plotted as a function of $\beta$ for $\alpha = 0.84$. The linear curve also shown in Fig. 3.11 was obtained from the work of MARKIEWICZ and EMMETT [3.13] who estimated losses due to the finite r. Their estimates are accurate only when the losses are relatively small ($\beta \lesssim 0.15$), but are more pessimistic than the exact results when $\beta \gtrsim 0.15$. Typical losses encountered in driving circuits amount to 5-15%.

It is obvious that it is possible to design an initially undamped pulse ($\alpha < 0.84$) so that addition of a finite r results in a pulse that is just critically damped. Another way of stating this is that solutions of (3.25) exist which represent values of ($\alpha, \beta$) such that the current pulse is just critically damped. The locus of all such points was first shown in [3.15]. As one

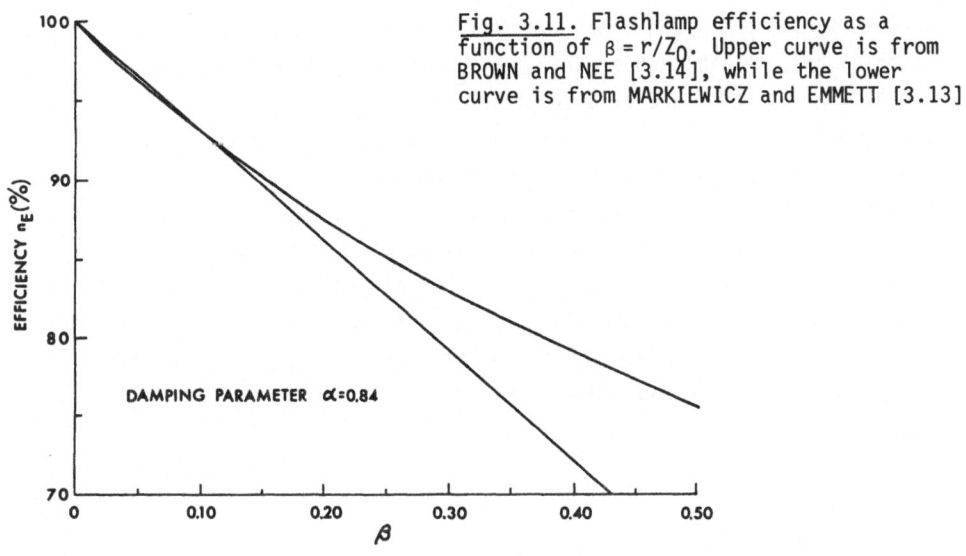

Fig. 3.11. Flashlamp efficiency as a function of $\beta = r/Z_0$. Upper curve is from BROWN and NEE [3.14], while the lower curve is from MARKIEWICZ and EMMETT [3.13]

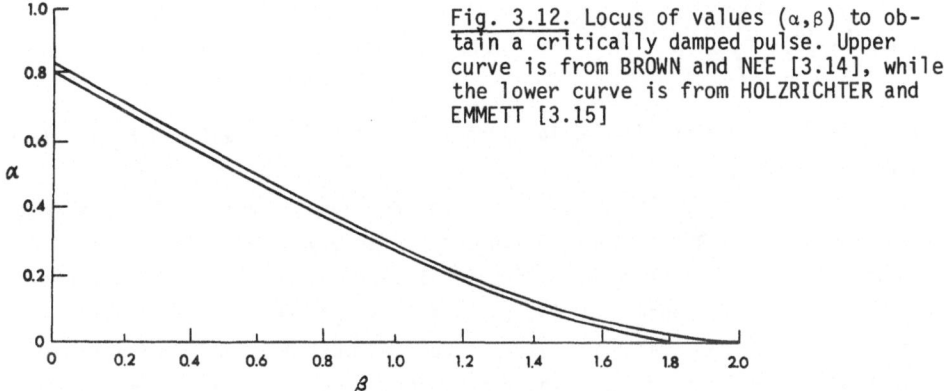

Fig. 3.12. Locus of values $(\alpha,\beta)$ to obtain a critically damped pulse. Upper curve is from BROWN and NEE [3.14], while the lower curve is from HOLZRICHTER and EMMETT [3.15]

extreme case, we have ($\beta=0,\alpha=0.84$) whereas at the other extreme $\alpha = 0$, a linear LRC circuit results; because $K_0 = 0$ it is necessary to solve

$$\frac{dI}{d\tau} + 2I + \int_0^\tau I\, d\bar{\tau} = 1 \quad , \tag{3.28}$$

where $\beta = 2$. In Fig.3.12, the more accurate upper curve is taken from [3.14], where for $\beta = 0$, $\alpha \simeq 0.84$, and for $\beta = 2$, $\alpha = 0$. The lower curve, reproduced from [3.15], was obtained by use of $\alpha \simeq 0.78$ for $\beta = 0$, whereas for $\alpha = 0$, $\beta \simeq 1.8$. In general, points $(\alpha,\beta)$ above the curve in Fig.3.12 are overdamped, whereas those below the curve are underdamped. The procedure for finding an optimal circuit to drive a given flashlamp is now given. We begin by calculating, using (3.21), the impedance parameter $K_0$, after L, D, and P have been chosen. Usually $E_0$, $\alpha$, and T are also specified. Calculations are done first for $\alpha \simeq$ 0.84, $\beta = 0$. The full width at 1/3 maximum $T_{1/3}$ of the current pulse is given approximately by $T_{1/3} \simeq 3T$; therefore the desired pulse width must be specified and T thus obtained. To find $E_0$, we must make some assumptions about the maximum amount of energy $E_L$ to be delivered to the lamp. This process will be described in Sect.3.6 in which the value of $E_L$ is determined by the fraction of explosion energy that must be used to obtain a particular desired lifetime from a lamp. After the parameter set $(K_0,\alpha,\beta,E_0,T)$ is obtained, it is possible to finish the PFN design. The initial energy $E_0$ stored in a capacitor C is

$$E_0 = \frac{1}{2} C V_0^2 \ . \tag{3.29}$$

Combining (3.29) with (3.24) yields C in terms of $K_0$, $E_0$, $\alpha$, and T,

$$C = \left( \frac{2E_0 \alpha^4 T^2}{K_0^4} \right)^{1/3} \ . \tag{3.30}$$

After finding C, we use $T = \sqrt{LC}$ to specify L. Finally, since C and $E_0$ are then known, we find $V_0$ from (3.29). We can then find the value of $\alpha$ for a real circuit in which $\beta \neq 0$. The resistance r can be found in a real PFN by direct measurement, using a resistance bridge or preferably by using the current ringing technique described by NEWELL [3.11]. We now have a set of parameters $(K_0,E_0,T,r)$ and we choose $\alpha < 0.84$. Using the design procedure mentioned for $(\alpha=0.84, \beta=0)$, we calculate $(C,L,V_0,Z_0)$. We then obtain $\beta = r/Z_0$ and determine from Fig.3.12 whether the pair intercepts the curve. If it does, then the trial $\alpha$ is the correct one; if not, a new $\alpha$ is specified and the process is repeated. It is easiest to do the design by use of a computer. Note that if the energy-transfer efficiency $\eta_E$ is desired, it should be calculated for the set of $(\alpha,\beta)$ obtained because Fig.3.11 is valid only for $\alpha = 0.84$.

It is fairly obvious to anyone working with Xe flashlamps that the arc does not fill the bore during the entire current pulse; as a consequence, the resistance of the arc is not constant. DISHINGTON [3.16] has developed an empirical theory in which the changing resistance R of the arc is taken into account by an equation of the form

$$R = \frac{k\ell}{\delta d i^{\frac{1}{2}}} \ , \tag{3.31}$$

where $\delta = d_a/d$ and $d_a$ is the inside diameter of the arc lamp and d is the instantaneous diameter of the arc. For most practical purposes, the added complication of this theory is not needed because the calculated peak current and pulse width usually agree closely with measured values. Hence, in flash-lamps common to large Nd:glass laser systems, variation of R is not an effect important in the design of driving circuits. This is apparently due to the long pulse durations (typically in the range of $300 \leq T_{1/3} \leq 600$ μs). A model that in all respects is equivalent to that of DISHINGTON [3.16] has been developed at LLL [3.17] and considers the voltage V across a lamp to vary according to

$$V = K_0 |i|^{\frac{1}{2}} + \frac{d}{dt} (Li) \quad , \tag{3.32}$$

where the first term is the same as (3.20) and the second is a time-dependent inductance that approximates the change of inductance in the lamp during the arc-growth process. The form

$$L = L_0 \, e^{-t/\tau} \tag{3.33}$$

was assumed; both L and $\tau$ are found by fitting experimental data from the lamp to be modeled. The accuracy may be judged from the result shown in Fig. 3.13, where calculated and measured values of V(t) are shown as functions of t.

From the foregoing it should be fairly obvious that a flashlamp pulse whose duration $T_{1/3}$ is comparable to the fluorescence decay time $\tau$ is not optimum,

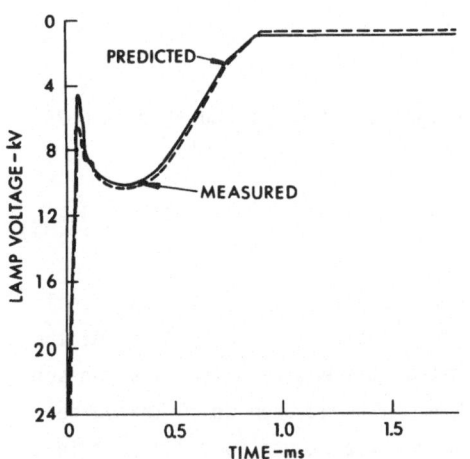

Fig. 3.13. Comparison of model-predicted and measured output of a flashlamp circuit [3.17]

primarily because decay losses reduce the stored-energy density to a value far less than the peak that would be obtained if $T_{1/3} \ll \tau$. With current flash-lamp designs, it is not possible to deliver the required amount of energy in a time short compared to $\tau$ because the explosion limits of the flashlamps prevent this (Sect.3.6 below). Instead, an alternative used recently by TRENHOLME [3.17] can be considered. He addressed the question as to which of all possible pump-pulse shapes is optimum for pumping an exponentially decaying process, if some relationship is assumed between the input power and the pump rate. That relationship, in a coupled case, is

$$\frac{dE_s}{dt} = \eta P(t) - \frac{E_s}{\tau} \quad , \tag{3.34}$$

where $E_s$ is the stored-energy density, $P(t)$ the pump power, and $\eta$ a transfer efficiency that depends upon the current density J. By considering a waveform $P(t)$ that dissipates an amount of energy

$$W = \int_{-\infty}^{+\infty} P(t)dt \quad , \tag{3.35}$$

and requiring that the pumping should terminate at the time of the peak gain because any further pumping is wasted, Trenholme used an Euler-LaGrange variational approach and obtained

$$P(t) = \frac{3W}{2Q\tau} e^{3t/2Q\tau} \quad , \tag{3.36}$$

where Q is an empirical parameter that characterizes the assumed variation of $\eta$ with J (or P), according to

$$\eta(J) \sim P^{-2/3Q} \quad . \tag{3.37}$$

For disk amplifiers modeled by TRENHOLME [3.17], it was found that $Q \simeq 0.75$. Hence, (3.36) becomes

$$P(t) = \frac{2W}{\tau} e^{2t/\tau} \quad . \tag{3.38}$$

Thus the optimum pulse for driving a laser amplifier is a rising exponential with a rise time approximately one half the fluorescence decay time. At the time of the peak $E_s$, the pulse is terminated; the entire pulse is shown schematically in Fig.3.14. This pulse, which has not yet been demonstrated because of the complexity of the required circuit elements, would result in a

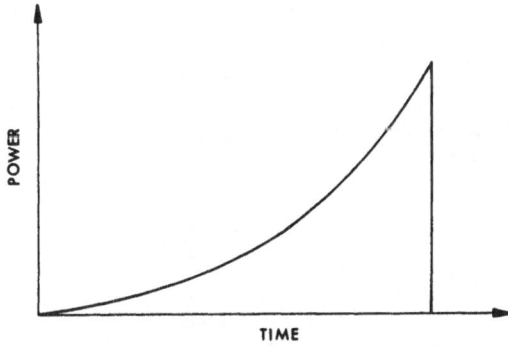

16% increase of the maximum value of $E_s$ over that obtained with a conventional single-mesh circuit. Equivalently, for a fixed gain 26% less pump energy [3.17] would be required. The importance of this analysis is that it establishes a limit on the improvement of pumping efficiency that can be obtained with conventional Xe flashlamps. To apply this analysis to other types of amplifiers, it is necessary to find the appropriate value of Q in (3.36).

Equation (3.34) is a standard first-order approximation of how $E_s$ varies in time while being pumped by some power P(t); it has been used by TRENHOLME [3.18], for instance, to investigate ASE. It is also useful to consider uncoupled performance, as in active-mirror amplifiers, for instance. Then since the relative transfer efficiency $\eta \sim 1$, we have

$$\frac{dE_s}{dt} = P(t) - \frac{E}{\tau} \quad . \tag{3.39}$$

Equation (3.39) may be solved by use of numerical techniques that employ experimentally determined values of P(t). For many first-order calculations, however, it is appropriate to seek analytical solutions; a number may easily be obtained. Here we consider only a half-sine pulse, although other more complicated solutions that involve pump pulses that vary as $\sin^2$ or $\sin^3$ are available. In particular, we assume that P(t) varies according to

$$P(t) = P_0 \sin\omega t \quad , \tag{3.40}$$

where $\omega = \pi/T$ and T is the total pulse length. Then the solution of (3.39) is

$$E_s(t) = \frac{E_0}{\omega} \left[ \omega(e^{-t/\tau} - \cos\omega t) + \frac{1}{\tau} \sin\omega t \right] \quad . \tag{3.41}$$

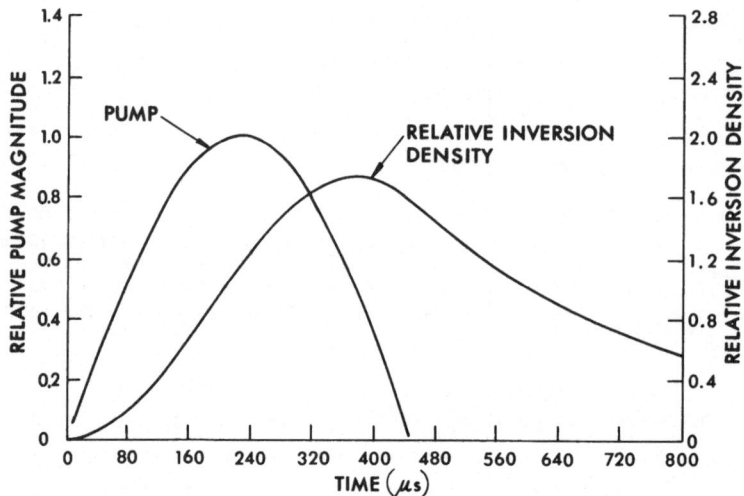

<u>Fig. 3.15.</u> Flashlamp pump and amplifier inversion density as a function of time using a half-sine pump approximation

Here the total energy $E_0$ stored initially is given by

$$E_0 = P_0 \omega / \left[ \omega^2 + \left( \frac{1}{\tau} \right)^2 \right] \quad .$$ (3.42)

Equation (3.41) can be parameterized according to the prescription

$$\left[ \eta(t) = \frac{E_s(t)}{E_0} \quad ; \quad \alpha = t/T \quad ; \quad \beta = \tau/T \right]$$ (3.43)

to obtain

$$\eta(t) = e^{-\alpha/\beta} - \cos\pi\alpha + \frac{1}{\pi\beta} \sin\pi\alpha \quad .$$ (3.44)

In Fig.3.15 we show an example of a half-sine pulse of duration $T \simeq 450$ μs pumping an amplifier with $\tau = 350$ μs. After the peak of $E_s$, which occurs $\simeq 100$ μs after the peak of the pump pulse, $E_s$ decays as $\exp(-t/\tau)$.

Although the optimum design of an amplifier involves the numerical solution of (3.39), a simple approximation for $P(t)$ such as (3.40) can be used to obtain a great deal of useful information, for instance, on how the maximum $E_s$ varies with T, or how $E_s$ is degraded if $\tau$ is reduced in an amplifier, owing to the presence of $H_2O$ impurity in the glass.

## 3.5 Slope Efficiency in Nd:Glass Amplifiers

It is an experimentally demonstrated fact that, in Nd:glass amplifiers pumped
by Xe flashlamps, if stored-energy density or gain is plotted as a function
of pump energy then the slope (efficiency) of the curve is not constant but
decreases as the pump energy increases. Examples of this can be found, for
instance, by consulting LLL annual reports (UCRL-50021-73 to UCRL-50021-76).
We discuss here the various mechanisms responsible for the rolloff and recent
unpublished results that make it possible to sort out the separate effects
unambiguously.

The first effect is heating or $i^2r$ loss in each flashlamp/PFN network. As
the applied voltage $V_0$ increases, so does the peak current, leading to significant losses due to resistive heating at the highest voltage. Physically,
proportionally less of the initial stored energy $E_0$ is being dissipated by
the flashlamp. The second effect is due to the opacity of the flashlamp. Most
amplifiers use some reflector geometry behind each flashlamp to redirect a
portion of the radiation towards the Nd:glass being pumped. Depending upon
the geometry, some fraction of the redirected light is passed through the Xe
plasma, where it is, depending upon wavelength, absorbed partially or totally.
The effect is to heat the plasma, resulting in a shift of the peak output of
the lamp towards the blue. If the effect is large enough, the pumping process
becomes less efficient. The third and final effect we consider here is due
to ASE or parasitics. If a parasitic oscillation occurs, an abrupt "clamping"
of the gain or stored-energy density $E_s$ may result so that further increases
in $V_0$ result in no increase of $E_s$. Parasitics do not always turn on in such
an abrupt manner, however, and it is sometimes impossible to tell physically
whether ASE or a parasitic oscillation is occurring, for example in an active-
mirror amplifier (Chap.4). Both effects lead in some cases to a gradual rolloff of the gain and may be partly responsible for any observed rolloff of
amplifier-slope efficiency.

To find which effects occur in any given pumping situation, KELLY et al.
[3.19] employed a fluorescence monitor. The fluorescence monitor was first
devised by HOLZRICHTER and DONICH [3.6] to obtain relative values of the
stored-energy density when ASE and parasitics do not occur because the incident flux is low. It also has the property that it samples only 1.06 µm
fluorescence and not 1.06 µm flashlamp radiation as well. Note that it is
assumed that fluorescence is proportional to stored-energy density which
should be the case when ASE or parasitics are not present, subject to small
(≈5%) corrections because of the fluorescence line shape [3.20]. Figure 3.16

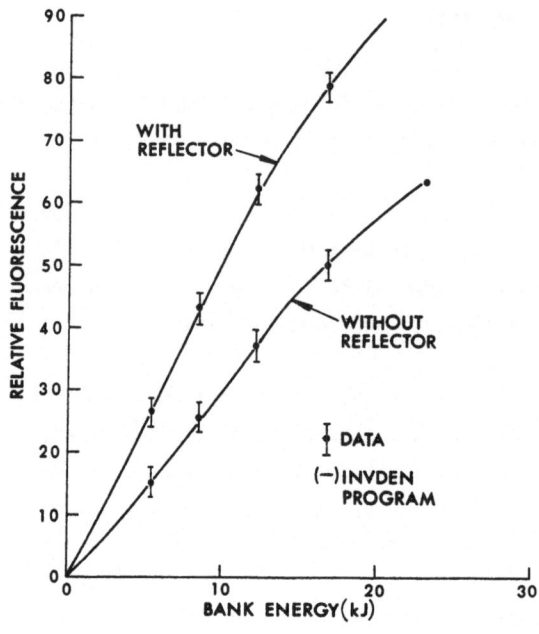

Fig. 3.16. Relative fluorescence as a function of PFN bank energy with and without a reflector, for LHG-8 laser glass [3.9,19]

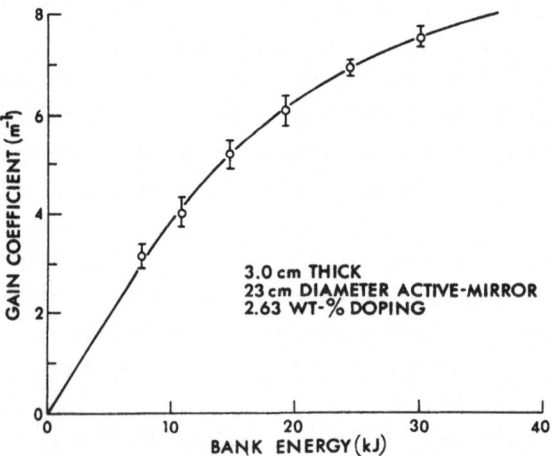

Fig. 3.17. Gain coefficient as a function of PFN bank energy for a 23 cm diameter active-mirror amplifier with an LHG-8 disk [3.9]

shows a result obtained by pumping a fluorescence monitor containing LHG-8 laser glass with a linear flat array of 13$\phi$ Xe flashlamps. Two curves were obtained, one with and the other without a flat reflector behind the array of the fluorescence voltage as a function of applied voltage. The slope of the lower curve is entirely due to the effects of $i^2r$ loss and the way in which the lamp output is convoluted with the glass absorption spectra. The upper curve, however, also includes the effect of lamp plasma opacity, which has been modeled by KELLY et al. [3.19] by running ≈89% of the flashlamp output back through a plasma plate 0.24 cm thick.

Upon completion of the above experiments, KELLY et al. [3.19] performed small-signal gain measurements on a 23 cm diameter active-mirror amplifier containing a 3 cm thick LHG-8 disk, pumped by the identical 13ϕ array of flashlamps. In Fig.3.17 we show the active-mirror small-signal gain as a function of bank energy. The solid curve was generated by the computer code INVDEN, described in Chap.5. A comparison of the data in Figs.3.16,17 shows that an additional rolloff is present in the active-mirror amplifier, from ASE, and that the loss due to ASE is ≈20%.

The combination of experiments performed using fluorescence monitors and amplifiers, along with a modest modeling capability, have allowed for the first time an unambiguous sorting out of the physics affecting slope efficiency in amplifiers.

## 3.6 Lifetime Limits of Xe Flashlamps

When large-bore Xe flashlamps are employed to pump Nd:glass amplifiers, it is advantageous to maximize the lifetime of the lamps. We explore here the limits on lamp lifetime and show how the design of an amplifier is thus influenced.

The first problem to be discussed is the explosion limit of Xe flashlamps. If a lamp of a given diameter is increasingly loaded (J/cm of arc length), a point will be reached where the lamp will catastrophically fail. Such failure normally results in damage to expensive optical components in an amplifier and thus should be minimized. The explosion lifetime L of a lamp can be calculated from [3.21]

$$L = \left(\frac{E_x}{E}\right)^{\beta} \quad , \tag{3.45}$$

where $E_x$ is the explosion enery, E the energy delivered to the lamp, and β a dimensionless exponent. Note that

$$E = \eta_E E_0 \quad , \tag{3.46}$$

where $\eta_E$ was calculated and shown previously (Fig.3.11) for a critically damped pulse. The explosion energy $E_x$ is given by [3.21]

$$E_x = a\ell DT^{1/2} = a\ell D(LC)^{1/4} \quad . \tag{3.47}$$

Here a is a constant, $\ell$ the arc length, D the inside diameter of the flash-lamp, and T the pulse-width defined in (3.24). The constants a and $\beta$ are normally determined experimentally and can be obtained for almost any size lamp from manufacturers data sheets. Accepted values for Xe flashlamps are [3.21]

$$\left\{ \begin{array}{l} a = 2.46 \times 10^4 \ [W(s)^{\frac{1}{2}}/cm^2] \\ \beta = 8.5 \end{array} \right\} \tag{3.48}$$

which are widely used to calculate $E_x$ and L. Recently, however, studies of $19\phi$ and $27\phi$ lamps have produced values of a and $\beta$ substantially different than those [3.22,23]. In [3.22] for example, tests on $19\phi$ lamps showed that $a \approx 1.51 \times 10^4$ [W(s)$^{\frac{1}{2}}$/cm$^2$] and $\beta \approx 15.6$ for Xe flashlamps that have clear fused-quartz envelopes. Significantly different values were obtained with lamps that had Germasil envelopes.

In all of the above, the values of a and $\beta$ are for Xe flashlamps operated in a $N_2$ atmosphere. To date, no similar studies have been done using $H_2O$ jackets around the lamp for active cooling. It is suspected, but not yet proven, that such an arrangement is beneficial to lamp lifetime since the $H_2O$ should, in effect, increase the thickness of the flashlamp envelope, and because scratches in the envelope which otherwise absorb and cause local heat-ing, would be index matched and, in effect, eliminated. Because the effect of the immediate environment on explosion energy has not been studied, it is standard conservative practice to calculate $E_x$ and L using the values of a and $\beta$ given in (3.48). For a reasonable lamp lifetime it is usual to run such Xe flashlamps in the range $0.30 \le (E_x/E)^{-1} \le 0.35$, giving L in the correspond-ing range $28000 \ge L \ge 7500$.

The explosion lifetime L places strict limits on the operation of Xe flash-lamps, if a reasonable flashlamp lifetime is expected. It is usually desir-able to minimize the number of flashlamps in a laser amplifier cavity because of cost and/or space constraints, thus it is necessary to maximize the J/cm of arc length. The lifetime L places a limit on that value, however, and it is necessary to increase the duration T of the pump pulse to limit L to a desired value when the lamp loading is increased. Because, however, a decay process with a finite lifetime $\tau$ is being pumped, increasing T beyond a cer-tain value is inefficient. Such effects were studied by HOLZRICHTER and DONICH [3.6] who compared experimental pumping with a simple theoretical model dev-

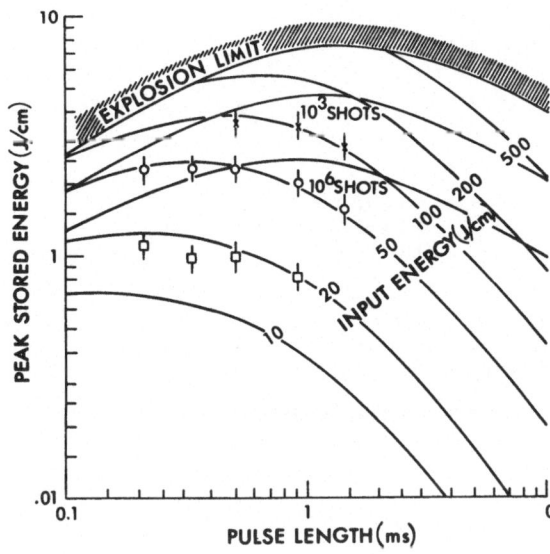

**Fig. 3.18.** Peak stored energy as a function of PFN pulse length for 1 cm diameter Xe flashlamps pumping a 1 cm ED-2 disk. Contours show flashlamp input energy (J/cm) and explosion limits [3.6]

eloped by Trenholme at LLL, utilizing the half-sine pump pulse described in Sect.3.4. Their results are shown in Fig.3.18, in which peak stored-energy density is plotted as a function of pulse duration for energy loadings of 10-500 J/cm. The Xe flashlamps were pumping ED-2 silicate laser glass with a fluorescence lifetime of 310 µs. To obtain a lifetime of $10^3$ shots, for instance, it is necessary to operate at a pulse duration greater than ≈600 µs if an energy loading of 100 J/cm is desired. A similar study was performed by NEWELL [3.11], who obtained numerous curves of relative fluorescence intensity as a function of dissipated energy (J/cm) for various fixed pulse durations. That study was also noteworthy because it investigated the effect of Xe fill pressure upon the output of the lamp; only a weak dependence was found.

In addition to explosive failure, other failure modes can be particularly severe in Xe flashlamps that are operated in a $H_2O$ environment. Coatings may appear on the quartz envelope that significantly reduce light output, if sufficiently pure $H_2O$ is not used [3.24]. The coating apparently results from corrosion of the Xe flashlamp. It can be eliminated by use of Ni plating and filtered high-purity $H_2O$. In systems that are conservatively designed for large L and in which attention has been paid to $H_2O$ quality, the normal mode of failure is lack of triggering. Recently, ABATE et al. [3.25] investigated the influence of large amounts of impurities on the output of a 19$\phi$ Xe flashlamp, as measured by a fluorescence monitor. In one experiment, large amounts

124

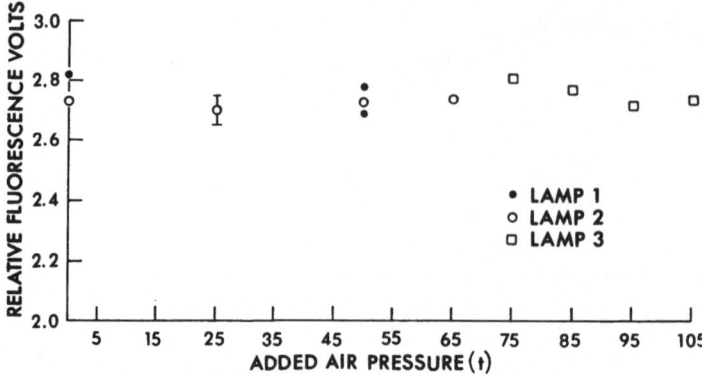

<u>Fig. 3.19.</u> Relative fluorescence as a function of added air pressure for three 19φ Xe flashlamps [3.25]

of air were added to a lamp that initially contained 300 t of pure Xe. In Fig.3.19, the relative fluorescence is shown for three lamps as a function of added air pressure. Triggering, in a standard arrangement, was observed up to 105 t of air or 35% of the initial fill. The fluorescence was unchanged in that range, indicating that Xe flashlamp performance is unaffected by the presence of such large amounts of impurity. The main effect is failure of the lamp to trigger. Flashlamps operated in $N_2$ or $H_2O$ environment are subject to leaks in seals or small micro-cracks in the envelope. At LLE, no explosive failure occurred in over 25,000 firings of 19φ lamps with 30.5 cm arc length, operated at ≈30% of their explosive energy. The failure of 12 lamps to trigger was investigated and in 11 out of 12 was found to be due to the presence of massive quantities of $H_2O$ inside the lamp [3.25], which indicates that the lamps were either initially defective or developed small leaks from the $H_2O$ environment.

## 3.7 Time-Resolved Xe Flashlamp Spectra

We have seen in Sect.1.1 examples of time integrated Xe flashlamp spectra. A simple model of the output spectra was discussed in Sect.3.3. Here we show recent results obtained by KELLY et al. [3.9] at the Laboratory for Laser Energetics of the time-resolved spectra of 13φ Xe flashlamps. Time-resolved spectra have been obtained previously by a number of workers, including HILL [3.26] and BAKER and STEED [3.27]. The most thorough attempt was by CHURCH

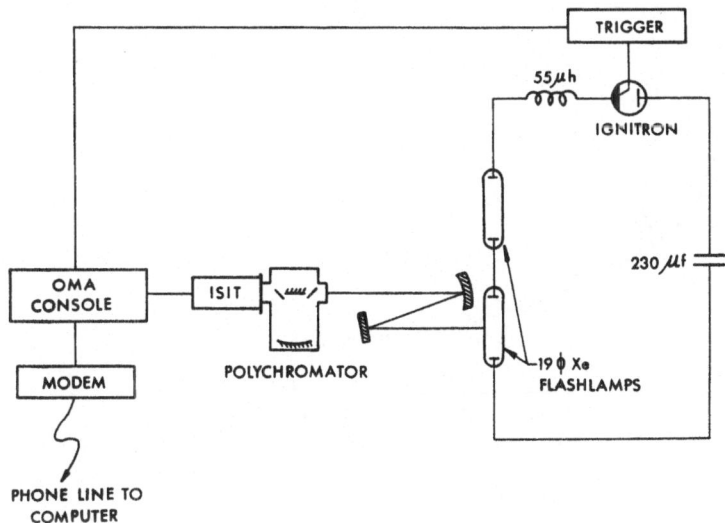

Fig. 3.20. Experimental configuration used to measure time-resolved spectra
of 13φ Xe flashlamps [3.9]

and co-workers [3.7,28-31], who developed rapid-scan spectrometers for that
purpose. Although high resolution was obtained and the entire spectral range
of interest covered, time resolution was not sufficient (50-100 μs/scan).
Time-resolved spectra were not published, but used to test the consistency
of an elaborate model mentioned previously [3.7].

To be useful in studies of Nd:glass laser amplifiers, the time-resolved
spectra obtained should cover the entire spectral range of interest (approximately 0.35-1 μm), should have a spectral resolution of ≈0.5 nm so that line
structure can be investigated, and should have a time resolution of ≈1 μs.
All of these requirements are met by use of an optical multichannel analyzer
(OMA) [3.32], using an ISIT (intensified silicon intensified target) head,
which may be gated. The device is placed in the focal plane of a polychrom-
ator, as shown in Fig.3.20. The signal is obtained by gating the ISIT with
respect to the Xe flashlamp current pulse; by adjusting the delay, a time-
resolved slice of the spectrum is obtained, and the resulting data are sent
to a computer, by which they are corrected for spectral response, ISIT distortions, etc. Two 19φ Xe flashlamps with a 30 cm arc length were driven in
series by a PFN with C = 230 μF and L = 55 μH. Data were obtained in the range
$0 \leq t \leq 620$ μs; the current FW1/3M was ≈500 μs. Examples of the data are shown
in Figs.3.21,22 where time-resolved spectra are displaced at t = 100 μs, 420 μs
into the flash. The absolute intensity was measured in the range 0.37-0.90 μm.

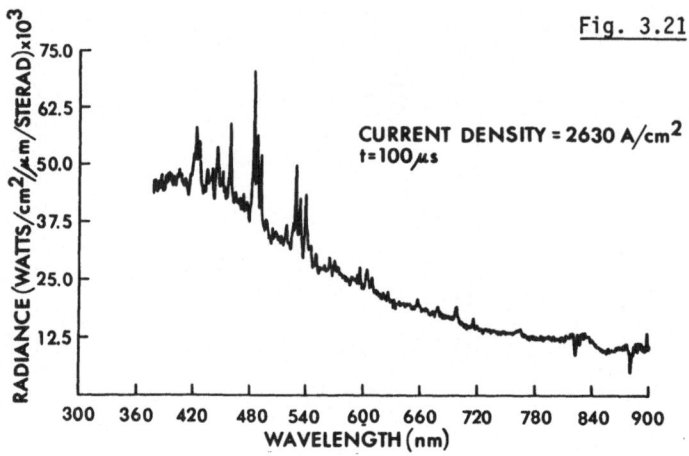

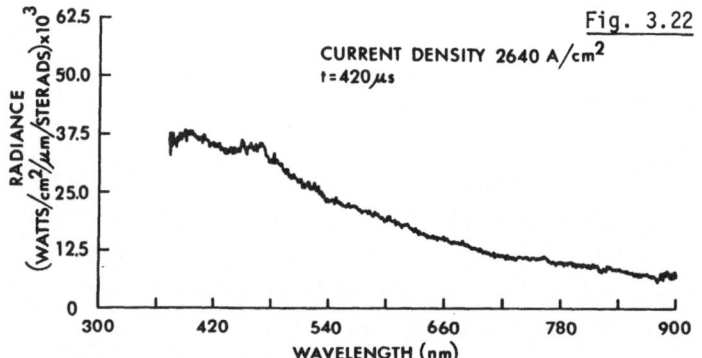

Fig. 3.21. Spectral radiance of a 13φ Xe flashlamp as a function of wavelength 100 μs after the beginning of the flash and a current density of 2.63 kA/cm² [3.9]

Fig. 3.22. Spectral radiance of a 13φ Xe flashlamp as a function of wavelength 420 μs after the beginning of the flash and a current density of 2.64 kA/cm² [3.9]

The results show that, in the beginning of the flash line spectra are prominent, superimposed upon the broad background, in agreement with earlier observations. As time advances, the background comes to dominate so that at t = 420 μs, for example, the line spectra are totally absent. Indeed, for identical current densities on the positive and negative slope of the current pulse, the spectra are substantially different since the line spectrum present on the positive side is almost totally absent on the negative. This is in disagreement with the Trenholme-Emmett flashlamp model discussed previously [3.8] which predicts that for a given current density J, the spectrum is always the same. Another difference is that different lines, line widths, and intensities are found for the most prominent Xe lines, also in disagreement with [3.8]. The absence of line spectra late in the flash is interpreted as being due to the greater degree of ionization later as compared to the beginning of the pulse, as well as pressure and Stark broadening.

These data on 13ϕ lamps have been extremely useful in constructing a time-dependent pumping model to be used in designing optimized active-mirror amplifiers. The entire model has used the experimentally derived flashlamp data and incorporates multi-phonon, $H_2O$ quenching, and radiative relaxation processes discussed in Chap.1. It will also be used to investigate time-dependent processes such as the use of organic dyes or other wavelength converters to make the pumping of Nd:glass by Xe flashlamps more efficient [3.19].

# 4. Amplified Spontaneous Emission and Parasitic Oscillations in Nd : Glass Amplifiers

In early research on Nd:glass amplifiers, it was discovered that scaling performance to larger diameters was often invalid; in some cases the expected values of certain parameters were not even approached. That experience demonstrated that there are certain limits to the size of large-aperture amplifiers, determined primarily by the complementary effects of amplified spontaneous emission (ASE) and parasitic oscillations (PO). These effects result in less energy storage or gain/cm, for example, when an amplifier is scaled up to a larger diameter. The diameter and gain of a stage are important in determining the X-factor or ultimate power capability for short pulses as we will investigate in Chap.7. When X-factor analysis, including both ASE and PO, is combined with cost, optimized system designs may be arrived at (Chap.8).

We treat here, first, the general problem of ASE, in Sect.4.1, followed by parasitic oscillations in disk and active-mirror amplifiers (Sect.4.2) and in rod amplifiers (Sect.4.3).

## 4.1 Amplified Spontaneous Emission in Nd:Glass

It is well known from the Einstein theory that the rate of spontaneous emission $A_{mn}$ and stimulated emission $B_{mn}$ between two energy levels m and n are intimately related. This is shown by the well-known relationship between the two given by

$$A_{mn}/B_{mn} = \frac{8\pi h \nu_{mn}^3}{c^3} , \tag{4.1}$$

where h is Planck's constant, $\nu$ the frequency of the transition, and c the speed of light. Thus, in any amplifier, spontaneous emission of photons is accompanied by the process of stimulated emission, whereby the initial photons of intensity $I_0$ are amplified according to

$$I = I_0 \, e^{\alpha \ell} \quad , \tag{4.2}$$

where I is the intensity after traveling through a medium of length $\ell$ that is characterized by a positive gain coefficient $\alpha$. The total emission rate $S_T$ is given by (assume $A_{mn} = A$, $B_{mn} = B$)

$$S_T = A + B = A\left(1 + \frac{B}{A}\right) \quad , \tag{4.3}$$

or, if we let $\xi = B/A$,

$$S_T = A(1+\xi) = MA \quad , \tag{4.4}$$

where

$$M = 1 + \xi \quad . \tag{4.5}$$

The lifetime $\tau$ of a transition has been seen to be related to A by

$$A = \frac{1}{\tau} \quad . \tag{4.6}$$

The total rate $S_T$ is characterized from (4.4) by a lifetime $\tau'$ given by

$$\tau' = \tau/M \quad , \tag{4.7}$$

that is, the lifetime $\tau$ in the absence of stimulated emission is decreased by a factor $(1/M)$ in its presence. In an amplifier in which ASE is important, a major effect is to reduce the fluorescence lifetime, which decreases the efficiency of pumping by a source whose pulse duration is constant and usually comparable to or greater than $\tau$. Chapter 3 showed that by use of a simple pumping model (3.39) the peak stored-energy density or gain can be calculated as a function of pump level or $\tau/T$ where T is the pump-pulse duration in the half-sine approximation. In the presence of ASE, (3.39) must be modified to take account of the new fluorescence decay time, according to

$$\frac{dE}{dt} = P(t) - \frac{E}{\tau'} = P(t) - \frac{ME}{\tau} \quad . \tag{4.8}$$

By use of this equation and its solution, it is easy to find how the peak stored-energy density is modified in the presence of ASE.

The general problem of ASE, and its effect upon optical amplifiers was considered early by TONKS [4.1], who investigated the rod-amplifier geometry. Using analytic techniques, he showed that the maximum attainable inversion was insensitive to rod aspect ratio and that total-internal reflections can be very important. SIBERT and TITTEL [4.2] used a Monte Carlo technique for the first study of ASE in large-aperture disk amplifiers. By far the most complete and careful study was done by TRENHOLME [4.3] who investigated both ASE and PO in spherical, circular-disk, and elliptical disk amplifiers.

The case of the sphere, which is of no practical interest, is amenable to analytic calculation and was the test case used to evaluate the Monte Carlo program used in this study. In all that follows, only single pass ASE is considered, and the boundary is assumed to be totally absorbing. The inversion density is assumed to be uniform throughout the sample, which is not the case in practice. If the material is characterized by a gain coefficient $\alpha$, it can be shown [4.3] that, for a sphere, the quantity $\xi$ is given by

$$\xi = \frac{3}{2\beta} \left[ \frac{2e^{\beta}}{\beta} \left( 1 - \frac{1}{\beta} \right) + \frac{2}{\beta^2} - 1 \right] - 1 \quad, \tag{4.9}$$

where $\beta = \alpha D$ and $D$ is the diameter of the sphere. The spontaneous emission in any laser material, as we have seen previously, is not monochromatic, but distributed with some effective bandwidth $\Delta\lambda_{eff}$. Although the actual profile is usually complicated, it is not difficult to evaluate $\xi$ by averaging over some standard line profile. As an example, for a Lorentzian profile with center wavelength $\lambda_0$, peak gain coefficient $\alpha_p$, and full-width W, $\alpha$ is given by

$$\alpha = \frac{\alpha_p}{1 + 4\left(\frac{\lambda - \lambda_0}{W}\right)^2} \quad. \tag{4.10}$$

By averaging $\xi$ over the Lorentzian profile, we can find the average ratio $\xi_L$ of stimulated to spontaneous emission for a Lorentzian [4.3]

$$\xi_L = \sum_{n=1}^{\infty} \left[ \frac{3\beta^n}{(n+3)(n+1)} \prod_{j=1}^{n} \left( \frac{2j-1}{2j} \right) \right] \quad, \tag{4.11}$$

where now $\beta = \alpha_p D$. Similar results were obtained for a flat-topped and Gaussian profile.

A computer program, known as ZAP, was developed to trace optical power flow in a large number of geometric systems. For this study, up to 2500 random rays were traced resulting in an accuracy of 2-3% in the calculation of $\xi$. In Fig.4.1, results are shown for both the analytical and ZAP-generated

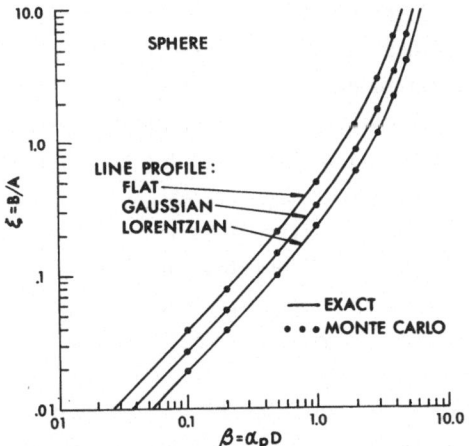

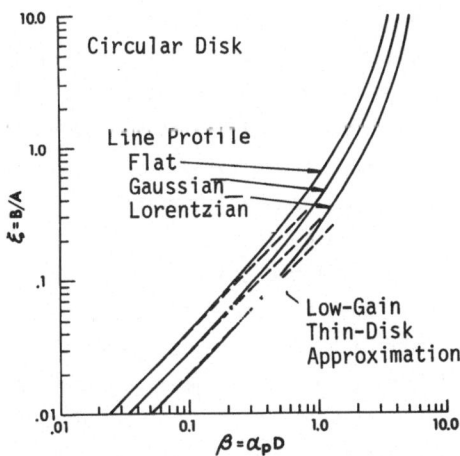

Fig. 4.1. Ratio of stimulated to spontaneous emission in a sphere with uniform inversion density, as a function of the peak gain-diameter product ($\beta=\alpha_pD$), as calculated by exact and Monte Carlo methods, for flat, Gaussian, and Lorentzian line profiles [4.3]

Fig. 4.2. Ratio of stimulated to spontaneous emission in a circular disk with uniform inversion density, as a function of the peak gain-diameter product ($\beta=\alpha_pD$), as calculated by exact and Monte Carlo methods, for flat, Gaussian, and Lorentzian line profiles [4.3]

cases ($\xi$ as a function of $\beta$), and for the flat, Lorentzian, and Gaussian profiles. The agreement is seen to be excellent, and shows that the Lorentzian profile is preferred, to reduce ASE. Note that $\xi$ is linear up to $\beta \sim 1$, but that for $\beta > 1$ pumping becomes more and more difficult.

This work was extended to the case of a circular disk, where again ZAP was used. Here reflections from the disk faces were taken into account, and averaged over the two principal polarizations. Uniform inversion density was also assumed. In Fig.4.2, we show the result for a circular disk of laser glass with an aspect ratio of 5:1 and $n_d = 1.56$. $\xi$ was evaluated for the usual three profiles. The results are qualitatively the same as in Fig.4.1 for the sphere. Curves obtained from an analytic, low-gain, thin-disk approximation [4.3] are also shown in Fig.4.2. The quantities $\xi$ and M are dependent not only upon the product $\alpha D$ but also to a lesser extent upon the disk aspect ratio and the ratio of the refractive index of the disk to that of the medium that is in contact with its faces. By use of the results presented in [4.3], it is easily possible to assess ASE effects in disk or active-mirror amplifiers with uniform energy-storage profiles. All disk and active-mirror devices, however, have rather nonuniform energy profiles, as was first shown by SOURES et al. [4.4]. Such effects will be considered in detail below (Sect.4.3) in connection with PO. Recently, BROWN et al. [4.5] showed that nonuniform inversion profiles in active-mirror and disk amplifiers cause thickness-dependent ASE losses. They used a phenomenological method to evaluate the effect. They also showed that the ASE loss was dependent upon glass composition; phosphates have much larger ratios of surface to average gain than silicates have. Fluorophosphates, amongst commercially available glasses, minimize ASE because the stored-energy density slopes are least in these glasses. This is shown, for example, by Table 2.3 in which values of $(dE/d\nu)$ are tabulated for various compositions. By use of Monte Carlo techniques with a spatially dependent gain coefficient $\alpha$ ASE in real disk or active-mirror amplifiers could be simulated, but no such studies have yet been reported.

In real amplifiers, a considerable amount of radiation is emitted by Xe flashlamps in the vicinity of 1.06 μm, as is shown by Fig.3.6. It has been suggested [4.6] that such radiation may significantly reduce pumping efficiency in disk amplifiers, because of amplification in the active laser material. Practically no pumping of $Nd^{3+}$ occurs below $\approx 0.9$ μm, hence it is worthwhile using filters or coatings to absorb the large amount of energy at longer wavelengths which otherwise produces only heat in the amplifier and increases the thermal-cycling time. Although $Nd^{3+}$ radiates predominantly by the transition $({}^4F_{3/2} \rightarrow {}^4I_{11/2})$ at 1.06 μm, the transition $({}^4F_{3/2} \rightarrow {}^4I_{13/2})$ at 1.35 μm may also produce a large ASE loss, in certain situations. In particular, although the cross-section and branching ratio for the latter are typically less than 1/4 and 1/3 of the former (see Chap.1), in amplifiers in which the average $\alpha D$ is great, the peak value due to nonuniform deposition can be at least six times as great in phosphate glasses [4.5]; the corresponding gain-diameter product at 1.35 μm can be great enough in some cases to warrant inclusion of it when pumping is modeled.

Finally, although we will not consider the process in detail here, ASE can be very important in large laser systems that are used to irradiate laser-fusion targets. If sufficient ASE arrives on a target prior to the main laser pulse, damage or complete destruction may result. For that reason, Pockels cells are placed at strategic locations when a system is staged. The amount of ASE that can be tolerated is normally dictated by the size and susceptibility of the targets whose use is anticipated; that in turn strongly influences the laser-system design. FOUNTAIN et al. [4.7] have calculated the ASE energy incident on a target in the LLL Cyclops system and showed good agreement with experimentally measured levels. Similar work by SAMPATH and LEWIS [4.8] on the GDL and OMEGA systems at the Laboratory for Laser Energetics has also yielded good agreement with experiments. KOSTOMETOV and ROZANOV [4.9] investigated ASE in disk amplifiers and calculated the angular distribution of radiation emitted by such devices.

## 4.2 Parasitic Oscillations in Disk and Active-Mirror Amplifiers

We have, thus far in this chapter, considered the general problem of ASE in disk and active-mirror amplifiers. A more specialized case of ASE is when feedback occurs, for instance, by specular or diffuse reflection at a disk edge. If a ray is reflected back upon itself in a direction in which the process may continue indefinitely, we encounter the second major effect that may limit energy storage in an amplifier, parasitic oscillations (PO). The condition for the occurrence of PO is that the round-trip gain just exceeds the losses. SWAIN et al. [4.10] and McMAHON et al. [4.11] observed PO in large-aperture disk amplifiers. The same phenomenon was encountered by McMAHON et al. [4.12] in rod amplifiers, which will be the subject of Sect.4.4.

We first consider lossless parasitic modes in a bare-disk geometry, as shown in Figs.4.3,4. TRENHOLME [4.3], in the first detailed analysis of PO, showed that lossless bulk modes could exist in disks in which all reflections are by total internal reflection. The dual condition

$$\left[\begin{array}{l} \theta > \theta_c = \sin^{-1}\left(\dfrac{n_1}{n_2}\right) \\[3mm] \phi > \phi_c = \sin^{-1}\left(\dfrac{n_3}{n_2}\right) \end{array}\right] \tag{4.12}$$

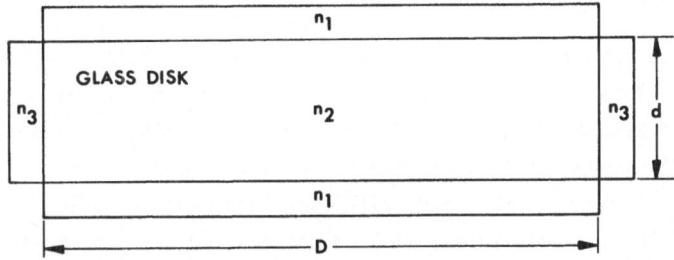

Fig. 4.3. Disk geometry showing the disk faces in contact with a medium of index $n_1$, edges in contact with a medium of index $n_3$, and having thickness and diameter d and D, respectively

134

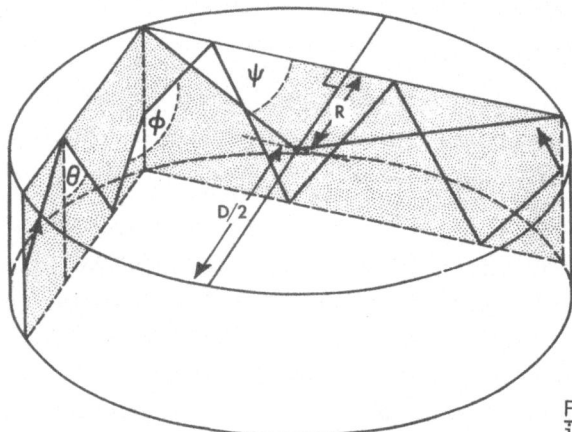

Fig. 4.4. Lossless bulk modes in a disk

must be satisfied for oscillation. $\theta$ and $\phi$ are the angles of incidence on the disk face and edge, respectively (Fig.4.4). $\psi$ is the angle between the ray path and the diameter, and satisfies

$$\cos\phi = \cos\psi \, \sin\theta \quad ,$$
(4.13)

so that

$$\cos\psi \geq \frac{\cos\phi_c}{\sin\theta_c} = \frac{n_2}{n_1}\left[1 - \left(\frac{n_3}{n_2}\right)^2\right]^{\frac{1}{2}}$$
(4.14)

for oscillation. There is a minimum value $\psi = \psi_0$ for which the threshold occurs. It is easy to see from Fig.4.4 that oscillation may occur outside the radius R given by

$$R = \frac{D}{2} \sin\psi_0 \quad .$$
(4.15)

Consequently, an annulus between $R \leq r \leq D/2$ may have its energy density reduced. To suppress this lossless mode, it is necessary that $n_3 > n_2$. In all cases of practical interest, $n_1$ is the index of air or $N_2$, hence $n_1 < n_2$ and total internal reflections may still take place at the disk faces. If the gain is great enough, a lossy bulk mode may still oscillate. This is principally because the reflection coefficient $R < 1$ at the disk edge may be great enough to allow oscillation. It can be shown [4.3] that the threshold condition for this lossy bulk PO is given by

$$R \exp\left[\frac{n_2}{n_1} (\alpha D)\cos\phi\right] = 1 \quad , \tag{4.16}$$

where R is the largest of Fresnel's reflection coefficients at the disk edge, D the disk diameter, and $\alpha$, the peak gain coefficient is given by (1.24).

It has been shown [4.3] that the lossy bulk mode oscillates first when $\phi = 0$, or the multiple bounces occur across the disk diameter D, hence

$$R \exp\left[\frac{n_2}{n_1} (\alpha D)\right] = 1 \quad . \tag{4.17}$$

Early analysis assumed a uniform energy-storage profile across the thickness of a disk. The assumption is poor, as was shown by SOURES et al. [4.4]. The ratio of peak to average stored-energy density in an active-mirror disk may be greater than 6 for some phosphate compositions, as was shown by BROWN et al. [4.5]. In that case, (4.17) is replaced by

$$R \exp\left[\frac{n_2}{n_1} (\bar{\alpha} D)\right] = 1 \quad , \tag{4.18}$$

where the average gain $\bar{\alpha}$ is calculated according to

$$\bar{\alpha} = \frac{1}{d} \int_0^d \alpha(x)dx \quad . \tag{4.19}$$

The way in which $\alpha(x)$ varies with thickness has been discussed previously in Chap.2. It is obvious that as $R \to 0$, large energy storage is possible. In recent years, much effort has gone into the development of solid edge claddings that absorb strongly at 1.06 µm, and reflect as little as possible of the incident radiation back into the disk [4.13]. Such claddings are subject to large heat loading in large disk amplifiers; matching of the cladding expansion coefficient to that of the glass bulk is desirable, if not necessary, to avoid fracture. Another solution to the PO problem is to use liquid edge claddings. This method, first proposed by DUBE and BOLING [4.14], was implemented by GUCH [4.15] on a large disk amplifier, and has resulted in the maximum ($\alpha D$) reported to date in disk amplifiers. It is difficult to implement on disk amplifiers, however, owing to the necessity of sealing an "o" ring around the entire disk periphery, materials problems, and cost. Active-mirror amplifiers, however, because of their geometry, are easily liquid-edge clad and operate that way routinely; the liquid also provides cooling.

Regardless of the value of $n_1$, it is also possible to have a ring mode, as shown in Fig.4.5. The mode path is in a plane parallel to the disk surface

136

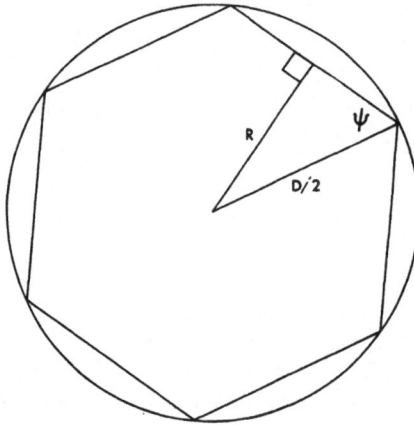

<u>Fig. 4.5.</u> Ring modes in a disk

and is sustained by total internal reflection from the disk edge. The mode
depletes an annular region outside the radius

$$R = \frac{D}{2} \left(\frac{n_3}{n_2}\right) \quad .$$  (4.20)

If $n_3 > n_2$, total internal reflection is impossible and this lossless mode is
suppressed.

An important PO mode is the so-called surface mode, which occurs in straight-
across paths along a diameter D. Such a mode may begin to oscillate before the
lossy bulk mode (if $n_3 > n_2$), because the gain at the disk face can be many times
the average [4.5]. BROWN [4.16] derived the threshold condition for the surface mode as

$$R_N \exp(\alpha_s D) = 1 \quad ,$$  (4.21)

where $R_N$ is the normal reflection coefficient at the disk edge and $\alpha_s$ is the
surface gain. It is clear that when the average stored-energy density increases so that the peak surface gain increases beyond the threshold value,
the mode may become a bulk mode, because the threshold condition is satisfied from the surface to a depth determined by the form of $E_s(x)$. The condition for the surface mode to predominate over the lossy bulk mode is

$$\alpha_s D > \frac{n_2}{n_1} (\bar{\alpha} D) + \ln\left(\frac{R}{R_N}\right) \quad .$$  (4.22)

It is of obvious interest in analyzing laser systems, to be able to evaluate the energy-storage capability of a disk in terms of the edge reflectance,

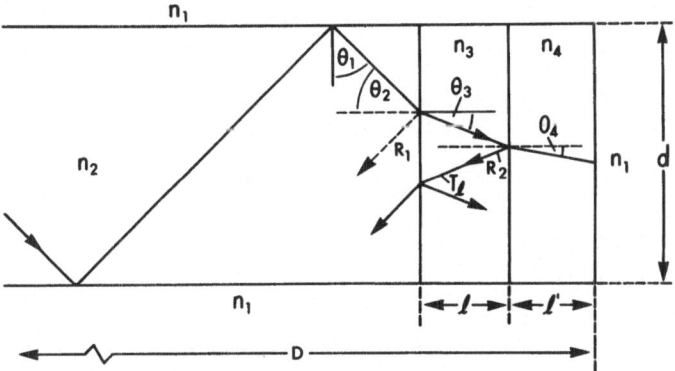

Fig. 4.6. General disk edge geometry

which normally results from a glass-glass or glass-liquid interface. Results
of such a study have been published by GLAZE et al. [4.17]. They obtained
parasitic limits on energy storage in ED-2 silicate laser glass, for both
the surface and bulk modes. A similar study was published by BROWN et al.
[4.5] in two phosphate compositions (EV-2, LHG-5) as well as ED-2. The edge
reflectance was calculated by use of a somewhat more general approach than •
that used by GLAZE et al. [4.17]. We refer the reader to Fig.4.6, which shows
the geometry considered. $n_1$ is normally the index of air, $n_2$ that of a glass
laser disk, $n_3$ that of a solid or liquid edge cladding, and $n_4$ the index of
whatever material is adjacent to the cladding. The disk thickness and diameter are d and D, respectively; $\ell$ and $\ell'$ are the thicknesses of the cladding and the adjacent material. A ray that is interreflected in the disk is
incident on the interface boundary $(n_1,n_2)$ at the angle $\theta_1$ and on the edge
$(n_2,n_3)$ at angle $\theta_2$. $\theta_3$ and $\theta_4$ are the angles at which the ray travels at in
$n_3$ and $n_4$, respectively, relative to the edge normal. A ray incident on
$(n_2,n_3)$ has a reflectance $R_1$, whereas at $(n_3,n_4)$ it is $R_2$. The transmittance
of the highly absorbing layer $(n_3)$ is $T_\ell$. It is not difficult to show that
the effective reflectance $R_B$ for the bulk parasitic mode is given by

$$R_B = R_1 + R_2(1-R_1)^2 T_\ell^2 \ ,$$ (4.23)

where

$$R_1 = \left[\frac{\sin(\theta_2-\theta_3)}{\sin(\theta_2+\theta_3)}\right]^2$$ (4.24)

$$R_2 = \left[\frac{\sin(\theta_3 - \theta_4)}{\sin(\theta_3 + \theta_4)}\right]^2 , \qquad (4.25)$$

and

$$T_\ell = \exp(-\alpha_\ell \ell \sec\theta_3) = (T)^{\sec\theta_3} . \qquad (4.26)$$

T, the one-way transmittance is

$$T = \exp(-\alpha_\ell \ell) . \qquad (4.27)$$

Here $\alpha_\ell$ is the absorption coefficient at 1.06 $\mu$ of the medium ($n_3$). Note that it is assumed that the absorption in ($n_4$) is very great so that no return is allowed from ($n_4, n_1$).

The following relationships can also be derived:

$$\theta_1 = \sin^{-1}\left(\frac{n_1}{n_2}\right) \quad \text{(total internal reflection)} \qquad (4.28)$$

$$\theta_2 = \frac{\pi}{2} - \theta_1 , \qquad (4.29)$$

so that together with (4.24-26)

$$R_1 = \left[\frac{(n_3^2 - n_2^2 + n_1^2)^{\frac{1}{2}} - n_1}{(n_3^2 - n_2^2 + n_1^2)^{\frac{1}{2}} + n_1}\right]^2 \qquad (4.30)$$

and

$$R_2 = \left[\frac{(n_3^2 - n_2^2 + n_1^2)^{\frac{1}{2}} - (n_4^2 - n_2^2 + n_1^2)^{\frac{1}{2}}}{(n_3^2 - n_2^2 + n_1^2)^{\frac{1}{2}} + (n_4^2 - n_2^2 + n_1^2)^{\frac{1}{2}}}\right] . \qquad (4.31)$$

For the surface mode, the effective reflectance $R_s$ is

$$R_s = R_1^1 + R_2^1(1 - R_1^1)^2 T^2 , \qquad (4.32)$$

where now $R_1^1$ and $R_2^1$ are simply

$$R_1^1 = \left(\frac{n_3 - n_2}{n_3 + n_2}\right)^2 \qquad (4.33)$$

$$R_2^1 = \left(\frac{n_4 - n_3}{n_4 + n_3}\right)^2 . \qquad (4.34)$$

Using (4.18,21), we can obtain

$$-(\bar{\alpha}D) = \frac{n_1}{n_2} \ln\left[R_1 + R_2(1-R_1)^2 T^{2\sec\theta_3}\right] \qquad (4.35)$$

and

$$-(\alpha_S D) = \ln\left[R_1^1 + R_2^1(1-R_1^1)^2 T^2\right] \quad . \qquad (4.36)$$

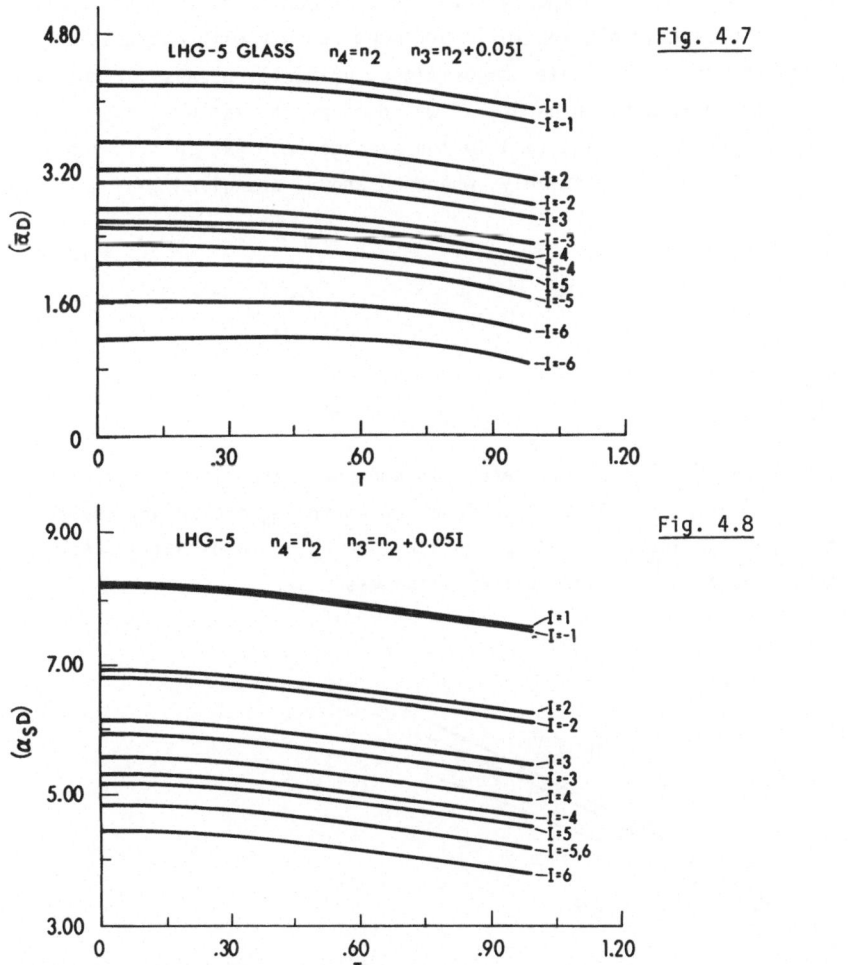

Fig. 4.7

Fig. 4.8

Fig. 4.7. Average gain–diameter product ($\bar{\alpha}D$) for an LHG-5 laser disk as a function of single pass transmission T, for $n_2 = n_4$

Fig. 4.8. Peak (surface) gain–diameter product ($\alpha_S D$) for an LHG-5 laser disk as a function of the one-way transmission T, for $n_2 = n_4$

In [4.5], (4.35,36) have been evaluated for EV-2 laser glass. Figures 4.7,8 show similar results obtained for LHG-5 phosphate glass. In Fig.4.7, $(\bar{\alpha}D)$ is shown as a function of T for the case $n_4 = n_2$, and $n_3 = n_2 + 0.05$ I, where I = ±1, ±2, ±3, ±4, ±5, and ±6. Similar curves are shown in Fig.4.8 for $(\alpha_S D)$. Study of such results yields the general conclusion that the maximum attainable $(\alpha D)$ is larger in phosphate than in silicate glasses, owing to the lower refractive index of phosphate compositions.

Another effect, shown to be important in evaluating laser glasses for large amplifiers is that of glass composition. Chapter 2 mentioned that the inversion density in phosphate glasses falls off more rapidly with penetration distance than it does in silicate compositions, primarily because of the narrower, more intense absorption bands. If we consider the ratio $n_D$ and $n_A$ of gain on the disk surface [peak] $(\alpha_S)$ to the average gain $(\bar{\alpha})$ in disk and active-mirror amplifiers, respectively, where $n_D$ and $n_A$ are given by

$$\eta_D = \frac{\alpha_{SD}}{\bar{\alpha}_D} \tag{4.37}$$

and

$$\eta_A = \frac{\alpha_{SA}}{\bar{\alpha}_A} , \tag{4.38}$$

we find that $n_D$ and $n_A$ are always larger in phosphate than in silicate glasses. This is shown in Figs.4.9,10 in which $n_D$ and $n_A$ are shown, respectively, for ED-2 (silicate), EV-2 (phosphate) and PHG-5 (phosphate) laser glasses [4.5], as a function of optical thickness.

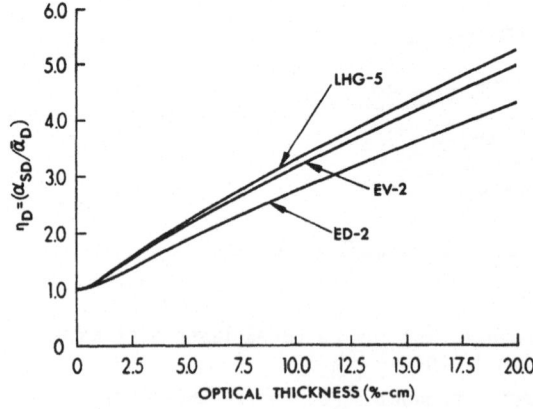

Fig. 4.9. Ratio of surface gain $(\alpha_S D)$ to average gain $(\bar{\alpha}D)$ for a disk amplifier for three laser glasses, as a function of optical thickness

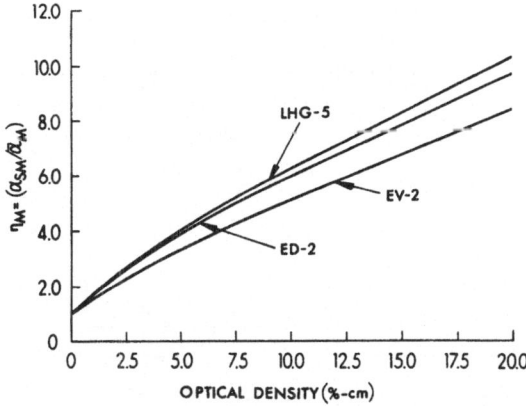

**Fig. 4.10.** Ratio of surface gain ($\alpha_{SM}$) to average gain ($\bar{\alpha}_M$) for an active-mirror amplifier, for three laser glasses, as a function of optical thickness

From these results, we may conclude that, when phosphate rather than silicate laser glass is employed, the surface parasitic is more likely to occur. This has been verified in Q-88 phosphate laser glass by McMAHON [4.18]. Another conclusion is that the surface parasitic is more likely to occur in active-mirror amplifiers than disk amplifiers.

This section has shown that, if $n_1$, $n_2$, $n_3$, $n_4$, T, and the stored-energy density distribution for a given laser glass are known, the limits of energy storage in disk or active-mirror amplifiers can be determined. All of the above is a ray-optics treatment; it ignores diffraction or scattering, which, of course, occurs in all real systems. Diffuse parasitics were investigated by TRENHOLME [4.19] but as of this writing the effect has not been demonstrated experimentally.

It has been noted [4.5] that ASE rather than PO is the major energy-storage limitation in active-mirror amplifiers. The same conclusion has been drawn recently in connection with disk amplifiers [4.20]. It is likely that PO do not exist in active-mirror amplifiers because of geometric discrimination [4.5]. If the disk edges and surfaces are leveled, which is easily done for active-mirror amplifiers, the major surface parasitic may be prevented. This will be discussed more fully in Chap.5.

## 4.3 Parasitic Oscillations in Rod Amplifiers

Until now in this chapter we have considered PO only in the disk geometry that is common to active-mirror and disk amplifiers. The other major type of

142

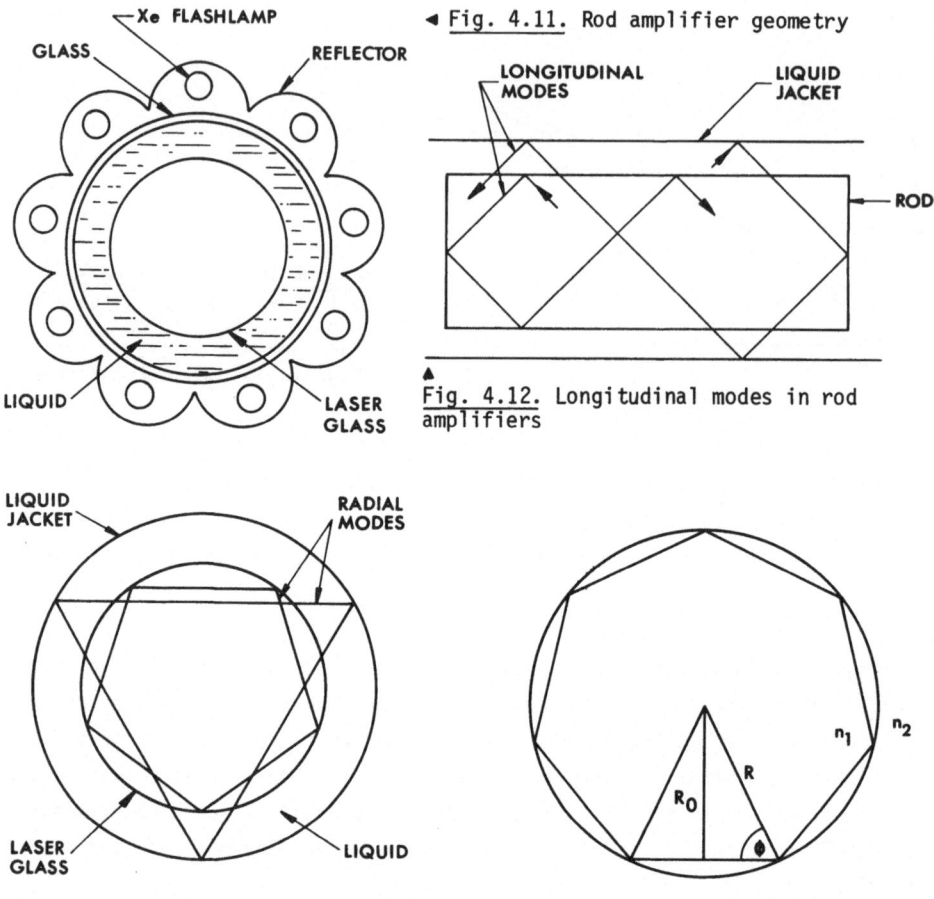

Fig. 4.11. Rod amplifier geometry

Fig. 4.12. Longitudinal modes in rod amplifiers

Fig. 4.13. Radial modes in rod amplifiers

Fig. 4.14. Radial mode geometry

amplifier in use today is the rod, or right-circular cylinder amplifier, which, of course, is a thick disk. An axial view of a typical rod amplifier is shown in Fig.4.11; it consists principally of a right cylinder of laser glass surrounded by a cooling liquid that is contained in a right cylinder of glass, normally Pyrex. The rod is pumped by an array of Xe flashlamps and reflectors arranged around the entire assembly. Two major types of parasitics are allowed; longitudinal and radial. The longitudinal mode, shown in Fig. 4.12, which is analogous to the lossy bulk modes in disks, is normally not present since the faces of the rod are either wedged or anti-reflection coated, effectively defeating it. Different than in disks, however, lossless reflection at or beyond the total-internal-reflected (TIR) angle occurs along the rod circumference or liquid jacket. Although the longitudinal mode is

normally absent in large-aperture rod amplifiers, LABUDDE et al. [4.21] observed it in small high-gain YAG rods. They used a novel scheme to suppress
it; grooves inscribed along the barrel length. The most important modes in
rod amplifiers are the radial modes, shown in Fig.4.13. We first consider the
radial modes in the rod itself and use the geometry shown in Fig.4.14. A radial mode may be either lossy or lossless, depending upon the angle of incidence $\phi$, and whether that angle is less than, or greater than or equal to $\phi_c$;
the TIR angle is given by

$$\sin\phi_c = \frac{n_2}{n_1} \ . \tag{4.39}$$

For $\phi < \phi_c$, although the reflection coefficient at the rod circumference is
<1, there still may be enough gain for a lossy mode to oscillate. A mode in-
cident at an angle $\phi$ on the boundary has associated with it at radius $R_0$
that is always less than or equal to the rod radius R,

$$R_0 = R \sin\phi \ . \tag{4.40}$$

It can easily be shown that the threshold condition for a mode that has m
bounces associated with it is given by

$$\alpha L_T = -m \ln(R) \ , \tag{4.41}$$

where $\alpha$ is the gain coefficient given by (1.24), $L_T$ the total path length,
and R the reflection coefficient at the angle $\phi$. By noting that the total
path length $L_T$ is m times the path length of a single leg $\ell$, we can rewrite
(4.41) as

$$\alpha\ell = -\ln(R) \ , \tag{4.42}$$

the result being independent of m. $\ell$ can be shown to be given by

$$\ell = 2R \cos\phi \ , \tag{4.43}$$

which, combined with (4.42), gives

$$\alpha D = - \frac{\ln(R)}{\cos\phi} \ . \tag{4.44}$$

Analogous to the radial mode in the rod there is a mode due to TIR at the liquid-jacket-air interface $(n_2, n_3)$. A threshold condition similar to (4.42) may be found and is given by

$$\alpha\ell_1 - 2\alpha'\ell_2 = -\ln(R) \quad . \tag{4.45}$$

Here, $\alpha'$ is the absorption coefficient at 1.06 μm in the liquid, $\ell_1$ the length in the glass and $\ell_2$ the length in the liquid. Both $\ell_1$ and $\ell_2$ refer to a single leg of a path that consists of m separate reflections.

There are a number of obvious and effective ways of suppressing radial parasitics. First, the radial mode due to TIR at the rod-liquid interface can be eliminated by matching the index of the laser glass to the liquid, a technique first employed by McMAHON et al. [4.12] using ED-2 silicate glass and mixtures of $ZnCl_2$ and $ZnBr_2$. Considerably better index matches may be obtained by use of phosphate glasses and easier-to-handle liquids such as ethylene glycol. It should be recalled that the linear indices of the phosphate compositions are significantly less (typically 1.50) than those of silicate glasses (typically 1.55). The consequence is that even though the index of the liquid is less than that of the glass, the radial mode is still allowed. It occupies only a small fraction of the rod radius, which is normally unused anyway, because a beam does not normally fill the entire rod because of a desire to minimize diffraction at the edge. To stabilize the radial mode due to TIR at the liquid-jacket-air interface, a number of methods are available. The first is to use a liquid such as $ZnCl_2$, which absorbs strongly at 1.06 μm but which transmits pump radiation [4.12]. The properties of such liquids for absorption and index matching have been recently presented by RINEFIERD et al. [4.22]. The second method, which has not yet been implemented, is to use an anti-reflection treatment pioneered recently by Corning Glass Works [4.23] to reduce the reflectance of the Pyrex liquid jacket. Recent measurements by JACOBS and ABATE [4.24] and reproduced in Fig.4.15, show that the treatment results in low reflectances, even for very large angles of incidence θ. TIR from an air-Pyrex interface normally occurs at $\theta \simeq 43°$. Another method for suppressing the radial jacket mode is to make the diameter $D_J$ of the liquid jacket significantly greater than that of the rod (D) such that no ray is allowed to intercept the gain medium. The lowest-order mode of interest is the one that corresponds to three bounces, m = 3. For m = 3, the angle of incidence φ on the Pyrex-air interface is 30°, much less than the critical angle. This mode is lossy if it is allowed to intercept a portion of the rod. The next mode (m=4) corresponds to $\phi = 45°$, greater

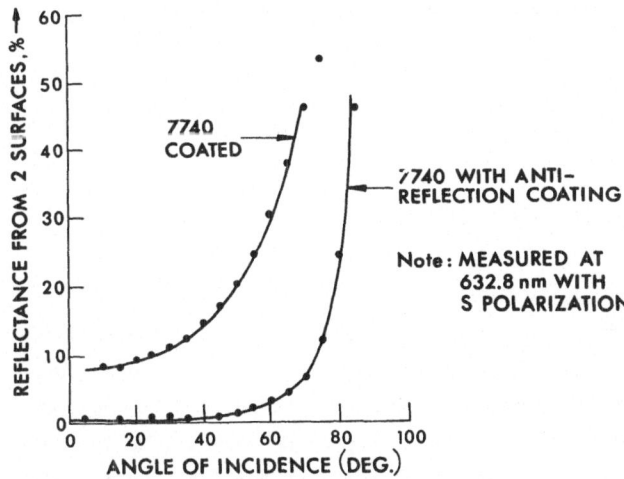

Fig. 4.15. % Reflectance from two surfaces as a function of angle of incidence for untreated Pyrex 7740 glass, and Corning anti-reflection treated
Pyrex 7740 glass, measured at 632.8 nm for S polarized light [4.24]

than $\phi_c$. In rod amplifiers, it is desirable to reduce the liquid-jacket diameter to the minimum value, because that results in the maximum coupling efficiency between the Xe flashlamps and the laser glass. The most-likely parasitic is that which corresponds to $\phi_c$, which, of course, does not require a
closed path. For that case, $\phi_c = 43°$ and the ratio of the rod diameter D to
the liquid jacket diameter $D_J$ is

$$\frac{D}{D_J} = \sin\phi_c = \frac{1}{n} \quad . \tag{4.46}$$

For Pyrex (n=1.46 at 1.06 μm) and quartz (n=1.45 at 1.06 μm), the ratio $D/D_J$
takes the value ≈0.69. In that case, it is possible that the lossy mode which
corresponds to m = 3 may oscillate. It can be shown that the depleted portion
of the rod diameter is 26%. The mode does not normally oscillate in practice,
because of the great reflection losses involved. The use of the AR-treated
liquid jacket, which has no substantial reflectance up to ≈70° angle of incidence provides a ratio $D/D_J \approx 0.94$.

The foregoing discussion shows that the analysis of PO in rod amplifiers,
while not as straightforward as that in disks, is perfectly tractable; it can
best be studied by developing a computer program to do the complete analysis.

# 5. Amplifiers for High-Peak Power Nd : Glass Laser Systems

The most important elements in any high-power laser system are the amplifiers
that are used to boost the initial oscillator power to the desired output
level. Optimization of the performance of such amplifiers leads to maximum
system performance at minimum cost. In this chapter we investigate the general characteristics and present operating conditions for disk, active-mirror
and rod amplifiers. It will be shown how it is possible to design an optimum
amplifier by varying such parameters as disk doping, flashlamp bank energy,
and pulsewidth. Also to be considered here are the present limits to achievable energy storage in amplifiers due to ASE or PO. We will also discuss the
thermal-cycling time and storage efficiency of all three types of amplifiers.
Short-pulse damage and nonlinear effects will not be discussed in this chapter but will be covered in detail in Chaps.6-7.


## 5.1 Disk Amplifiers

We begin by investigating the disk amplifier, a schematic diagram of which
is shown in Fig.5.1. Nd:glass disks are placed in the optical path of a beam
at Brewster's angle, resulting in no reflection loss for p (horizontal) polarization. Brewster's angle ($\theta_B$) is given by

$$\theta_B = \tan^{-1}(n) \quad , \tag{5.1}$$

where n is the linear index of the laser glass at 1.06 μm. One consequence
of Brewster's angle is that the area A of a disk is $\approx$n times that for a disk
placed at $0^\circ$ angle of incidence, which is important in considerations of nonlinear and damage effects. In particular, if D and D' represent the major
axis of the disk at normal incidence and Brewster's angle, respectively, it
can be shown that they are related through

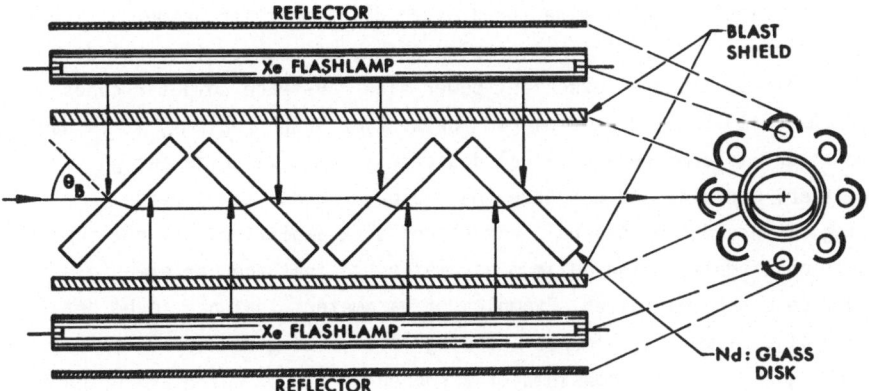

Fig. 5.1. Schematic diagram of a typical disk amplifier

$$\frac{D'}{D} = \sqrt{1+n^2} \simeq n \quad . \tag{5.2}$$

Generally, an even number of disks is used in an amplifier since beam translation occurs due to refraction in a single disk. Nonlinear beam steering in disks can be minimized if the disks are placed in an alternate (Fig.5.1) rather than parallel arrangement. The lateral displacement $\delta$ is given by

$$\delta = t \sin\theta_B \left(1 - \frac{\cos\theta_B}{\sqrt{n^2-\sin^2\theta_B}}\right) \quad , \tag{5.3}$$

where t is the disk thickness. The entire periphery of each disk is generally solid edge clad; liquid claddings are occasionally used for suppression of PO (Chap.4). To protect such claddings from the effect of flashlamp radiation and to provide a means of mounting, the disk edge is generally surrounded by a metallic sheath or holder. A blast shield is normally placed between the flashlamp array and the disks. Generally fabricated from quartz, its purpose is twofold. First, in the case of the catastrophic explosion of a flashlamp, it serves to protect the disks, which are costly to refinish or replace. In some cases, however, the blast shield itself is also destroyed. Second, the blast shield makes it possible to completely encase the disk assembly. That is desirable because of the susceptibility of disk surfaces to accumulation of particles, which form sites of damage from the great incident flashlamp flux. More will be said later about damage in disk amplifiers. Surrounding the blast shield is a layer of Xe flashlamps, which optically pump the Nd-doped disks. The flashlamps are located as close as possible to the blast shield, to increase coupling efficiency. The diameters and number of flashlamps are difficult to determine for a given disk-amplifier geometry. The reflector design is an additional problem which, of course, is coupled to those parameters. Although excellent power-flow ray-trace computer codes such as ZAP [5.1] have been developed for work on such problems, it is usually necessary to combine analytical design with experience to design such an amplifier, aided by the guidance that computer codes can provide.

Having discussed some general features of disk amplifiers, we now turn to the question of determining optimum parameters so that the performance of a given device can be maximized. Examples of parameters that are to be determined are disk diameter, thickness, doping level, and glass type. The process of designing an amplifier consists of a number of steps, which are often time consuming and expensive. By far the most difficult step is to decide the number and size of Xe flashlamps to be used to pump the Nd:glass, at what fraction of the explosion energy they can prudently be run, on what diameter circle the lamps should be placed, and what reflector geometry is optimum. A number of effects should be considered. The first is the fact that the Xe flashlamp spectral efficiency varies according to the diameter of the lamp, as discussed in Chap.3. The diameter of the circle of lamp centers is generally chosen so that the lamps can be placed as close as possible to the disks. That consideration is usually complicated by the reflector design. A large fraction of the Xe flashlamp radiation is emitted in directions different than the locations of the disks. Some fraction of the lost radiation can be redirected towards the disks by the use of a reflector placed behind or wrapped around a lamp. Radiation that is reflected through the lamp is partially absorbed, which reduces the effectiveness of the reflector. The absorption coefficient in the lamp depends upon the wavelength and current density (Chap.3). Radiation absorbed at one wavelength may be re-emitted (into $4\pi$ steradians) at the same or different wavelengths. The net effect of the integrated absorption of radiation in the lamp is to heat the plasma further; a resulting problem is that the J/cm of energy dissipated by the flashlamp, normally chosen to be $\approx$30-40% of the explosion energy, is really much higher. It is obvious from this brief discussion that to analyze this problem to find the optimum design is not easy. As mentioned previously, Monte Carlo programs such as ZAP [5.1] have been developed. Such sophisticated codes are only first-order models of processes that are not amenable to analysis; they include a large number of geometries and require large computers. Furthermore, it is necessary to "normalize" the code to the amplifier being modeled. Most laboratories rely upon a combination of experiments, modeling, and experience to achieve optimal amplifier performance. Bread-board arrangements, in which

the number and size of lamps, type of reflectors, etc. are varied, are generally built and their characteristics measured and compared to the computer
models.

Sophisticated computer codes are used to simulate the performance of Nd:
glass amplifiers. We describe here one such code developed at LLL for modeling disk amplifiers; a similar one developed for active-mirrors at LLE will
be described in the next section. The model, a version of which is known as
VODAC, begins by solving the flashlamp pumping equation discussed in Chap.3.
By assuming a constant conversion of lamp-dissipated electrical energy into
photons (typically 75-80%), the current and power dissipated by the lamp are
found by solving (3.25) numerically. It is necessary initially to specify the
pulse-forming-network parameters: lamp-impedance parameter and fill pressure,
initial voltage, inductance, capacitance, damping parameter, and dc resistance. Only a fraction of the lamp output is actually absorbed in the Nd:
glass, the remainder is dissipated in reflectors or other elements in the
optical cavity, or escapes out the open ends. Furthermore, because Xe flashlamps become more opaque at higher current densities, the fraction of the
photons reaching the glass is not constant but a function of the current density J. Radiation collected by reflectors and directed towards the Nd:glass
is partially absorbed and later re-emitted by the flashlamp. TRENHOLME [5.2],
using the ZAP [5.1] laser analysis program and fitting experimental data,
found that the cavity transfer efficiency $\eta(J)$ can be modeled by a function
of the form

$$\eta(J) = \frac{A(1-e^{-BJ})}{J^C} \quad , \tag{5.4}$$

where A, B, C are constants. The way in which $\eta(J)$ varies with J is shown in
Fig.5.2. The way in which the Xe flashlamp output is absorbed by the laser
glass was discussed in Chap.2 where it was shown that the energy density obtained is a nonexponential function of optical thickness. A way of describing
the absorption $A_T(x)$ as a function of the optical depth x is to use the so-
called A,B,C,D function described previously and given by (2.14). The four
constants are obtained by a nonlinear least-squares fitting program. The pump
power arriving in the upper laser level is then given by

$$P_U(t) = P_L\eta(J)A_T(x) \quad , \tag{5.5}$$

where $P_L$ is the lamp power. Equation (5.5) allows us to find the gain coefficient $\alpha$ [cm$^{-1}$] from

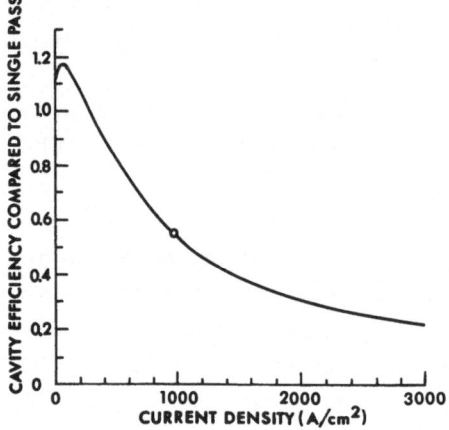

Fig. 5.2. Cavity transfer efficiency in a disk amplifier as a function of lamp current density. The rapid decrease at high current density is due to an increase of lamp opacity, which causes the lamps to absorb more and more of the cavity photons [5.2]

$$\alpha = \frac{\alpha_0 P_u(t)}{V} \quad , \tag{5.6}$$

where $V$ is the laser-glass volume and $\alpha_0$ the specific-gain coefficient defined previously. Although the upper ($^4F_{3/2}$) laser level is populated by Xe flashlamp pumping, the level undergoes loss from a number of mechanisms. The first is normal fluorescence-decay losses; as mentioned in Chap.1, the decay is nonexponential and dependent upon the Nd doping level. The more highly doped glasses display increasingly nonlinear fluorescence decay. To take account of the nonexponential decay and concentration quenching, TRENHOLME [5.2] found

$$\tau_{eq}(\rho) = \frac{\tau_R}{1+\left(\frac{\rho}{\rho_Q}\right)^p} \tag{5.7}$$

useful for finding the equivalent lifetime $\tau_{eq}$ as a function of Nd doping $\rho$ [ions/cm$^3$]. $\tau_R$ is the zero-doping lifetime, and $\rho_Q$ and $p$ constants determined by a nonlinear least-squares fitting program. Amplifier simulations that used the actual nonlinear decay were re-run using a single exponential decay to find the equivalent lifetime $\tau_{eq}$ that would result in the same gain. Equation (5.7) is plotted in Fig.5.3 for five laser glasses, along with experimental data.

Two other mechanisms that deplete the upper laser level are PO and ASE, both of which were treated in Chap.4. The effects have been treated semi-phenomenologically in a manner that is subsequently outlined.

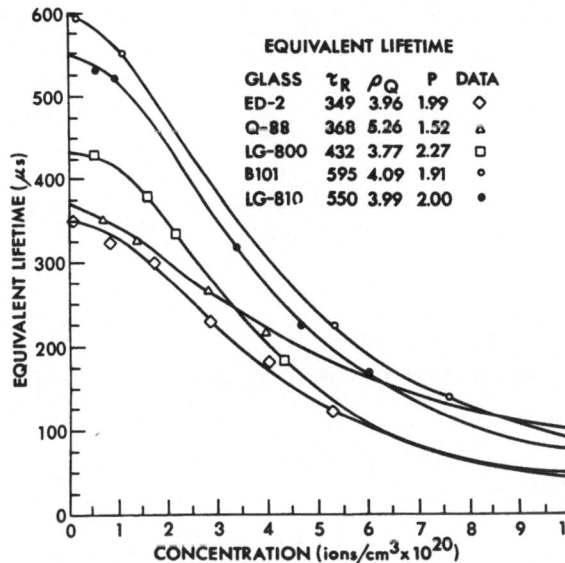

Fig. 5.3. Equivalent single-exponential decays as a function of doping for a number of glasses. The points are found by matching the nonexponential decay of samples with single exponentials that give the same gain [5.2]

Chapter 4 showed that ASE reduces the fluorescence lifetime by a divisor M (4.7). TRENHOLME [5.2] showed that the decay in the presence of ASE can be represented by the rate

$$R = - \frac{\alpha}{\tau_{eq}} \left[ (1-b)+b \ \exp\left(c_1\beta+c_2\beta^2+c_3\beta^3\right)\right] \ , \tag{5.8}$$

where $\alpha$ and $\tau_{eq}$ are as defined previously, b is the branching ratio, $c_1$, $c_2$ and $c_3$ are constants and $\beta = \alpha L$ where L is the major axis of the disk. The divisor M in (4.7) is determined by fitting (5.8) to the ZAP Monte Carlo results mentioned in Sect.4.2. For this case, an elliptical disk with uniform inversion and a Gaussian fluorescence profile is assumed; the reader interested in the details may consult [5.3]. The other effect of interest, PO, may be simulated by rapidly increasing the decay rate R as the quantity nears the parasitic limit. That is effected by multiplying (5.8) by the factor K where

$$K = 1 + \frac{W}{1-(\beta/\beta_p)} \ . \tag{5.9}$$

In (5.9), W is the relative width of the parasitic onset and $\beta_p$ is the parasitic limit, defined by

$$\beta_p = \min(\beta_B,\beta_S) \ , \tag{5.10}$$

where the bulk $\beta_B$ and surface limits $\beta_S$ are found from (4.18,21), respectively. Then

$$\beta_B = -\frac{n_1}{n_2} \ln(R) \quad , \tag{5.11}$$

and

$$\beta_S = -\ln(R_N) \quad , \tag{5.12}$$

where $n_1$, $n_2$, R, and $R_N$ have been defined previously.

Finally, the total differential equation that must be solved, which has all of the relevant physics, is

$$\frac{d\alpha}{dt} = \frac{\alpha_0 P_L \eta A_T}{V} - \frac{\alpha}{\tau_{eq}} [(1-b)+bS]K \quad . \tag{5.13}$$

The fluorescence amplification factor S can be shown, from (5.8) to equal

$$S = \exp\left[c_1\beta+c_2\beta^2+c_3\beta^3\right] \quad ; \tag{5.14}$$

all of the other quantities have been defined previously. The general pumping equation (5.13) represents the difference of two terms; the first represents ions that arrive in the $^4F_{3/2}$ metastable level, the second represents the losses that result from fluorescence decay, ASE, and PO. An additional loss term due to $H_2O$ quenching, which is found to be important in phosphate laser glasses, will be discussed in Sect.5.3.

Having arrived at (5.13), we are now in a position to optimize amplifiers. Using a computer to vary the glass doping and PFN pulsewidth, the representative contour plot shown in Fig.5.4 can easily be generated; there stored-energy density or gain contours are shown as a function of wt-% doping and PFN pulse duration $(3\sqrt{LC})$. For a constant pulse duration, a doping maximum occurs that represents the optimum operating point [5.2]. It is physically due to two competing processes: 1) concentration quenching for large dopings, and 2) low absorption for small dopings. Similarly, for a given doping there is a maximum when the pulse duration is varied. For short pulse duration lamp opacity increases owing to the high current density in the lamps; for pump pulses long compared to the fluorescence lifetime, fluorescence decay depletes the gain. For any disk of chosen thickness, it is possible to generate quantitative data for each glass type used. That information, combined with cost information about disks and other parameters that characterize long-

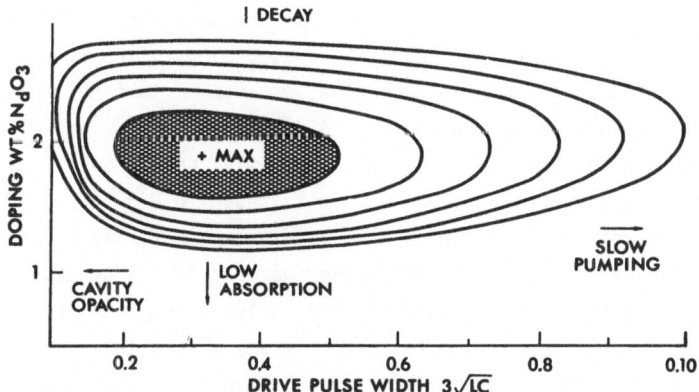

Fig. 5.4. Stored-energy density or gain contours as functions of drive pulse duration ($3\sqrt{LC}$) and $Nd^{3+}$ doping. The processes which reduce gain if we deviate from the optimum are shown on the figure [5.2]

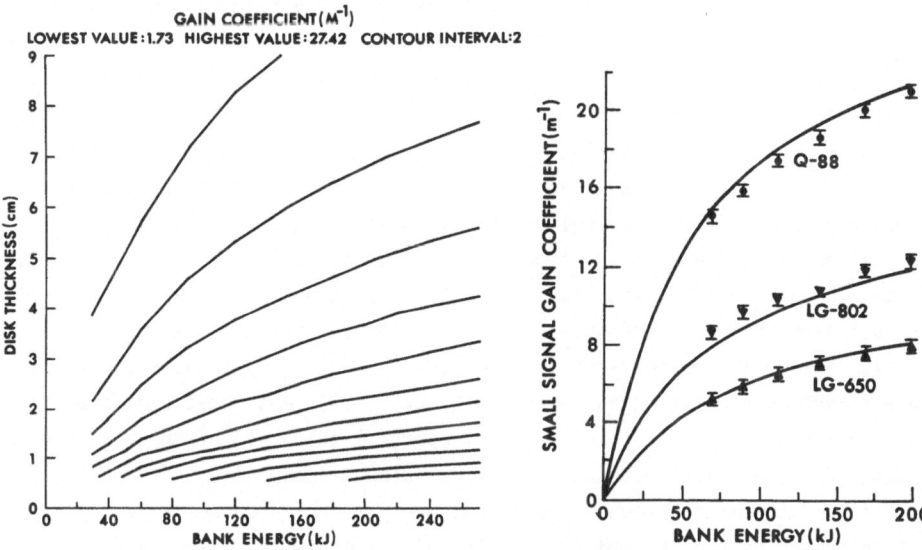

Fig. 5.5. Maximum gain coefficient as a function of disk thickness and bank energy. At each point, the doping and pump duration have been optimized as shown in Fig.5.4. For thin disks and large banks, the gain is limited by ASE [5.2]

Fig. 5.6. Gain coefficient as a function of bank energy for a LLL disk amplifier using Q-88, LG-802, and LG-650 laser glasses, showing experimental data and the model predictions [5.4]

and short-pulse performance, permits design of the most cost-optimal amplifier for a desired application from only spectroscopic data on laboratory samples. That procedure will be fully discussed in Chap.8. The thickness and bank energy can be varied to generate the data shown in Fig.5.5. Peak-gain coefficient contours are plotted as functions of disk thickness and bank energy; each point on the contours has been optimized according to the foregoing procedure in which doping and pulse duration were varied to find the peak gain. For thin disks and large bank energy, ASE dominates.

The model described in the foregoing has been carefully checked by use of experimental data at LLL; a representative example is shown in Fig.5.6. Experimental data represented by the discrete data points were obtained by use of a standard disk head at LLL [5.4]. Three glasses were used; a high-gain phosphate (Q-88), a low-gain silicate (LG-650), and a fluorophosphate (LG-802). The model had been previously normalized to ED-2 silicate glass. The computer simulations, shown by solid lines in Fig.5.6, were obtained by changing only the glass type. The agreement with the measured gain is excellent.

An important characteristic of large-aperture amplifiers is thermal-cycling time, defined as the time required after the amplifier is fired, before the optical elements return to a relaxed state so that a beam passed through the amplifier will suffer no loss or optical distortion. Owing to the rather low efficiency associated with amplifiers operating at peak gain, typically at most 1 to 1.5%, the remaining energy ends up as heat, mainly in the disks or flashlamp array. In typical disk amplifiers in which a quartz blast shield totally encloses the disk assembly, forced active cooling is not possible; instead convection must be relied upon to transfer heat to the blast shield, resulting in thermal-cycling times of at least an hour and sometimes greater than two or three hours for large-aperture amplifiers. Such long cycling times result even when cool nitrogen gas is circulated along the outside of the blast shield. The thermal-cycling time may be reduced by the use of the "box amplifier", currently being developed by LLL [5.5]. A schematic diagram of the device is shown in Fig.5.7. In it, approximately half of the Xe flashlamps have been omitted because the lamps directly opposite the disk faces contribute the major amount of pumping, whereas those around the periphery of the disks do not. By omitting twelve of the forty flashlamps usually used with a 30-cm-aperture (D type) disk amplifier and arranging the remaining twenty eight lamps as shown in Fig.5.7, a storage efficiency (total peak-stored energy density/bank energy) at peak gain of $\approx 0.85\%$ was achieved, an improvement of approximately a factor of 1.5-1.6 over the older design. Gain performance as a function of total bank energy is shown in Fig.5.8 for the standard D amplifier and the rectangular-box amplifier. An advantage of this scheme is that it effectively opens up two sides of the amplifier, now used only as reflectors, to be used in an active-cooling scheme, probably by flowing cool nitrogen or some other gas along the disk faces. Further development

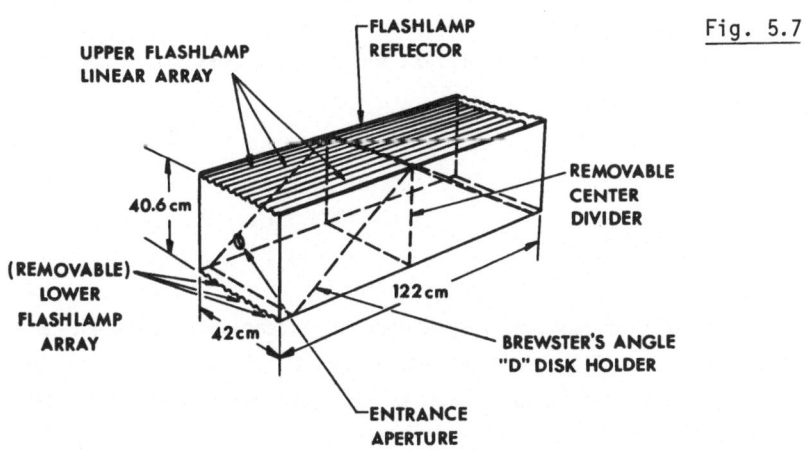

Fig. 5.7

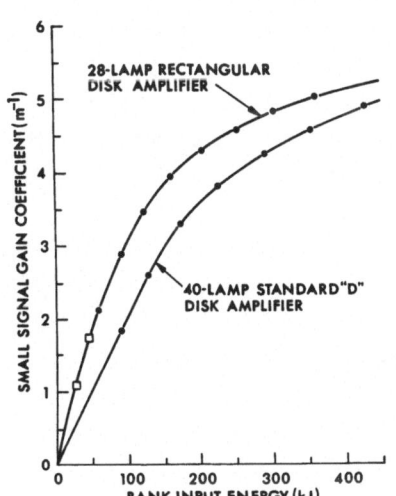

Fig. 5.7. Schematic diagram of rectangular flashlamp pumping fixture (box-amplifier) [5.5]

Fig. 5.8. Comparisons of small-signal gain coefficients as a function of bank energy for 28-lamp rectangular disk amplifier and standard 40-lamp cylindrical "D" disk amplifiers [5.5]

of this amplifier using active cooling will likely result in a much-decreased thermal-cycling time, which at present is determined by the convection rate and not by heat transfer in the disk itself. In active-mirror and rod amplifiers, as will be seen, heat transfer is limited by the heat-diffusion time in the glass itself.

Thermal distortions have been measured in a disk amplifier, by use of time-resolved holography [5.6]. In that work, two major effects were observed. The first was a general decline of fringe visibility starting after the peak of the pump pulse. That effect was attributed to the intrusion of hot gas into the volume between adjacent disks; it may be eliminated by completely sealing the disks from the flashlamps. The second effect was thermal lensing,

which changes the beam divergence by ≃20 µrad. A twenty minute thermal-cycling time was determined for the amplifier, which had a very small aperture
(3.5 cm).

## 5.2 Active-Mirror Amplifiers

The active-mirror amplifier has recently undergone extensive development at
LLE. Originally conceived, like the disk amplifier, at the General Electric
Company [5.7], its use in high-power-laser systems was delayed by the slow
development of damage-resistent multilayer dielectric coatings. The device
is shown in Fig.5.9. The front-face coating is antireflective for the wavelength that is to be amplified, but may be reflective for certain flashlamp
wavelengths that are not absorbed on the first pass. The rear coating is designed to be highly reflective for the amplified beam and also to pass as
large a fraction of the pump light as possible. The disk is supported by the
use of uniform-size glass beads which provide stress-free mounting. The

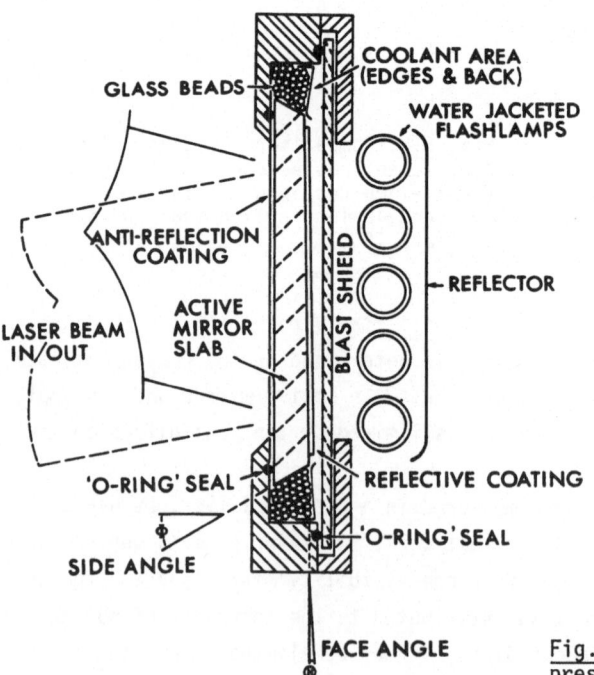

Fig. 5.9. Key features of the
present active-mirror design

glass beads are Cu doped, so as to absorb strongly at 1.06 μm and suppress
ASE or PO. The refractive index of the beads is matched as closely as possible
to the index of the disk, and to the index of refraction of the cooling solution which is usually a mixture of ethylene glycol and $H_2O$ or just ethylene
glycol. The coolant flows along the rear face of the disk, which actively
cools the surface where most of the heat is produced by flashlamp pumping.
The coolant also serves to protect the rear multilayer dielectric coating.
The normals to the edge of the disk are not parallel to the surface, because
a bevel has been ground on the entire periphery. Its purpose is to discriminate geometrically against PO. The surface mode along the rear face is particularly troublesome because the face gain is typically six times the average
for a phosphate glass. A blast shield is placed between the flashlamp array
and the disk to protect the disk against catastrophic explosions and to form
a channel for the cooling liquid. The flashlamp array which can be closely
coupled to the mirror, consists of a number of 13- or 19-mm bore Xe flashlamps
whose arc lengths are chosen, individually, to match the chord of the circular
disk where they are mounted, to increase the pumping efficiency. The flashlamps are $H_2O$ jacketed to speed thermal recovery and to reduce explosion haz-
ards. A reflector is placed behind the array to direct a portion of the flashlamp radiation that is emitted in that direction back towards the active-
mirror disk. Both the blast shield and the $H_2O$ jackets have been treated with
the Corning antireflection treatment mentioned in Chap.4 and summarized in
Fig.4.15; this results in substantial increases of stored-energy density over
earlier designs.

   Active-mirror amplifiers possess a number of advantages over conventional
disk amplifiers. Among these are the ability to propagate circularly polarized light, resulting in a nonlinear index lower by a factor of 1.44 than
that achievable with linear polarization (see Chap.1), and rather great storage efficiencies, typically 1.5% at the peak gain. Also, because of the double
pass, small-signal gain is two times that of an equivalent disk for short
pulses with which extraction occurs. However, difficulties are also associated
with active mirrors. In systems staged in such a manner, they are run at the
damage limit to minimize cost. Active mirrors always require larger apertures
than disks, because the damage threshold for multilayer dielectric coatings
is always substantially less than for bare surfaces, which are used in disk
amplifiers. Furthermore, because the surfaces of disk amplifiers are inclined
at Brewster's angle to the axis, their area is increased by a factor of $\sqrt{n^2+1}$,
where n is the linear index. An additional problem for active mirrors is that
for long pulses ($\simeq 1$ ns), there is a substantial overlap of the incoming and

exiting waves in the mirror coating. Therefore the coating may have two to
four times greater loading than a surface that is traversed only once.

We now turn to the pumping model (INVDEN) developed at LLE to model the
performance of active-mirror amplifiers. INVDEN [5.8], although very similar
to VODAC and other codes developed at LLL, contains a number of important
differences. First, the model has been thoroughly tested by using actual time-
resolved data from 13-mm-bore Xe flashlamps, as discussed in Chap.3. The
Trenholme-Emmett flashlamp model [3.8] has been suitably modified according
to that spectroscopy and is now used to drive the model, because less computer
time is involved. The basic equation solved numerically by INVDEN is

$$\frac{dE_S(z,t)}{dt} = \beta(\lambda,J) \int_{\lambda_1}^{\lambda_2} \eta(\alpha)I(z,\lambda,t) \frac{\lambda}{\lambda_L} \alpha(\lambda)d\lambda - \frac{E(z,t)}{\tau} \quad , \tag{5.15}$$

where $E_S$ is the stored-energy density and is a function of disk thickness z
and time t. $\lambda_1$ and $\lambda_2$ are the wavelength limits over which the flashlamp output $I(z,\lambda,t)$ is convolved with the glass-absorption data $\alpha(\lambda)$, obtained by
spectrophotometry of small laboratory samples. The factor $(\lambda/\lambda_L)$ represents
the quantum-defect loss of energy to multiphoton relaxation (heat) in the
glass, and $\eta(\lambda)$ is the quantum efficiency from an excited state to the upper
$(^4F_{3/2})$ laser level, which is assumed to be independent of wavelength and
equal to unity for all glasses. The second loss term, which involves the total
loss rate $(1/\tau)$ given by

$$\frac{1}{\tau} = \frac{1}{\tau_{CQ}} + \frac{1}{\tau_{H_2O}} + \frac{1}{\tau_R} \quad , \tag{5.16}$$

has contributions from three separate effects. The first is the concentration-
quenching lifetime $\tau_{CQ}$ given by Equation (5.7). As before, the constants $\rho_Q$
and p in (5.7) are gotten from a nonlinear least-squares fit to experimental
data. The second rate, involving the quantity $\tau_{H_2O}$ is required because $H_2O$,
which is present in certain glasses, most notably phosphates, can cause severe quenching of the $^4F_{3/2}$ level. This nonradiative loss is due to high-energy phonon vibrations of the O-H radical associated with the glass forming
network. Based upon measurements obtained by JACOBS at LLE [5.9], it has been
found that $\tau_{H_2O}$ may be represented by

$$\frac{1}{\tau_{H_2O}} = \frac{1}{\tau_R} \left[ 1 + \left(\frac{O.D.}{\alpha}\right)^\beta \right] \quad , \tag{5.17}$$

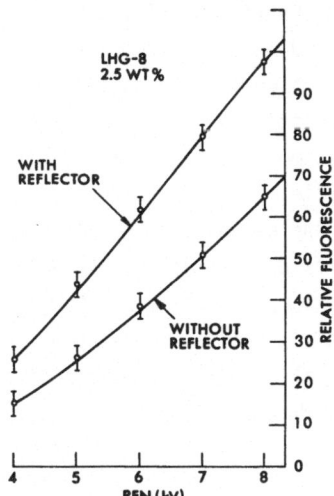

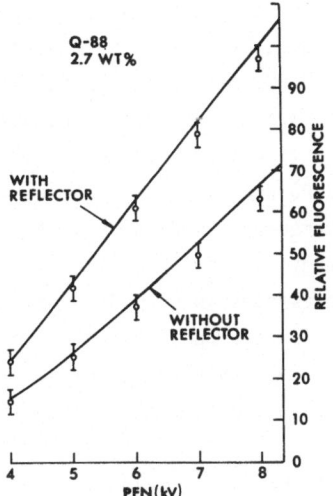

Fig. 5.10. Fluorescence voltage as a function of PFN voltage for LHG-8 laser glass, with and without a reflector on the flashlamp array [5.8]

Fig. 5.11. Fluorescence voltage as a function of PFN voltage for Q-88 laser glass, with and without a reflector on the flashlamp array [5.8]

where $\alpha$ and $\beta$ are constants determined by nonlinear-least-squares fitting experimental data, and O.D. is the optical density/cm measured at 2.2 $\mu$m, the location of a strong $H_2O$ absorption in glass. As before, $\tau_R$ is the radiative lifetime obtained for very low ($\rightarrow 0$) doping and negligible $H_2O$ impurity. The coupling constant $\beta(J)$ is an adjustable parameter that is used to normalize the code to experimental results. It is a function of the lamp-current density J as well as the wavelength $\lambda$, and is used to simulate opacity effects in active-mirror amplifiers. The formula

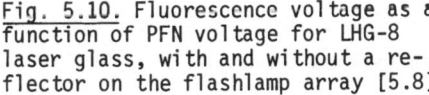

$$\beta(\lambda,J) = K\left[1+C_1 \exp\left\{-C_2 J^2 [C_3 \exp(-C_4(\lambda-0.7)^2)+C_5]\right\}\right] \qquad (5.18)$$

gives excellent agreement with experiments. Typically $C_1 = 0.894$, $C_2 = 0.24$, $C_3 = 0.15$, $C_4 = 8$, $C_5 = 0.01$. This result was derived from experiments performed with fluorescence monitors rather than active-mirror data, with and without a reflector. The constant K is adjusted to give agreement with experiment. A typical example is shown in Fig.5.9. Fluorescence monitors were used to obtain data because the level of stored-energy density present is so low that ASE or PO do not occur. In Fig.5.10, the fluorescence voltage from a photodiode is shown as a function of PFN voltage for LHG-8 laser glass. The data without reflector were used to normalize the code with $\beta(\lambda,J) = 1$. The constants K, $C_1$, $C_2$, $C_3$, $C_4$, and $C_5$ were then obtained by fitting $\beta(\lambda,J)$ to the reflector data. By use of the same normalization, code predictions were obtained for Q-88 laser glass substituted in the same fluorescence monitor. Figure 5.11 shows excellent agreement with measurements.

Actual active-mirror data did not agree with the model unless the effects of ASE and PO were included. Unlike the disk amplifier in which the peak-to-average stored-energy density is typically 2-3:1, the assumption of uniform stored-energy density in active-mirror amplifiers is not valid, and the ASE analysis presented in Chap.4 is not applicable. Rather than use Monte Carlo techniques for a nonuniform inversion, the problem was modeled by using a parasitic clamp. By fitting the model to experimental data on a single active-mirror disk, a limit on the across-disk gain G(z) given by

$$G(z) = [\alpha_0 E_S(z)_{max} - \alpha_L]D \quad , \tag{5.19}$$

where D is the disk diameter and $E_S(z)_{max}$ the peak stored-energy density, was obtained. In effect, when the gain G(z) reaches the limit $G_{max}$, no further increase of stored-energy density can occur at that location. A depletion layer is formed beginning at the edge of the disk nearest the flashlamps, which moves inward as the average stored-energy density increases. By use of this technique, excellent agreement has been obtained with experimental data. From the parasitic limit determined from the LHG-8 data shown in Fig.5.12, the Q-88 curve was predicted, which is in close agreement with the data. The maximum disagreement is less than 8%, which is excellent considering all of the assumptions used in the model.

A major application of the model has been to calculations that are concerned with the front-face coating. To obtain maximum stored-energy density,

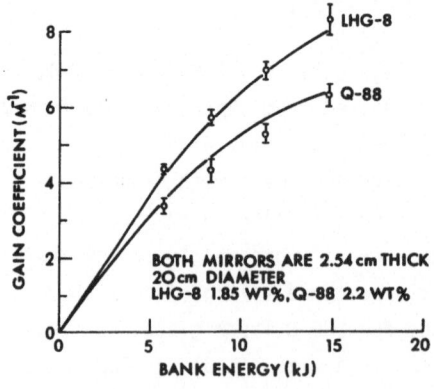

Fig. 5.12. Inversion density (J/cm$^3$) in a 20 cm diameter active-mirror amplifier using LHG-8 and Q-88 laser glass, as a function of PFN voltage and including ASE [5.8]

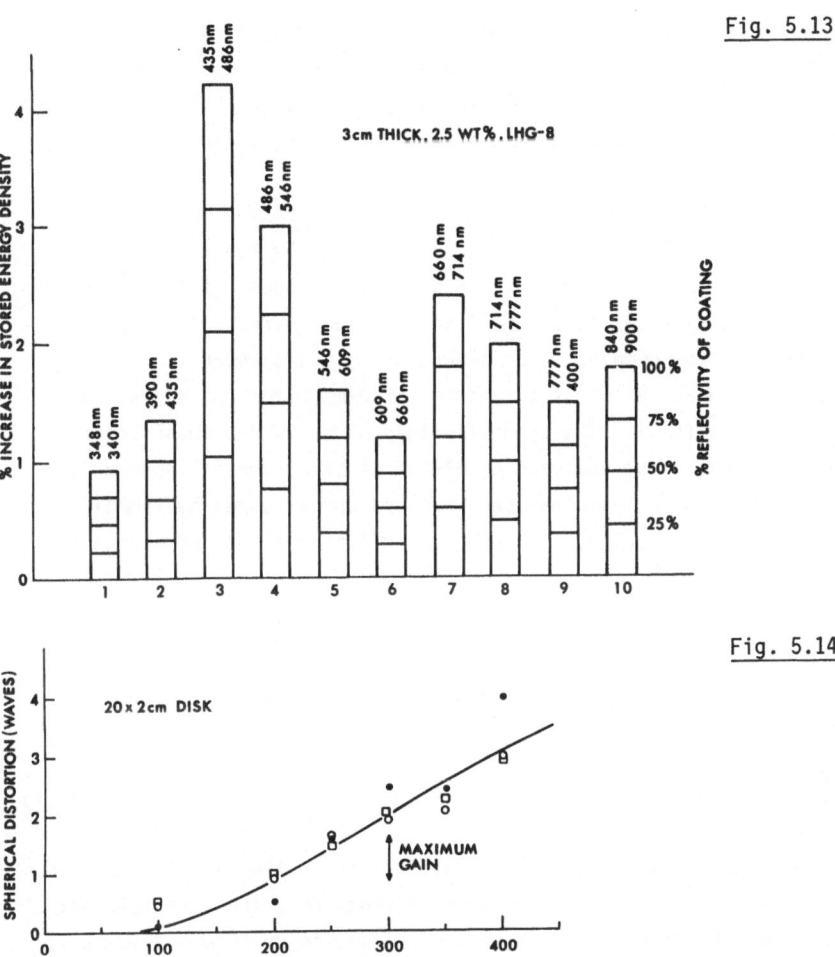

Fig. 5.13

Fig. 5.14

Fig. 5.13. Percent increase in inversion density for an active-mirror amplifier using a front-face coating which has 25, 50, 75, 100% reflectance at the location of 10 major $Nd^{3+}$ bands [5.8]

Fig. 5.14. Spherical-wavefront distortion as a function of time after the beginning of the flashlamp pulse, for a 20-cm diameter, 2-cm thick active-mirror amplifier. Also shown is the location of peak amplifier gain [5.11]

it is desirable to maximize the reflectance for certain pump bands, notably those that are weak. An evaluation of this for an active-mirror disk is shown in Fig. 5.13, where the % increase of stored-energy density is shown for each major $Nd^{3+}$ band, for 0, 25, 50, 75, and 100% reflectance across the wavelength band. This figure shows that the weakest bands are the major contributors, whereas the maximum increase to be expected from a front-face coating is 1.20%.

The model has also been used to calculate the efficiency of a split-disk active mirror design, in which a single disk is replaced by two with different dopings, the disk nearest the flashlamps having the lower. Substantial ($\gtrsim 20\%$) increases in efficiency are predicted because of the partial elimination of ASE [5.10].

We now turn to thermal effects in the active mirror. The major effect is the so-called thermal lensing, which occurs because the heat is deposited nonuniformly within the thickness of the disk. The induced distortion causes the mirror to distort so as to become like a concave mirror; that occurs during the peak of the pump pulse as shown in Fig.5.14, where the number of waves of distortion are shown as a function of time after the beginning of the flashlamp pulse [5.11]. The thermally induced curvature in an active-mirror amplifier was first analyzed by MARTIN at the General Electric Co. [5.12]; recently that analysis has been refined and extended by SAMPATH at the LLE [5.13]. In particular, it can be shown that the induced radius of curvature R is given by

$$R = (-B)^{-1} , \tag{5.20}$$

where B is

$$B = \frac{12\beta}{\ell^3} \int_0^\ell T(z)z \, dz - \frac{6\beta}{\ell^2} \int_0^\ell T(z)dz . \tag{5.21}$$

Here the temperature distribution $T(z)$ varies only in the direction z of disk thickness. $\beta$ is the thermal-expansion coefficient and $\ell$ the disk thickness. If the disk lies in the x-y plane, the temperature is uniform and there is only one radius of curvature, an assumption that is supported by experiments. Note that if the temperature $T(z) = $ constant, $B = 0$ and R is infinite, as would be expected. To minimize B (maximize R), $\beta$ should be as small as possible, whereas the thickness should be maximized. Clearly, it is also desirable to minimize the thermal gradient, which may be accomplished by using a front-face coating to make the profile more uniform, or by glass selection, as discussed in Chap.2.

The thermal-cycling time is also of interest; we shown in Fig.5.15, data for a 2 cm thick disk as a function of PFN voltage. It is noteworthy that at the 8-9 kV maximum operating point, the mirror is fully recovered after less than eight minutes. It can be shown [5.14] that the thermal-decay-time constant $\tau$ for a slab of thickness $\ell$ follows (2.46) and may be used as a rough guide in scaling the data from Fig.5.15. The classical heat-flow equation

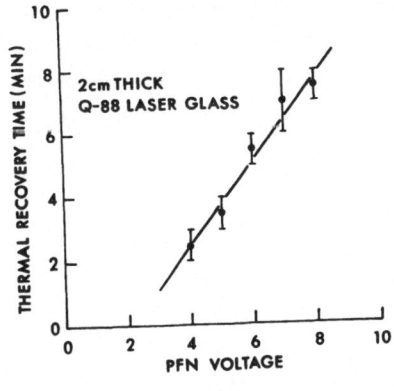

Fig. 5.15. Thermal recovery time for a 2-cm-thick active-mirror amplifier as a function of PFN voltage [5.11]

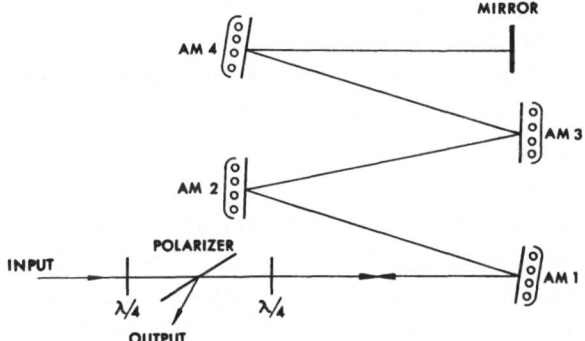

Fig. 5.16. Double-pass active-mirror system [5.16]

with initial temperature profile $T(z)$ and the appropriate boundary conditions should be solved to find $\tau$. Such a study has been made recently [5.15] and shows that the problem is not very sensitive to the surface coefficient of heat transfer because the heat diffuses so slowly in glass.

It may also be shown that stress-birefringence effects vanish in active mirrors because they are compensated by the double pass for any angle of incidence [5.12,13]. This is important for avoiding losses associated with propagating circular polarization.

A unique advantage of active-mirror amplifiers is the ability to operate them in a double-pass mode. This system, invented by BROWN [5.16], and shown schematically in Fig.5.16, consists of a circularly polarized beam which is passed through a polarizer by means of two quarter-wave plates. The beam makes a single pass of all active mirrors, is reflected by a mirror, and is further amplified on a second pass. If the number of reflections after the polarizer is odd, then right- or left-circular polarization is converted to left or right, respectively, and rejection at the polarizer results. Two ob-

vious advantages of this scheme are that no birefringence losses occur due to
the polarizer, and increased energy extraction occurs in the active mirrors.
This system has been tested experimentally and found to agree very well with
computer simulations [5.17]. It may be used with rod amplifiers if birefringence losses are minimized, but not with disk amplifiers.

## 5.3  Rod Amplifiers

The last major type of amplifier to be discussed here is the rod amplifier.
That device was shown schematically in Chap.4 (Fig.4.11). A right-circular
cylinder of laser glass is mounted coaxially with a surrounding liquid jacket
normally fabricated from Pyrex or quartz, whose purpose is to enclose a layer
of liquid coolant that actively cools most of the pumped length of the laser
rod. The ratio of the radius of the laser rod to that of the liquid jacket
is chosen according to rules discussed in Chap.4, to suppress radial-mode
parasitic oscillations. Wrapped around the entire periphery of the laser rod
is an array of Xe flashlamps, which pump most of the length of the rod. Behind each flashlamp is usually a reflector, used to redirect radiation emitted
in directions other than towards the rod, back towards the active volume. Because stored-energy density levels in rod amplifiers normally do not reach
values for which ASE is important, and because effective techniques exist to
suppress parasitic oscillations, rod amplifiers are pump-limited and every
effort is made to maximize the flashlamp flux incident upon the rod.

At present, there is no comprehensive rod-amplifier pumping model that
includes all of the effects discussed for disk and active-mirror amplifiers.
Although models such as VODAC or INVDEN could be modified to be applicable
to rod amplifiers, such a program has not been undertaken. Nevertheless, excellent performance has been obtained from amplifiers developed, for example,
for the OMEGA laser at LLE, as we will see.

Rod amplifiers differ in many respects from active-mirror and disk amplifiers. Because, on the average, each wavelength of flashlamp radiation is
absorbed exponentially in the radial direction, the total sum of which gives
rise to a nonexponential radial inversion profile, three effects occur. The
first is that, because the radial gain profile is not constant, important
modifications of propagating beams occur that do not affect disk or active-
mirror amplifiers, as will be discussed in Chap.8. The second effect is thermally induced birefringence which arises from the radial temperature gradient.

Disk amplifiers display little birefringence because propagation is approximately in the same direction as the temperature profile, whereas in active mirrors it does not occur owing to the compensating double pass. The third effect, also of thermal origin, is thermally induced lensing and also results from the radial temperature gradient. Both thermal effects will be discussed in detail below.

A large number of papers [5.18-20] have appeared that propose to derive analytically the stored-energy density or gain in a rod-type amplifier. To model such an amplifier properly, the following important effects should be taken into account:

a) concentration quenching;
b) $H_2O$ quenching;
c) focusing effects associated with the reflector;
d) reabsorption of emitted radiation by the flashlamp plasma, with subsequent re-emission of radiation at the same and other wavelengths;
e) attenuation of broadband Xe flashlamp radiation by the discrete $Nd^{3+}$ bands;
f) laser-glass impurities, such as platinum and iron;
g) reflection and absorption of radiation by the flashlamp envelope, $H_2O$ and $H_2O$ jacket, liquid coolant and jacket, laser glass, and reflector.

Although all of these could be taken into account in a complicated computer model, the problem is difficult because, in rod amplifiers, it is also necessary to take properly into account the fact that each point in the laser rod is illuminated to a considerable extent by the entire length and number of flashlamps, within the constraints imposed by total internal reflection in the rod and other optical elements. The best way to study the problem is to use a computer code such as ZAP, described in Chap.4, but which would have to be modified considerably. In that way, the number and diameter of the lamps, PFN, the diameter of the circle on which the lamp centers are located, and the reflector shape could be varied to determine an optimal design. In practice, it has been easier to vary those parameters and reflector shapes in laboratory experiments, resulting in amplifiers with excellent performance. The small-signal-gain performance of rod amplifiers used in the OMEGA system at LLE is shown in Table 5.1; it is typical for phosphate-glass rod amplifiers

Table 5.1. OMEGA rod-amplifier performance

| Rod diameter [mm] × length [cm] | Glass type | Average small-signal gain |
|---|---|---|
| 16 × 36.5 | LHG-8 | 200 |
| 30 × 36.5 | LHG-8 | 50 |
| 40 × 36.5 | LHG-8 | 25 |
| 64 × 36.5 | LHG-8 | 18 |
| 90 × 36.5 | LHG-8 | 8 |

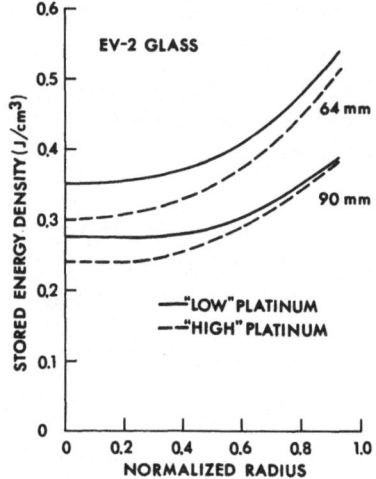

Fig. 5.17. Stored-energy density as a function of normalized radius for a 64- and 90-mm-diameter rod amplifier, using "low" and "high" platinum containing EV-2 laser glass

with the indicated parameters. The effect that reflector design can have upon the radial stored-energy density distribution, as well as the effect of impurities, have been described previously [5.21,22]. The major effect of iron impurity is to attenuate radiation in the vicinity of 1.06 μm; it should be minimized to have the lowest passive absorption at that wavelength. Platinum, on the other hand, absorbs flashlamp radiation in the vicinity of 350 nm, one of the weaker $Nd^{3+}$ absorption bands (Chap.2). The effect in rod amplifiers, is to reduce considerably the stored-energy in the rod center as shown in Fig.5.17. There, stored-energy density is plotted as a function of normalized radius ($r/r_0$) for 64- and 90-mm rod amplifiers using EV-2 phosphate glass, for both low (less than 20 ppm) and high (greater than 100 ppm) platinum impurity. The main effect is to reduce the stored-energy density at the rod center, but also to make the temperature profile greater, which has a deleterious effect on beam propagation.

We now turn to thermal effects in rod amplifiers, as analyzed for instance by QUELLE [5.23], or BROWN and JACOBS [5.24]. In general, a ray traveling along the rod at any radial position may be broken up into a radial (r) and a tangential (θ) component. The change in pathlength $P_r$ due to thermal effects for r polarization is

$$P_r = L\left\{\left[\beta + \frac{2\alpha E}{(1-\nu)} B_\perp\right]T(r) + \frac{\alpha E}{(1-\nu)}(B_{11} - B_\perp)R \right.$$
$$\left. + \left[2\alpha(n-1) - \frac{\alpha E}{(1-\nu)}(3B_\perp + B_{11})\right]F\right\}$$

(5.22)

and for $\theta$ polarization, $P_\theta$ is

$$P_\theta = L\left\{\left[\beta + \frac{\alpha E}{(1-\nu)}\ (B_\perp + B_{11})\right]T + \frac{\alpha E}{(1-\nu)}\ (B_\perp - B_{11})R\right.$$
$$\left. + \left[2\alpha(n-1) - \frac{\alpha E}{(1-\nu)}\ (3B_\perp + B_{11})\right]F\right\} \quad . \tag{5.23}$$

In (5.22,23), $\beta = -dn/dT$ as defined in Chap.2, $\alpha$ is the coefficient of thermal expansion, E is Young's modulus, $\nu$ Poisson's ratio, and T(r) the radial temperature distribution. $B_\perp$ and $B_{11}$ are the stress optic coefficients

$$B_\perp = \frac{n}{E}\left[(1-\nu)\ \frac{p}{v_0} - \nu\ \frac{q}{v_0}\right] \tag{5.24}$$

and

$$B_{11} = \frac{n}{E}\left[\frac{q}{v_0} - 2\nu\ \frac{p}{v_0}\right] \quad . \tag{5.25}$$

Here, p and q are piezo-electric coefficients which must be determined experimentally, and $v_0$ the speed of light. The stress-optic coefficient B, given by (2.53) or the difference between (5.24,25), has already been discussed and the results presented in Chap.2 for various laser glasses. In (5.22,23), R and F are

$$R = \frac{1}{r^2}\int_0^r T(r)r\ dr \tag{5.26}$$

and

$$F = \frac{1}{a^2}\int_0^a T(r)r\ dr \quad . \tag{5.27}$$

Because $\Delta P_r \neq \Delta P_\theta$ for any except the most trivial temperature profile, one component is retarded relative to the other, and birefringence results, which can lead to severe losses in a system, for example, a polarizer, if the effect is not minimized. The difference $|\Delta P_r - \Delta P_\theta| = \Delta P$, when evaluated for temperature profile of the form $T(r) \sim r^n$ results in

$$\Delta P = |\Delta P_r - \Delta P_\theta| \sim B \quad , \tag{5.28}$$

for which the path difference is directly proportional to the stress-optic coefficient. That is why, for the rod-amplifier discussion of FOM in Chap.2,

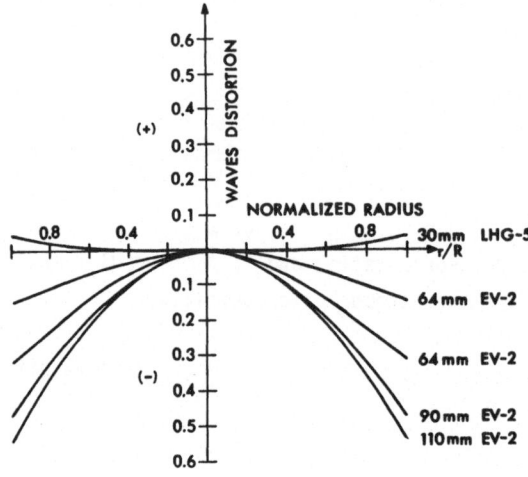

Fig. 5.18. Number of waves distortion as a function of normalized radius for 30-, 64-, 90-, and 110-mm rod amplifiers showing defocusing in EV-2 laser glass and focusing in LHG-5 glass [5.24]

it was pointed out that it is desirable to minimize B. A measure of the average wavefront distortion in a rod amplifier is

$$\overline{\Delta P} = \frac{\Delta P_r + \Delta P_\theta}{2} \quad , \tag{5.29}$$

which may be positive, negative, or zero. It has been evaluated for a number of laser amplifiers [5.24] in use at LLE with the result shown in Fig.5.18 where the average number of waves distorted is shown as a function of normalized radius. The 64-, 90- and 110-mm amplifiers all show a negative lensing or defocusing for EV-2 laser glass, whereas the 30-mm rod displays positive lensing or focusing using LHG-5 glass. Thus, by alternating suitable glass types, thermal lensing could be at least partially compensated in rod-amplifier chains.

Unlike disk amplifiers, and in common with active-mirror amplifiers, rods are actively cooled with a liquid circulated around the barrel. For a cylinder in which the temperature T(r) is a function of r only, T(r) must satisfy

$$\frac{d^2T}{dr^2} + \frac{1}{r}\frac{dT}{dr} + \alpha^2 T = 0 \quad , \tag{5.30}$$

which is Bessel's equation and $\alpha$ is a root of it. The general solution of (5.30) is

$$T'(r,t) = \sum_{n=1}^{\infty} A_n J_0(\alpha_n r) \exp(-\gamma\alpha_n^2 t) \quad , \tag{5.31}$$

where the $A_n$ are constants, $\alpha_n$ Bessel roots, $J_0$ the zero-order Bessel function and $\gamma$ the thermal diffusivity. For a number of assumed temperature profiles and boundary conditions, the solution of (5.30) yields the result that the thermal recovery time $\tau$ for rod amplifiers to first order, scales according to (2.46), or directly proportional to $\gamma$ and the square of the rod radius. In fact this relationship has been verified [5.22] for a number of rod amplifiers. Rod amplifiers with diameters less than 90 mm were shown to have thermal-recovery times of less than 30 minutes, a major reason why they were chosen for use in the OMEGA system at LLE.

Rod amplifiers up to 90 mm diameter possess a number of advantages over other types of amplifiers. Among these are low cost and ease of construction and maintenance. They are also relatively efficient (up to $\approx 1\%$), thus minimizing bank costs, and have medium repetition rate (1 shot every 30 minutes for 90 mm diameter). Their disadvantages include the necessity to grow, with excellent optical quality, pieces of laser glass larger than 9 cm diameter, and a large B-integral contribution relative to disk amplifiers, which reduces their ultimate power-handling capability for short-pulse propagation. The major reason for this is the decrease of gain or stored-energy density as the diameter is increased. The pump or number of flashlamps can increase only as the radius of the amplifier whereas the volume of laser glass increases as the radius squared. Additional disadvantages are the radial dependence of stored-energy density and temperature profile, which have beam-propagation and thermal effects that are absent or minimal in disk or active-mirror amplifiers.

# 6. Damage Effects in High-Peak-Power Nd : Glass Laser Systems

We now turn our attention to a subject that is interesting from a fundamental-
physics standpoint as well as extremely important in the design of large Nd:
glass laser systems, namely optically induced damage to dielectric materials.
The subject has attained considerable importance because, if permanent irreversible damage occurs in such systems, it is expensive to replace the damaged
component and may lead to (if not replaced) a gradual degradation of performance. Furthermore, certain staging schemes, particularly for long pulses, are
intended to run at as large a flux as possible, which is limited by the damage threshold of the component in question. The aperture or diameter of a
stage is often determined by the damage threshold of the limiting component
in the stage; because the cost of a component is a strong function of the diameter it is obviously of interest to obtain the maximum damage threshold to
minimize the diameter. Although the general subject of optically induced damage has matured considerably in recent years, progress has been slow owing
to the experimental and theoretical complexities. At the present time, workers
in the field cannot even agree on the basic physical cause of optical damage;
it is likely that a number of separate mechanisms are involved. Such disagree-
ments have not impeded progress, however; much of the work in this area is
experimental and empirical and has yielded a number of practical results of
immediate importance, particularly in the inertial-confinement fusion community.

We are concerned here with transparent dielectric materials that have
linear absorption coefficients in the range of $10^{-1}$-$10^{-5}$ cm$^{-1}$. In the present picture of damage, an intense electromagnetic wave may generate the initiating electrons through the physical mechanisms of multi-photon ionization
and/or avalanche ionization. After the liberation of the initiating electrons,
an absorbing plasma is produced, which may transfer a damaging amount of energy to the surrounding-material lattice. Damage may occur at surfaces, in
the material bulk, or in multi-layer dielectric coatings, each of which will
be described. The experimental fact that bare-surface damage thresholds are

always significantly lower than the bulk has led to a number of investigations; at present the reasons for this are still controversial.

We will attempt to outline the major developments in the separate areas of bulk, surface, and dielectric-coatings damage, in that order. Because a survey such as this cannot be exhaustive, the interested reader is advised to consult the excellent published volumes that recount the annual conferences on Laser Induced Damage in Boulder, Colorado [6.1-10].

## 6.1 Bulk Damage to Optical Materials

The first type of damage mechanism to which we turn is bulk damage in transparent dielectric materials. The subject is of obvious practical interest, but more importantly, in the bulk damage processes may be studied without the additional complications that are encountered on bare or dielectric-coated surfaces. We begin by first reviewing the currently favored electron-avalanche theory, pioneered by BLOEMBERGEN, YABLONOVITCH and co-workers at Harvard University and reviewed recently by SMITH [6.11]. To begin an electron avalanche, it is necessary to have a number, $N_0$, of initiating electrons. Such electrons are generally agreed to be supplied from shallow traps, present to some degree in all optical materials, and arising from structural defects, voids, color centers, etc. The liberation of electrons from such shallow traps may take place through linear or multi-photon absorption processes, and $N_0$ may be written

$$N_0(I) = \alpha I + \beta I^2 + \gamma I^3 + \ldots , \tag{6.1}$$

where $\alpha$, $\beta$, $\gamma$ are constants that correspond to linear, two-photon, and three-photon absorption, respectively, and I is the laser-pulse intensity. As pointed out by SMITH [6.11], $N_0$ will then depend upon the laser pulse width, the single-photon energy, the band gap, extrinsic-impurity levels, and the ambient conditions. Linear absorption is present in all optical materials of interest; two-photon absorption can be significant if the condition

$$2\hbar\omega \gtrsim E_B \tag{6.2}$$

is met, where $\omega$ is the laser frequency, $\hbar$ Planck's constant divided by $2\pi$, and $E_B$ the bandgap energy. Processes higher than the second order have not been detected owing to optical breakdown in the samples. Estimates of the value of $N_0$ typical for optical materials had been in the range of $10^8$-$10^{11}$ cm$^{-3}$ [6.12]; to date the only direct measurement of $N_0$ was by SMITH et al. [6.13]. 30 ps, 1.06 μm laser pulses were incident on a sample of NaCl; at threshold a number of spatially distinct microsites were observed. Individual damage sites did not overlap because their growth was limited by the short-duration of the beam pulses. On the assumption that each microsite was caused by a single (or a few) initiating electron, a density $N_0 \sim 7 \times 10^9$ cm$^{-3}$ at an intensity level of $\sim 2 \times 10^{11}$ W/cm$^2$ was determined; $N_0$ was observed to increase rapidly as the intensity was increased.

Once the initiating electron density $N_0$ has been established, the question arises as to how energy is transferred initially to form a plasma site, and how it ultimately arrives in the adjacent lattice to cause material damage. The view that an electron avalanche is responsible was based initially upon the important work of YABLONOVITCH [6.14] who investigated damage threshold in the alkali halides using a $CO_2$ laser. Surprisingly, he found that the data obtained showed the same trends as that found by VON HIPPEL [6.15] who measured dielectric breakdown in alkali halides at zero (dc) frequency. In Fig. 6.1 we show the measured bulk damage threshold for Na halides ($E_B^{XTAL}$), normalized to the field ($E_B^{NaCl}$) for NaCl for each data set, for the dc, $CO_2$, and more recent measurements at various wavelengths [6.11]. These data led to the unmistakable conclusion that the breakdown strength is independent of frequency in the range 0-$10^{15}$ Hz and for pulse durations in the range $10^{-11}$-$10^{-6}$ s. Based upon earlier work in electrical discharges, it was concluded that the operative mechanism in solids for the dc case was electron-avalanche multiplication. The measurements of YABLONOVITCH [6.14] led to the conclusion that electron-avalanche multiplication was the dominant mechanism at optical

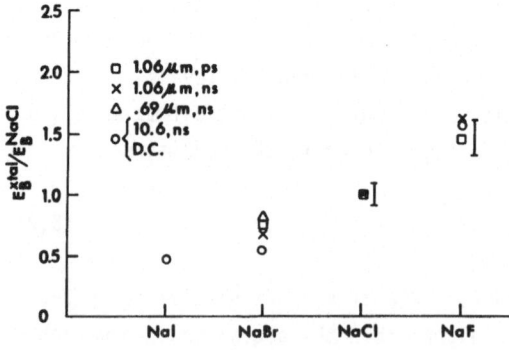

Fig. 6.1. Experimental trend of breakdown threshold electric fields through the Na-halides. Each data set is normalized to the field for NaCl of that data set [6.11]

frequencies as well. In that theory it is assumed that the rate of increase of the electron density N during breakdown is

$$\frac{\partial N}{\partial t} = \eta(F)N + \left(\frac{\partial N}{\partial t}\right)_{tun} - \left(\frac{\partial N}{\partial t}\right)_{loss} \quad , \tag{6.3}$$

where $\eta(E)$ is the probability per unit time that an ionizing collision will occur, averaged over the entire electron distribution, and E is the applied electric field, which is in turn a function of time. The second term is the rate of electron production due to tunneling or, at higher frequencies, multiphoton ionization; it is ignored because it is small compared to the first term. The last term represents the loss of electrons due to diffusion out of the discharge region, recombination, etc. For short ($<10^{-8}$ s) pulses of interest here, this term is also negligible and (6.3) may be integrated to obtain

$$N(t) = N_0 \exp\left[\int_0^t \eta E(t')dt'\right] \quad , \tag{6.4}$$

which represents the exponential growth of the plasma density by an avalanche-ionization coefficient $\eta$, a function of the time-averaged (rms) electric field E. In this process, an oscillating electron that has been liberated to the conduction band gains sufficient energy from the strong optical field to produce a secondary electron by impact ionization of a lattice atom. The resulting two electrons repeat the process to form four, then eight, etc. An important part of this process is the need for the electron to undergo rapid collisions with the lattice (at a rate $\sim 10^{15}$ s$^{-1}$). Without such collisions, an electron would oscillate at the "quiver energy" indefinitely, neither gaining nor losing energy, with its velocity in quadrature with the optical field vector. Through the intervention of rapid collisions, however, dephasing occurs and an electron may continue to gain energy until impact ionization occurs at the bandgap energy.

After the formation of the plasma predominantly near the peak of the laser pulse, continued heating of the electron gas occurs by the absorption mechanism of inverse bremsstrahlung; the absorption rate is given by [6.14]

$$\frac{d\varepsilon}{dt} = \frac{e^2\tau_c}{m^*} \frac{|E(t)|^2 N(t)}{(1+\omega^2\tau_c^2)} \quad . \tag{6.5}$$

Here $\tau_c$ is the characteristic electron-collision time determined primarily by collision with the lattice, $\omega$ the laser angular frequency, and e and $m^*$ the electron charge and effective mass, respectively.

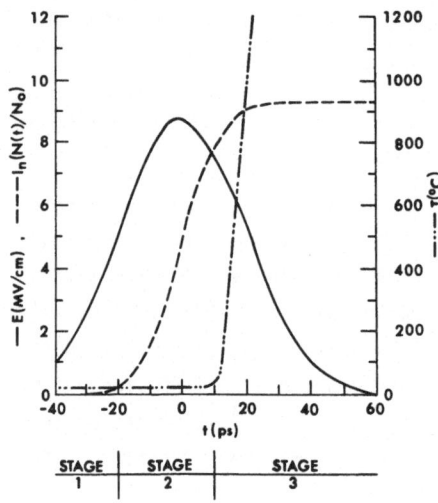

Fig. 6.2. Time dependence of the plasma growth $[N(t)/N_0]$ and lattice temperature $[T(t)]$ during breakdown induced by a picosecond laser pulse with a Gaussian electric field $E(t)$ [6.11]

YABLONOVITCH [6.14] has shown, by using (6.5), that the electron heating in an ac field is the same as that in a dc field when

$$E_{rms}(\omega) = \sqrt{1+\omega^2\tau_c^2} \; E_{dc} \quad , \tag{6.6}$$

where $E_{rms}$ is the rms ac field that gives the same heating as the dc electric field $E_{dc}$. It can be shown [6.14] that for 1.06 μm laser light, $\omega\tau_c < 1$ and thus $E_{rms} \sim E_{dc}$; hence it should be expected that the dc data that describe the breakdown in the alkali halides should also describe the threshold behavior at 1.06 μm.

SMITH [6.11] has temporally modeled the events that lead to damage, with the result shown in Fig.6.2. There, the rms electric field and the plasma growth, described by $\ln[N(t)/N_0]$, are shown as functions of time. The laser pulse modeled was a Gaussian with a wavelength of 1.06 μm focused into a NaCl crystal. Also shown in Fig.6.2 is the time evolution of the temperature $T(t)$ at the focus of the lens. If thermal diffusion and radiation are assumed negligible during the laser pulse, then T is given simply by

$$T(r,z,t) = \frac{1}{c} \int_0^t \left(\frac{d\varepsilon}{dt'}\right)dt' \quad , \tag{6.7}$$

where T is, in general, a function of the radial parameter r, the propagation direction z and the time t, c is the heat capacity of the material, and $d\varepsilon/dt$ is given by (6.5). The temperature shown in Fig.6.2 is valid up to the sample melting temperature of $\approx 800°C$; it is of interest that the rise time for the temperature is less than ten ps. Similarly, it can be seen that the plasma growth is extremely rapid; it reaches its peak in ~20 ps. In a recent publication, ANTHES and BASS [6.16] have monitored the laser-light transmission through the breakdown region in silica, with an ultrafast streak camera. Using 25 ps pulses at 0.53 μm, they observed a plasma buildup time of ~10 ps. In fact, the process may be faster, because the streak-camera resolution was ≈6 ps. These measurements are in agreement with the simple model of Fig.6.2 where the plasma density increases ≈7 orders of magnitude in ≈20 ps.

An important part of the electron avalanche model and its application is how the ionization rate $\eta(E)$ depends upon the rms electric-field strength E. This question was first addressed by YABLONOVITCH and BLOEMBERGEN [6.17] who obtained the results shown in Fig.6.3 for NaCl. There we observe the rate of ionization $\eta$ and the coefficient of ionization $\alpha$ as a function of E. The two coefficients are related through

$$\eta = \alpha v \quad , \tag{6.8}$$

where v is the electron speed. The three curves were determined by using the experimental dependence of the breakdown field on sample thickness [6.18-20] previously known from dc measurements on NaCl. In Fig.6.3 the reader will observe that two separate curves are obtained for $\eta(E)$ and represent upper and lower limits determined, respectively, by the low field mobility μ for electrons in NaCl, or by the increase of electron scattering expected when the energy $\frac{1}{2}mv^2$ of an electron becomes so great that it exceeds the energy $\hbar\omega_L$ of a longitudinal optical phonon. In Fig.6.4, ionization-rate data obtained by a number of workers [6.21-23] have been plotted by SMITH [6.11] along with a theoretical model fitted to the NaCl data based upon the work, discussed above, of YABLONOVITCH and BLOEMBERGEN [6.17]. The agreement is excellent. Note that the pulse duration $\tau_p$ has been plotted as a function of E. A relationship between the pulse duration $\tau_p$ and the ionization rate $\eta$ may be obtained by observing from (6.4) that

$$\ln\left(\frac{N}{N_0}\right) \simeq \int \eta(E)dt \sim \eta\tau_p = K \quad . \tag{6.9}$$

Values of $N_0$ were discussed previously ($N_0 \sim 10^{10}$ cm$^{-3}$); experimentally [6.24] it has been observed that in BSC-2 glass $N \sim 10^{18}$-$10^8$ cm$^{-3}$, hence $\eta\tau_p \sim 18$.

Finally, BETTIS et al. [6.25] have established theoretically that a $\tau_p^{\frac{1}{2}}$ dependence of the bulk-damage threshold on pulse width would be expected. Diffusive behavior scales as $(D\tau_p)^{\frac{1}{2}}$ where D is the appropriate diffusion co-

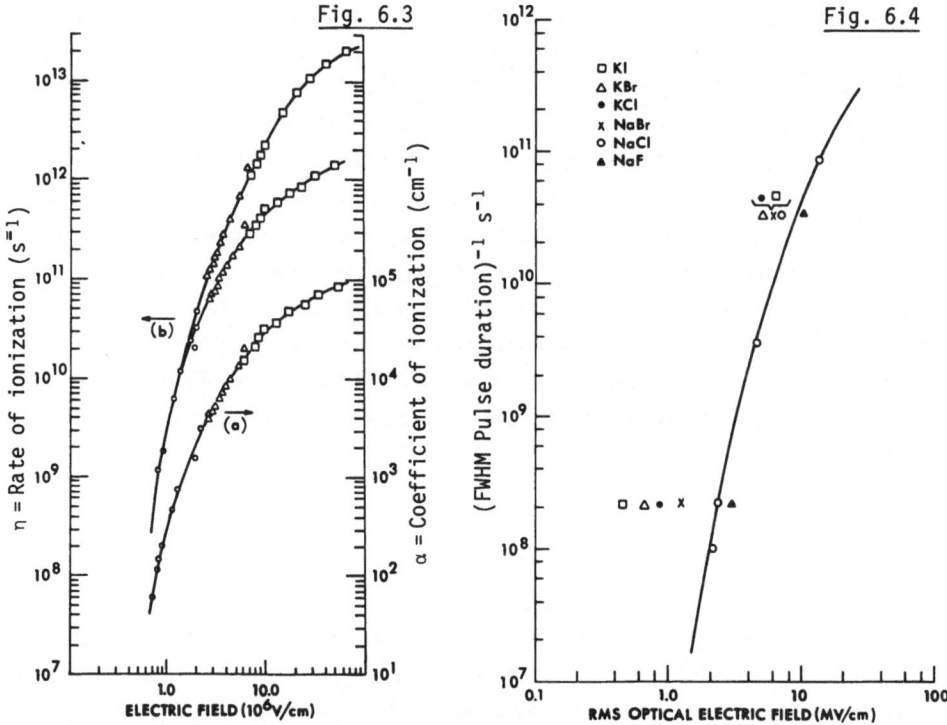

Fig. 6.3. The lower curve a is the coefficient of ionization with the scale on the right. The points are determined from the empirical thickness dependence of the breakdown strength in NaCl. The upper curve b is the ratio of ionization with the scale on the left. At high fields the curve has two branches due to the uncertainty in the drift velocity v. The actual value lies somewhere between the two branches [6.17]

Fig. 6.4. Experimental relationship between the breakdown threshold electric-field strength and the laser-pulse duration for alkali halides. The solid curve is a fit to experimental NaCl data and is of the form predicted by Yablonovitch and Bloembergen (from SMITH [6.11])

efficient. For long ($\gtrsim$ ns) pulses, D would represent thermal transport away from the breakdown region, whereas for short (ps) pulses D involves the diffusion of electrons away from the initiating electron location.

An important scientific and engineering question is how the breakdown electric field $E_{th}$ or intensity $I_{th}$ vary as a function of pulse width $\tau_p$. It has been addressed by BETTIS et al. [6.25] who studied both the pulse width and the spot-size dependence of the threshold electric field. Based upon considerations of laser plasma heating by inverse bremsstrahlung and measurements of electron density and temperature, they arrived at the relationships

$$E_{th}\sqrt{d} = K \qquad\qquad\qquad (6.10)$$

and

$$E_{th}\tau_p^{\frac{1}{4}} = \text{const} \quad , \qquad\qquad\qquad (6.11)$$

which predict that the threshold electric field varies inversely as the square root of the spot size d and inversely as the fourth root of $\tau_p$. Thus the intensity $I_{th} \sim E_{th}^2$ is expected to scale inversely as the square root of $\tau_p$. This result is contradicted by the work of SPARKS [6.26] who predicted a $\tau_p^{-1/6}$ variation of the threshold electric field; nevertheless experimental studies [6.25] suggest that a reciprocal square-root dependence is more appropriate. Although data supporting (6.10) were reported in [6.25], recently it has been found to be violated [6.27].

Table 6.1(a). Bulk breakdown in transparent solids [6.11]

| Substance | $\lambda$ [$\mu$] | $\tau_p$ [ns] | $I_B$ [$10^{10}$ W/cm$^2$] |
|---|---|---|---|
| $Al_2O_3$ | 1.064 | 4.7 | 18 |
| | 1.064 | 0.030 | 77 |
| BSC-2 glass | 1.064 | 4.7 | 8.8 |
| C (diamond) | 10.6 | $10^9$ (1 s) | >0.0001 |
| $CaF_2$ | 1.064 | 0.030 | 78 |
| | 0.532 | 0.021 | 240 |
| | 0.355 | ~0.020 | 120-420 |
| ED-4 glass | 1.064 | 0.030 | 40 |
| $KH_2PO_4$ (KDP) | 1.064 | 0.030 | 200 |
| | 0.532 | 0.021 | 220 |
| | 0.355 | ~0.020 | 480-2340 |
| LHG-5,-6 glasses | 1.064 | 0.030 | 69 |
| $La_2Be_2P_5$ (BEL):Nd | 1.064 | 0.030 | 130 |
| $SiO_2$ (silica) | 1.064 | 4.7 | 11 |
| | 1.064 | 0.030 | 53 |
| | 0.532 | 0.021 | 140 |
| | 0.355 | ~0.020 | 250-940 |
| $Y_3Al_5O_{12}$ (YAG):Nd | 1.064 | 4.7 | 0.64 |
| | 1.064 | 0.030 | 47 |
| $YAlO_3$ | 1.064 | 4.7 | 7.1 |
| ZnSe | 10.6 | ~92 | 0.04-0.05 |
| | 1.064 | 4.7 | 0.12 |

Table 6.1(b). Bulk breakdown in alkali halides [6.11]

| Substance | $\lambda$ [$\mu$] | $\tau_p$ [ns] | $I_B$ [$10^{10}$ W/cm$^2$] |
|---|---|---|---|
| LiF | 1.064 | 0.030 | 55 |
|  | 0.532 | 0.020 | 260 |
|  | 0.355 | ~0.020 | 460-1840 |
| NaF | 1.064 | 4.7 | 0.90 |
|  | 1.064 | 0.030 | 41 |
|  | 0.694 | 14.0 | 12 |
|  | 0.532 | 0.021 | 130 |
|  | 0.355 | ~0.020 | 570-3550 |
| NaCl | 10.6 | ~600 | 0.46-1.6 |
|  | 10.6 | ~92 | 0.39-1.4 |
|  | 10.6 | ~80 | 1.55 |
|  | 1.064 | 10.3 | 1.8 |
|  | 1.064 | 4.7 | 2.2 |
|  | 1.064 | 0.030 | 9.0 |
|  | 1.064 | 0.030 | 22 |
|  | 1.064 | 0.015 | 63 |
|  | 0.694 | 14.0 | 2.0 |
|  | 0.532 | 0.021 | 63 |
| NaBr | 10.6 | ~80 | 0.35 |
|  | 1.064 | 4.7 | 0.93 |
|  | 1.064 | 0.030 | 14 |
|  | 0.694 | 14.0 | 1.4 |
| NaI | 10.6 | ~80 | 0.27 |
| KF | 10.6 | ~80 | 2.1 |
|  | 1.064 | 0.030 | 25 |
| KCl | 10.6 | ~600 | 0.05-1.4 |
|  | 10.6 | ~92 | 0.065-0.26 |
|  | 10.6 | ~80 | 0.74 |
|  | 1.064 | 4.7 | 0.66 |
|  | 1.064 | 0.030 | 14 |
|  | 0.694 | 14.0 | 1.3 |
| KBr | 10.6 | ~600 | 0.1-0.17 |
|  | 10.6 | ~80 | 0.34 |
|  | 1.064 | 4.7 | 0.29 |
|  | 1.064 | 0.030 | 12 |
| KI | 10.6 | ~80 | 0.22 |
|  | 1.064 | 4.7 | 0.15 |
|  | 1.064 | 0.030 | 15 |
| RbCl | 10.6 | ~80 | 0.34 |
|  | 1.064 | 4.7 | 0.85 |
| RbBr | 10.6 | ~80 | 0.24 |
|  | 1.064 | 4.7 | 0.59 |
| RbI | 10.6 | ~80 | 0.17 |
|  | 1.064 | 4.7 | 0.33 |
|  | 1.064 | 0.030 | 5.0 |

Alternative theories or modifications of the electron-avalanche model discussed above have been proposed by SPARKS [6.26] and SCHMID et al. [6.28] but will not be discussed here. At the time of this writing it is fair to say that although the electron-avalanche theory has successfully accounted for a large body of experimental data, it is not predictive in the sense discussed by SPARKS [6.26], and experimental verification is still lacking. In particular, there has not yet been a direct demonstration of electron multiplication; more-precise measurements of the initial electron density $N_0$ and associated plasma parameters are required.

SMITH [6.11] has tabulated experimentally determined values of the bulk breakdown intensity $I_B$ for a large number of transparent solids; those data have been reproduced here in Table 6.1 for transparent solids and alkali halides. Bulk-damage thresholds in fluoride laser glasses have been measured by STOKOWSKI et al. [6.29]; other values may be found in the damage literature [6.1-10].

Finally, we have not considered the importance of two-photon absorption here except as regards the generation of initiating electrons (6.1). Although it is generally considered to be not an important process in causing damage at 1.06 μm, the recent interest in frequency doubling and tripling (0.53, 0.35 μm) will make the subject much more important in high-power laser design. It is expected that in addition to degrading damage thresholds in the visible and ultraviolet, the value of the nonlinear index will increase significantly and self-focusing will become even more important than at 1.06 μm.

## 6.2 Surface Damage to Optical Materials

In the previous section, we examined the phenomena of bulk damage to optical materials and reviewed the current understanding through the presently favored model, namely, electron-avalanche ionization. It is, however, an experimentally determined fact that optical surfaces always show a lower damage threshold than that characteristic of the bulk, typically by a factor of 3-5. Thus the problem is considerably more complicated, and an understanding of the phenomena requires consideration of a number of problems peculiar to such surfaces.

If a surface were "perfect", meaning free of cracks, inclusions, and absorption sites, it is expected that the damage threshold would approach that of the bulk and that, furthermore, the same limiting mechanism would be operable, electron-avalanche ionization. Because that limit has never been reached and there does not exist, at present, any theory that comprehensively explains all aspects of surface damage, we limit ourselves to only the major developments. For detailed information, the appropriate literature [6.1-10] should be consulted.

We begin by examining one of the first effects observed, namely that if a laser beam is normally incident on a piece of glass that has two parallel sides through which the beam propagates, it was observed [6.30] that the ratio of exit to entrance damage thresholds was not unity; the exit surface displayed a lower threshold than the entrance. This observation was explained by BOLING et al. [6.31] who reasoned that at the entrance surface the reflected wave is $\pi$ radians out of phase with the incident wave which results in an electric field that is less than that of the laser beam itself. At the exit surface, however, because there is no phase change, the reflected wave is in phase with the incident beam resulting in an increase of the electric field there. It was shown that the ratio of the output ($I_{EXIT}$) to input ($I_{ENT}$) intensities is given by [6.31]

$$\frac{I_{EXIT}}{I_{ENT}} = \frac{(E_{EXIT})^2}{(E_{ENT})^2} = \frac{4n^2}{(n+1)^2} \quad , \tag{6.12}$$

where n is the linear index and $E_{EXIT}$ and $E_{ENT}$ the electric fields at the exit and entrance faces, respectively. To test this conjecture, samples were prepared [6.30] with both faces at Brewster's angle. Because no reflected waves occur in that case, it was expected and confirmed [6.30] experimentally that the ratio of the damage thresholds was unity. The effect is important because for most laser glasses, the ratio $I_{EXIT}/I_{ENT} \sim 1.4$. Standing-wave interference patterns are not, however, the only electric-field effects to be expected at surfaces. In another landmark study, BLOEMBERGEN [6.32] assessed what effect local cracks, pores, and inclusions might have on the damage threshold. It was shown that the field inside a cavity on a dielectric surface is given by

$$E_{IN} = \frac{E_0}{[1+(1-n^2/n^2)L]} \quad , \tag{6.13}$$

where $E_0$ is the field inside the dielectric, n the linear index and L a depolarization factor that takes the values of L = 1/2, 1/2 for the case of a spherical pore or a cylindrical groove, respectively. The crack gives L = 1 - ($\pi$/2)(c/a), where c and a are the width and depth, respectively. Calculation using typical values of n for optical glasses and alkali halides shows that an enhancement of the local intensity leads to an expected decrease of the damage threshold by a factor ranging from 2-5.

It was obvious from the above work that a correlation should exist between surface roughness, or quality, and damage threshold. A good deal of work has been reported on the subject [6.33,34]; a comprehensive study was reported by HOUSE et al. [6.35] who damage-tested surfaces prepared by conventional polishing, ion-polishing, flame polishing, acid etching and bowl-feed polishing. At a pulse duration of 40 ns, they observed that the damage threshold increased as surface quality improved. Another conclusion was that for a given surface roughness, techniques that cleaned the surface, such as ion or acid etching and flame polishing, significantly improved the damage threshold, presumably due to the removal of surface dirt or impurities, which are easily ionized. One measurable surface characteristic is the rms surface roughness $\sigma$ determined by profilometry [6.36]. A relationship has been reported between $\sigma$ and the breakdown electric field $E_b$ [6.35]. It may be written as

$$E_b \, \sigma^m = c \quad , \tag{6.14}$$

where a typical value of the exponent m is ~0.5 and c is a constant. In Fig. 6.5, we show data obtained from [6.35] for fused silica as well as measurements reported by MILAM et al. [6.37] at a pulse length of 150 ps. Although the form of (6.14) is valid for the 40 ns data (m≈0.61), for the short-pulse data the correlation is not conclusive indicating that techniques used to improve damage resistance at long pulses (ns) may not be applicable at reduced pulse widths (ps). MILAM [6.38] has reported a careful study of damage to bare dielectric surfaces that were either routinely polished or had surface-

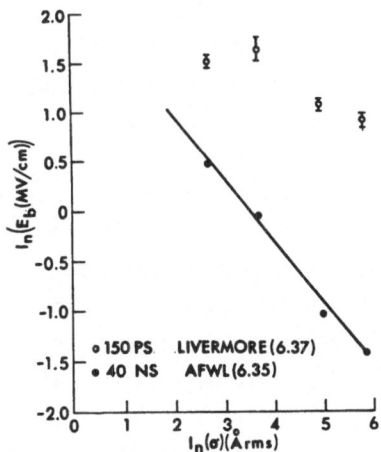

Fig. 6.5. Breakdown-electric-field strength of four fused silica surfaces as a function of rms surface roughness for two different pulse durations [6.37]

Table 6.2. 1064-nm 125-ps damage thresholds of dielectric surfaces [6.38]

| Surface | Surface finish | Beam incidence angle | Treshold [J/cm$^2$] |
|---|---|---|---|
| O.I. ED-2, sample A | Polished | 0$^\circ$ | 13.8 ± 2.0 |
| O.I. ED-2, sample B | Polished | 0$^\circ$ | 13.3 ± 2.0 |
| O.I. ED-2, sample C | Polished | 0$^\circ$ | 11.7 ± 1.5 |
| O.I. ED-2, sample D | Polished | 0$^\circ$ | 11.3 ± 1.5 |
| O.I. ED-2, sample E | Polished | 0$^\circ$ | 12.0 ± 1.5 |
| Kigre Nd-phosphate laser glass | Polished | 0$^\circ$ | 13.7 ± 2.0 |
| Hoya Nd-phosphate laser glass (several samples) | Polished | 0$^\circ$ | 13.5 ± 2.0 |
| Hoya LSG-95 | Polished | 0$^\circ$ | 12.3 ± 2.0 |
| Hoya LSG-91H | Polished | 0$^\circ$ | 13.1 ± 2.0 |
| Hoya FR-5, sample A | Polished | 0$^\circ$ | 11.8 ± 1.5 |
| Hoya FR-5, sample B | Polished | 0$^\circ$ | 8.0 ± 1.5 |
| Corning 7059 electronic substrate | Extruded/ flame | 56$^\circ$, p-state | 9.7 ± 1.5 |
| Corning 0211 Microsheet | Extruded/ flame | 56$^\circ$, p-state | 10.5 ± 1.5 |
| Hoya BK-7 | Polished | 0$^\circ$ | 9.2 ± 1.5 |
| Corning 7940 fused silica | Polished | 0$^\circ$ | 10.0 ± 1.0 |
| Schott WG-280 | Polished | 56$^\circ$, p-state | 8.1 ± 1.0 |
| Corning 7740 Pyrex | Polished | 56$^\circ$, p-state | 8.0 ± 2.0 |
| Schott FK-51, sample A | Polished | 0$^\circ$ | 10.0 ± 1.5 |
| Schott FK-51, sample B | Polished | 0$^\circ$ | 7.2 ± 1.1 |

leaching or flame-polishing treatment, at a pulse width of 125 ps. The results, reproduced in Table 6.2, show that at that pulse width no significant
increase of damage threshold results from surface treatment, which indicates
that there is no connection between surface cleanliness or roughness and
damage threshold at that pulse width. In a more recent study, MILAM [6.39]
reported measurements of damage threshold in BK-7 glass as a function of surface roughness and pulse width. He stated a number of important conclusions,
among them that for surfaces with an rms roughness of <30-40 Å, some other
variables, probably including surface absorption, are more important at short
pulse widths. Routinely polished samples actually displayed a higher damage
threshold than the smoothest surfaces. For long pulses (3.2 ns), particulate
contamination can decrease the threshold by 50%. For the investigated pulse
widths (0.17, 1.0, 1.6, 3.2 ns) the thresholds were approximately proportional

to the square root of the pulse width, as one would expect for bulk damage (see Sect.6.1). At the present time, the situation is confused; although for very rough (> 100 Å) surfaces the correlation between surface-damage threshold and roughness holds for short (0.15 ns) and long pulses (40 ns); for surfaces with roughness < 40 Å, the correlation holds only for long pulses.

Currently, much work is being done to evaluate the absorption characteristics of surfaces and to investigate correlations with damage threshold. Although such absorption is usually much greater than the absorption of the bulk material, the surface thickness is so small that the absorption spectrum cannot be measured using conventional techniques. Two techniques, in particular, have begun to yield interesting information. The first is attenuated-total-reflection spectroscopy used by BURDICK [6.40] to investigate surface absorption in ZnSe, and the technique of photo-acoustic spectroscopy [6.41] which may be used to measure very weak absorption in both the bulk and the surface, and which has been used as well to measure thin-film absorption.

The subject of surface damage to optical materials has received much attention recently and, although there are contradictory data, it is clear that considerable progress in understanding a very complicated problem has been achieved. Further careful studies of surface damage, with full characterization of all relevant parameters, should allow the construction of a model that accounts for all aspects of such damage.

## 6.3 Thin-Film Damage

Although it has been observed that surface-damage thresholds are always less than that of the bulk, it is equally true that damage thresholds for multi-layer dielectric coatings have been found to be significantly less than for the bare surface upon which they are deposited. Such coatings often determine the aperture of a given stage in a system design because they are the elements that are limited first by damage. The reader will notice from what follows that although bare-surface damage is not very well understood, the knowledge of thin-film damage is even more primitive. In this section, we will attempt to review the major developments obtained to date. The literature of thin-film damage is extensive; no attempt will be made here to be comprehensive; as with surface and bulk damage, [6.1-10] should be consulted for more detailed information.

One of the most difficult aspects of the damage problem in thin films is the lack of reproducible results, probably caused in large part by the number of variables involved. In the deposition process alone, it is necessary to characterize the starting-material purity, electron-beam characteristics, substrate temperature and surface purity, deposition rate, background impurity pressure and composition, and other parameters. The difficulty is that monitoring each of the above is complicated and expensive; it is not even clear at present that all of the critical process variables have been identified; hence this field is replete with largely empirical results.

Early studies of damage concentrated upon investigations of the effect of stress, defects, adhesion, and substrate roughness. The influence of stress upon damage threshold was the subject of an early study by AUSTIN et al. [6.42], who showed that the use of vapor-phase mixtures to relieve stress led to greater damage thresholds. The role of defects was investigated by DE SHAZER et al. [6.43], who showed that damage thresholds increased as the irradiating-beam radius decreased, indicating that the distribution and type of defect was important in thin-film damage. Correlations of film adhesion with damage threshold has suffered, until recently, by the lack of a reproducible adhesion-measurement system. Methods for improving this situation were described recently by SCHEELE and BERGSTROM [6.44]. The influence of surface roughness was reported by BETTIS et al. [6.45], following up on the previous work on bare surfaces [6.35], and showing that increased surface roughness led to lower damage thresholds for long pulses. Later work by SMITH et al. [6.46] showed, however, that for short pulses ($\sim 150$ ps) the damage threshold of anti-reflection films was not correlated with substrate-surface roughness. An interesting study has been made of the correlation between the $OH^-$ content of the fused-silica substrate and the damage threshold for both short [6.46] $\simeq 150$ ps pulses and long [6.47] $\simeq 40$ ns pulses. Although the latter study concluded that higher $OH^-$ concentration resulted in lower damage thresholds, in the former no correlation was found. As with many other facets of thin-film damage, correlations found in one pulse-width regime are not found to occur at another. To explain this apparent contradiction, the authors [6.46] invoke an argument involving a bottleneck to the breakdown process. It is proposed that the role of the $OH^-$ concentration is to assist in the liberation of carriers in the earliest stages of breakdown; this is particularly true for nanosecond pulses for which it is assumed that the bottleneck occurs in the initial phase. Consequently, damage thresholds would be expected to be correlated with $OH^-$ concentration only for long pulses. It has been conjectured that short pulses are more effective in providing the

needed field strengths to liberate electrons early in the pulse [6.48]. In
the same study [6.46], damage thresholds of AR coatings were measured for
designs which utilized, for example, either layers of $ZrO_2/SiO_2$ or $ThF_4/MgF_4$.
It was found that, by utilizing the lower indices appropriate to fluoride
materials, a significant (as much as 100%) improvement of the damage threshold could be obtained, a result that is in agreement with the observation
that damage in thin films most often takes place at the damage threshold of
the highest-index material.

Advances in the understanding and design of damage-resistant multi-layer
dielectric coatings has been made recently through the analysis of standing-
wave effects. Standing waves are quasi-steady-state interference patterns
produced by the addition of traveling waves reflected from the various parallel index discontinuities in the thin-film stack. The importance of such
effects was appreciated early by TURNER [6.49], who reported that half-wave
(high-reflectance) films had lower damage than quarter-wave (anti-reflection)
stacks. Until recently, standing-wave effects in thin films were prevented
from being unambiguously identified by the presence of defects. Using a 30 ps
pulse length at 1.06 μm, NEWMAN et al. [6.50] investigated the effect of
film thickness on damage threshold. For $TiO_2$ films, it was evident that films
with odd quarter-wave thicknesses have appreciably greater damage thresholds
than those that have even multiples, which correlated well with the calculated
maximum internal electric fields. The effect was less pronounced in $ZrO_2$ films
and absent in $SiO_2$ films, again in agreement with calculations. An additional
interesting point in this study was the equality of damage thresholds for
linearly or circularly polarized light. A comprehensive investigation of the
role of the electric-field distributions in damage to thin-film stacks has
been reported by APFEL et al. [6.51], who prepared four different multilayer
coatings under identical conditions (at the same time and in the same coater)
and damage-tested them with a 30 ps, 1.06 μm pulsed beam. A major conclusion
was that damage is best correlated with the peak values of the electric field
and that damage is most likely to occur in the high-index layers. Additionally,
it was pointed out that the first high-index layer deposited on the substrate
may have a lowered threshold. In experimental studies of damage to high-reflection and anti-reflection coatings using a 125 ps laser source at 1.06 μm,
MILAM [6.38] found that anti-reflection coatings failed at generally lower
intensities than did high-reflection coatings. Analysis of the electric fields
in such designs [6.51] would yield the opposite conclusion if the limiting
mechanism is due to the peak electric-field distribution. In an attempt to
explain this discrepancy, APFEL [6.52] prepared a high-reflection, anti-re-

flection, and combination coatings, and measured the damage thresholds. In agreement with MILAM's results [6.38], the anti-reflection coating was found to fail at less than half the flux required to damage the high-reflection coating; the combination was damaged midway between the two. It was shown that susceptibility of the coating/substrate to damage was the likely cause of the anomalous result, which is in agreement with the calculated peak electric-field strength, which shows that only the anti-reflection coating has a large value at that location. In a test of this hypothesis, APFEL et al. [6.53] deposited silica and coatings on glass with a barrier or protective layer of silica adjacent to the substrate. Damage testing was performed with a 150 ps, 1.06 μm beam source. The result was to increase substantially the damage threshold (in excess of 50%) of coatings with the barrier layer compared to a coating without it, prepared under identical conditions. Further investigation of this effect has been reported by CARNIGLIA et al. [6.54], who reported that the damage to the film/substrate interface is probably caused by thermal expansion of the material close to or at the boundary, and that it is likely that the absorption of the polishing or cleaning medium is responsible. At present it is not clear why the silica layer is so effective in raising the damage threshold; definitive answers await further study.

A great deal of progress has been made recently in measurements of the optical properties of thin films and interfaces with the substrate or air. Such measurements are necessary if damage in thin films is to be understood, once extrinsic effects such as stress and roughness have been eliminated. We now briefly review the variety of techniques which have become available.

Fairly simple techniques for measuring the bulk properties of thin films have been developed by BURDICK [6.55] using a wedged-film technique; an advantage is that surface effects are normalized out and the method is easily adapted to measure two-photon absorption coefficients as well. The absorption at the thin-film/air surface or at the thin-film/substrate surface can often be much larger than the bulk. Recently, a technique that can distinguish between bulk and surface absorption at the thin-film/air, thin-film/substrate surfaces was pioneered by TEMPLE et al. [6.55] using a wedged-film technique and a scanning adiabatic calorimeter. An advantage of the technique is that is can distinguish between absorption due to impurities either in the bulk or on the surface of thin films (e.g. $H_2O$). The measurement of bulk and interface absorptions in coatings as a function of wavelength and intensity is extremely important in trying to understand the processes that lead to thin-film damage. A technique that is also able to distinguish between surface and bulk absorption is photoacoustic spectroscopy; it will probably be widely used in damage investigations [6.57,58].

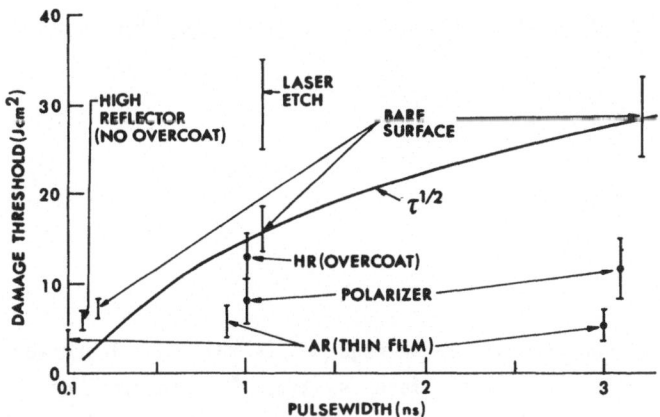

Fig. 6.6. Damage threshold as a function of pulse duration for selected optical components [6.59]

As has been seen above, the choice of coating material and the presence of a protective barrier layer can have a drastic effect upon the damage threshold of dielectric coatings. In mostly unpublished work, it has been found that the use of graded-index coatings [6.59] can lead to substantial improvements of the damage threshold (typically by a factor ~2.5) of anti-reflection coatings. As of this writing, this subject is so new that no comprehensive data are available. LOWDERMILK and MILAM [6.59] have summarized the current status of optical damage; we reproduce their results in Fig.6.6. There, we show damage threshold as a function of pulse width measured at 1.06 μm, for bare surfaces (which scale close to $\tau_p^{1/2}$), thin-film and graded-index anti-reflection coatings (AR) and high-reflector coatings (HR). Also included in Fig.6.6 are surfaces which have been laser-etched or annealed. This technique has also been used recently to significantly increase the threshold of surfaces, but is so new that virtually no data are available.

# 7. Nonlinear Effects in High-Peak Power Nd: Glass Laser Systems

We now begin an investigation of what may be regarded as the central problem
in the design of high-peak power Nd:glass laser systems. The phenomena of
self-focusing, discovered when high-power lasers first became available, presently limits the output of every laser system designed to operate at short
(50-200 ps) pulse widths. Although the problem has yielded to a number of improvements in laser materials, most notably the development of low-nonlinear-
index glasses (Chap.2), it is fair to say that the major developments leading
to present-day laser systems that deliver multi-terawatt performance per
beamline are due to the excellent work of colleagues at the Lawrence Livermore
Laboratory. In particular, they deserve credit for such concepts as spatial
filtering and imaging, to be treated in detail below; but, more importantly,
the major advancement is in our *understanding* of the nonlinear effects, well
known to the experimentalist in this area. Although the basics of self-focusing theory (Sect.7.2-4) were known for some time previous to the advent of
the construction of large Nd:glass laser systems, the major contribution here
is the development of semi-phenomenological theories that enable the laser
designer to proceed with a great degree of confidence and reliably predict
the operating characteristics of such a system. As is usually the case with
systems of such complexity, it is impossible to simulate the exact physics in
detail because it would call for a computational capability and expense that
is beyond the resources of any single laboratory. Consequently, what we present in Sects.7.2-6 below is the current phenomenological understanding that
has led to multi-terawatt performance of a number of laser systems in this
country and abroad.

In Sect.7.1, we review the fundamental notions of self-focusing (SF) theory that are relevant to the discussion to follow. In Sect.7.2, we specialize
the discussion to small-scale-self-focusing (SSSF), the major effect that
limits the output of Nd:glass laser systems. An associated phenomenon, whole-
beam self-focusing (WBSF) which may also limit system performance, is investigated in Sect.7.3; the experimental consequences of WBSF and SSSF are re-

viewed in Sect.7.4. The techniques of spatial filtering and imaging are investigated in Sect.7.5. A major tool needed in the staging of high-peak-power laser systems, X-factor analysis, is covered in detail in Sect.7.6. Finally, in Sect.7.7 we review the simulation and modeling techniques used to predict the performance of real systems.

## 7.1 Self-Focusing Theory

The main consideration in this chapter is that of SF. Its effects in high-peak-power laser systems can be due to the combined influence of WBSF and SSSF. Before treating these subjects in detail, we consider the general phenomena of SF. Excellent reviews of the subject have been written by MARBURGER [7.1], SVELTO [7.2] and SODHA et al. [7.3]. In solid optical materials, while they are propagating high-peak laser intensities ($\gtrsim$ 5-10 GW/cm$^2$), a situation such as that shown in Fig.7.1 arises. When an initially transverse Gaussian beam propagates through a medium with linear index $n_0$, the intensity-dependent nonlinear index causes a distortion of the wave. In particular, the high-intensity peak at the center of the pulse induces an index change $\Delta n$, whereas the change is virtually zero at the low intensity edge. Owing to the higher index at the center, that portion of the wave is retarded relative to the edge and the beam begins to come to a focus. The process of SF would continue in an infinite medium until the peak intensity of the wave becomes sufficiently high to damage the material. The distance at which a given wave will self-focus, of course, depends upon many variables, including the values of the nonlinear index, the beam intensity, etc. It should be realized that the SF distance $Z_f$ also is a function of time. The minimum distance occurs at the peak of the pulse and thus $Z_f$ zooms from infinity to a minimum and then back to infinity. We consider here only the cubic nonlinearity as discussed previously in Chap.1. Furthermore, the medium response is assumed to be fast compared to any pulse length we would consider here.

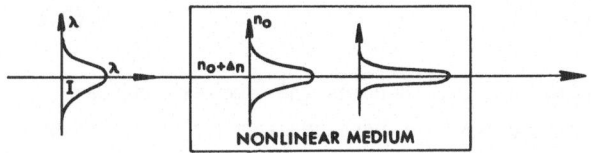

Fig. 7.1. Schematic representation of self-focusing of a wave in a medium in which a cubic nonlinear index is operative

In a real situation, in addition to SF, diffraction also has to be considered. For a sufficiently self-focused beam, diffraction spreading will become comparable and it is possible for a beam to propagate unchanged. This situation has been observed experimentally and is known as self-trapping [7.2]. SVELTO [7.2] has shown that the condition that self-focusing be compensated by diffraction leads, to first order, to the self-trapping condition when the critical intensity $I_{CR}$ is given by

$$I_{CR} = \frac{(1.22)^2}{32\pi D^2}\left(\frac{\lambda_0^2}{cn_2}\right) , \qquad (7.1)$$

where D is the beam diameter, $\lambda_0$ the wavelength, c the speed of light and $n_2$ the nonlinear index. The critical power $P_{CR}$ is

$$P_{CR} = \frac{(1.22)^2}{32}\left(\frac{\lambda_0^2}{cn_2}\right) \qquad (7.2)$$

and is independent of the beam radius. If the critical power is exceeded, then the beam will overcome diffraction and come to a focus. In that case, the focal distance $Z_f$ is described, to first order, by [7.2]

$$Z_f = \frac{r_0^2 n_0}{\sqrt{2cn_2}}\left(\frac{1}{P^{\frac{1}{2}}-P_{CR}^{\frac{1}{2}}}\right) , \qquad (7.3)$$

where $r_0$ is the beam radius; this relationship shows that $Z_f$ is proportional to $r_0^2$ and inversely proportional to $P^{\frac{1}{2}}$. The situation described by this simple formula is accurate for beams a long distance from focus. To investigate the focal region, a true propagation calculation must be performed, as we will describe now. We begin with Maxwell's equations with a linear and nonlinear source term, given as

$$\vec{\nabla} \times (\vec{\nabla}\times\vec{E}) = k^2\left(n_0+n_2|E|^2\right)^2\vec{E} \qquad (7.4)$$

in the steady state, where $k = 2\pi/\lambda$. In the quasi-optical approximation, we can write $\vec{E}$ as

$$\vec{E} = \frac{1}{2}\,\vec{e}\, A\, \exp[i(\omega t-kz)] + c.c. \quad , \qquad (7.5)$$

where $\vec{e}$ is the polarization vector and c.c. denotes the complex conjugate. In this approximation, it is assumed that the envelope function A varies by a small amount in a wavelength; substitution of (7.5) in (7.4) results in

$$\nabla_\perp^2 A - 2ik \frac{\partial A}{\partial k} + \frac{2k^2 n_2}{n_0} |A|^2 A = 0 \quad , \tag{7.6}$$

where we have neglected second derivatives along the propagation direction $|d^2A/dz^2| \ll k(dA/dz)$, and the transverse Laplacian $\nabla_\perp^2$ is

$$\nabla_\perp^2 = \frac{\partial^2}{\partial x^2} + \frac{\partial^2}{\partial y^2} \quad . \tag{7.7}$$

Equation (7.6) is a nonlinear Schrödinger equation that reduces for $n_2 = 0$ to the normal propagation equation of classical physical optics. If the beam is radially symmetric, (7.6) becomes

$$\frac{1}{r} \frac{\partial A}{\partial r} - 2ik \frac{\partial A}{\partial z} + 2k^2 \frac{n_2}{n_0} |A|^2 A = 0 \tag{7.8}$$

if we assume $r^{-1}(\partial A/\partial r) \gg \partial^2 A/\partial r^2$. Although various approximate solutions of (7.6,8) are available and give results virtually identical with (7.2,3), in physically realistic situations the equations are unsolvable analytically, and numerical methods must be employed. Analytic solutions were discussed by MARBURGER [7.1], and SODHA et al. [7.3].

Equations (7.6,8) are valid only in passive media. Here we are interested in laser media that exhibit some gain per unit length $\alpha$; in that case (7.6) becomes

$$2ik \frac{\partial A}{\partial z} - \nabla_\perp^2 A - \left(2k^2 \frac{n_2}{n_0} |A|^2 + ik\alpha\right) A = 0 \quad . \tag{7.9}$$

An elegant method for solving (7.9) numerically has been given by FLECK et al. [7.4] who represent the solution formally as

$$A(x,y,\Delta z) = \exp\left(-\frac{i\Delta z}{2k} \nabla_\perp^2 - i\Delta z \Phi\right) A(x,y,0) \quad , \tag{7.10}$$

where $\Phi$ is

$$\Phi = \frac{1}{2k} \left(\frac{2k^2 n_2}{n_0} |A|^2 + ik\alpha\right) \quad . \tag{7.11}$$

There, (7.10) is replaced to second order accuracy by the symmetrized split-operator form [7.5]

$$A(x,y,\Delta z) = \left(e^{-(i\Delta z/4k)\nabla_\perp^2}\right)\left(e^{-i\Delta z\Phi}\right)\left(e^{-(i\Delta z/4k)\nabla_\perp^2}\right) A(x,y,0) \quad . \tag{7.12}$$

The beam is initially propagated through a distance $\Delta z/2$, then alternately has its phase incremented by the value $-\Delta z\Phi$; it is then propagated a distance $\Delta z$ in a passive homogeneous medium with linear index $n_0$. Such propagation is governed by (7.9) with $\left[(2k^2 n_2/n_0)|A|^2 + ik\alpha\right] = 0$, or

$$2ik\ \frac{\partial A}{\partial z} = \nabla^2_\perp A \quad . \tag{7.13}$$

Equation (7.13) can be efficiently solved by use of finite Fourier-transform methods; such algorithms are available on all large computing machines. The importance of this and similar techniques is that the evolution of a pulse in a real laser system can be accurately calculated and, for example, the effects of WBSF can be assessed.

## 7.2  Small-Scale Self-Focusing

In Sect.3.2 we saw how when the critical power $P_C$ is exceeded by a beam, SF occurs in an optical material. Typical values for $P_C$ are in the range of 1-10 MW. Modern Nd:glass laser systems, on the other hand, operate with powers typically in the range of 0.1-2 TW, exceeding the critical power by factors of $10^4$-$10^5$. Under such conditions a new phenomenon occurs, namely SSSF which is the predominant limitation to the output of short-pulse, high-peak-power Nd:glass laser systems. We completely ignore here SF due to any effects other than the cubic nonlinear index, which has been found to be dominant in high-peak-power Nd:glass laser systems. In particular, coherent effects, such as those treated by MATTAR [7.6] are neglected. It has been shown [7.7] that small perturbations of the amplitude or phase under such conditions lead to the growth of such perturbations, ultimately leading to catastrophic localized damage once the intensity reaches a level sufficient to cause an electron avalanche. Perturbations arise in real systems from particulate matter on surfaces, bubbles or absorbing inclusions in the material, turbulence, initial fluctuations in the oscillator output, or other causes.

Following the initial work of BESPALOV and TALANOV [7.7], SSSF has been analyzed by a number of workers, among them GLASS [7.8], MARBURGER et al. [7.9] and TRENHOLME [7.10,11]. Here we initially follow the work of SUYDAM [7.12] who analyzed ripple growth (SSSF) in passive and active media. The analysis, known as linear-instability theory, has also been used by BRUECKNER and JORNA [7.13] to analyze instability phenomena in liquids and gases. We begin with the paraxial equation (7.9) which we write as

$$2ik \frac{\partial A'}{\partial z} = \nabla_\perp^2 A' + \left( \frac{2k^2 n_2}{n_0} |A'|^2 + ik\alpha \right) A' \quad , \tag{7.14}$$

where, as before $\nabla_\perp^2$ is the transverse Laplacian, $\alpha$ the gain coefficient, and $A'$ the complex field amplitude. We make the substitution

$$A' = A e^{i\phi} \tag{7.15}$$

and after some manipulation find the two coupled equations

$$2k \frac{\partial A}{\partial z} = A \nabla_\perp^2 \phi + 2(\vec{\nabla}_\perp A \cdot \vec{\nabla}_\perp \phi) + k\alpha A \tag{7.16}$$

$$2kA \frac{\partial \phi}{\partial z} = A(\nabla_\perp \phi)^2 - \nabla_\perp^2 A - 2k^2 \mu A \quad , \tag{7.17}$$

where we have separated real and imaginary parts and used the vector identity

$$\nabla_\perp^2 (A e^{i\phi}) = e^{i\phi}(\nabla_\perp^2 A) + A(\nabla_\perp^2 e^{i\phi}) + 2(\vec{\nabla} A \cdot \vec{\nabla} e^{i\phi}) \quad . \tag{7.18}$$

The fractional increase of the index of refraction $\mu$ is given by

$$\mu = \frac{\Delta n}{n_0} = \frac{n - n_0}{n_0} = \frac{n_2 |A_0|^2}{n_0} \quad . \tag{7.19}$$

SUYDAM [7.12] has found solutions of (7.16,17) which are

$$A = A_0(z) \quad , \tag{7.20}$$

along with the auxiliary condition

$$\frac{dA}{dz} = \frac{1}{2} \alpha A \quad , \tag{7.21}$$

and the phase

$$\phi = \phi_0 = K - \frac{8\pi^2 c n_2}{n_0 \lambda} \int_0^L I(z) dz \quad . \tag{7.22}$$

The second term in (7.22) is the familiar B integral so that it may be written

$$\phi = K - B \quad . \tag{7.23}$$

We now perturb the amplitude and phase by amounts $A_1$ and $\phi_1$, respectively, according to the prescription

$$A = A_0 + A_1 \quad , \quad (A_1 << A_0) \tag{7.24}$$

and

$$\phi = \phi_0 + \phi_1 \quad , \quad (\phi_1 << \phi_0) \tag{7.25}$$

and substitute into (7.16,17). Neglecting square and product terms, we obtain

$$2K \frac{\partial A_1}{\partial z} = A_0 \nabla_\perp^2 \phi_1 + k\alpha A_1 \tag{7.26}$$

and

$$-2kA_0 \frac{\partial \phi_1}{\partial z} = \nabla_\perp^2 A_1 + 4k^2 \mu A_1 \quad . \tag{7.27}$$

Equations (7.26,27) are solved [7.12] by setting

$$A_1 = p \, e^{\beta z} \, \psi_k(x,y) \tag{7.28}$$

and

$$\phi_1 = g \, e^{\beta z} \, \psi_k(x,y) \quad , \tag{7.29}$$

where $\psi_k(x,y)$ are solutions of the eigenvalue equation

$$\nabla_\perp^2 \psi_k = -K^2 \psi_k \quad . \tag{7.30}$$

Here K is the eigenvalue (transverse wavenumber) and the $\psi_k$ form a complete set. It is easy to show that

$$\frac{p}{g} = - \frac{A_0 K^2}{2\beta k} \quad . \tag{7.31}$$

Solutions of the nontrivial case are

$$\beta^2 = \frac{K^2}{4} \left[ 4\mu - \left(\frac{K}{k}\right)^2 \right] \quad . \tag{7.32}$$

The initial perturbation $\delta$ is given by

$$p = \delta A_0 \quad , \tag{7.33}$$

which by (7.32) gives

$$g = -\delta\left(\frac{\beta k}{K^2}\right) \quad . \tag{7.34}$$

Equation (7.32) is the central result of the linearized-instability analysis; it predicts that ripple intensity R grows according to

$$R = R_0 \, e^{\beta L} \quad , \tag{7.35}$$

where $R_0$ is the ripple intensity for $z = 0$, L the total path length, and no saturation occurs. Following TRENHOLME [7.10] we now define the weak-beam angle $\theta$ as

$$\theta = \frac{K}{k} = \frac{\lambda}{\lambda_R} \quad , \tag{7.36}$$

where $\lambda$ is the laser transition wavelength and $\lambda_R$ is the ripple wavelength. Physically, $\theta$ represents the angle at which a ripple of wavelength $\lambda_R$ travels relative to the main beam. Rewriting (7.32), we have

$$\beta^2 = k^2\theta^2\left(\mu - \frac{\theta^2}{4}\right) \quad . \tag{7.37}$$

To find the maximum growth rate $\beta_m$ which occurs at any angle $\theta_m$, we calculate $d\beta/d\theta = 0$ and obtain the important result

$$\theta_m = \sqrt{2\mu} = \left(\frac{2n_2|A_0|^2}{n_0}\right)^{\frac{1}{2}} \quad . \tag{7.38}$$

the corresponding maximum gain-coefficient $\beta_m$ is

$$\beta_m = \mu k = \frac{kn_2|A_0|^2}{n_0} \quad . \tag{7.39}$$

Substituting in (7.35), we see that in a passive medium of length L,

$$\ln(G_m) = \beta_m L = k\mu L \quad , \tag{7.40}$$

or in general that

$$\ln(G_m) = B \quad , \tag{7.41}$$

where B is the B integral. This is precisely why B is an important measure
of nonlinear effects in high-peak-power Nd:glass laser systems. Besides re-
presenting the amount of nonlinear phase delay of the beam important in WBSF,
it also represents the rate of growth of the worst ripples in a system. For
these reasons, B has become the standard way of measuring beam degradation
due to nonlinear effects in high-peak-power laser systems.

The growth of ripples, according to (7.35), leads to rapid self-focusing.
SUYDAM [7.12] has extended the above analysis to the case where the initial
perturbation becomes large enough to interact with itself. In that case the
growth is larger than exponential, and leads to the approximate condition

$$\delta \exp(\beta_m z_f) \simeq b \quad , \tag{7.42}$$

where $z_f$ is the self-focusing length, or

$$z_f = \frac{1}{\beta_m} \ln\left(\frac{b}{\delta}\right) \quad . \tag{7.43}$$

This relationship has been verified experimentally by CAMPILLO et al. [7.14]
for a ruby laser in $CS_2$, showing good agreement with the theoretical value
$b = 3$ [7.12].

The predictions of the Bespalov-Talanov theory have been verified by BLISS
et al. [7.15] who measured growth rates of interference fringes in passive
ED-2 laser glass. The gain coefficient as a function of wavenumber K was found
to agree well with the results of the linear-instability theory, modified to
take account of the nonexponential ripple growth and the spatial relationship
between the phase and amplitude modulations generated by a shear plate.

Until now, we have treated only passive media in which $\alpha = 0$. What happens
when gain is introduced into the analysis? This question has been answered
by SUYDAM [7.12] and others, but is most illuminating in the form given by
TRENHOLME [7.10]. The calculation begins by defining a critical relative index change $\mu_c$ given by

$$\mu_c = \frac{\theta^2}{4} \quad , \tag{7.44}$$

which by (7.37) represents the stage at which growth changes from oscillatory
to exponential behavior. By using (7.37), it is easy to show that

$$\beta = 2k\mu_c\left(\frac{\mu}{\mu_c} - 1\right)^{\frac{1}{2}} \quad . \tag{7.45}$$

Ripple intensity R grows in the adiabatic approximation with no saturation, according to

$$\frac{dR}{dz} - \beta R \quad, \tag{7.46}$$

or

$$R = R_0 \exp\left(\int_0^L \beta \, dz\right) \quad, \tag{7.47}$$

of which (7.35) is a special case. We also assume that, in an amplifier with no saturation, the relative index $\mu$ changes according to

$$\mu = \mu_i \, e^{\alpha z} \quad, \tag{7.48}$$

where $\mu_i$ is the relative index at the amplifier input (z=0), and $\alpha$ the gain coefficient. Substituting (7.45,48) into (7.47) results in

$$\ln\left(\frac{R}{R_0}\right) = \int_0^L 2k\mu_c\left(\frac{\mu_i}{\mu_c}\, e^{\alpha z} - 1\right)^{\frac{1}{2}} dz \quad. \tag{7.49}$$

If we make the variable change,

$$X = \frac{\mu_i}{\mu_c}\, e^{\alpha z} - 1 \quad, \tag{7.50}$$

we find

$$\ln\left(\frac{R}{R_0}\right) = \frac{2k\mu_c}{\alpha} \int_{\frac{\mu_i}{\mu_c}}^{\frac{\mu_0}{\mu_c}-1} \frac{X^{\frac{1}{2}}}{(X+1)}\, dx \quad. \tag{7.51}$$

Here $\mu_0$ is the relative index of the amplifier output (z=L) and is related to $\mu_i$ by

$$\mu_0 = G\mu_i \quad, \tag{7.52}$$

where G is the gain of the amplifier,

$$G = e^{\alpha L} \quad. \tag{7.53}$$

Standard evaluation of the integral in (7.51) results in

$$\ln\left(\frac{R}{R_0}\right) = \frac{4k\mu_c}{\alpha}\left[\left(\frac{\mu_0}{\mu_c}-1\right)^{\frac{1}{2}}-\left(\frac{\mu_i}{\mu_c}-1\right)^{\frac{1}{2}}-\tan^{-1}\left(\frac{\mu_0}{\mu_c}-1\right)^{\frac{1}{2}}-\tan^{-1}\left(\frac{\mu_i}{\mu_c}-1\right)^{\frac{1}{2}}\right] \quad . \quad (7.54)$$

It is obvious that if $\mu_0 < \mu_c$ or $\mu_i < \mu_c$, the gain is imaginary and oscillating solutions result, and no ripple gain is obtained. Then we should use

$$M_0 = \max\left(\frac{\mu_0}{\mu_c}-1,0\right) \qquad (7.55)$$

and

$$M_i = \max\left(\frac{\mu_i}{\mu_c}-1,0\right) \qquad (7.56)$$

in evaluating the gain. The trigonometric identity

$$\tan^{-1}A - \tan^{-1}B = \cos^{-1}\left\{\frac{1+AB}{[(1+A^2)(1+B^2)]^{\frac{1}{2}}}\right\} \qquad (7.57)$$

may be used to transform (7.54) into

$$\ln\left(\frac{R}{R_0}\right) = \frac{4k\mu_c}{\alpha}\left\{M_0^{\frac{1}{2}}-M_i^{\frac{1}{2}}-\cos^{-1}\left[\frac{1+(M_0M_i)^{\frac{1}{2}}}{[(M_0+1)(M_i+1)]^{\frac{1}{2}}}\right]\right\} = g \quad . \quad (7.58)$$

This expression for the amplifier gain g is the result we have been seeking; it will be seen to yield a number of interesting results. We now determine the angle $\theta$, or ripple size, at which the maximum gain occurs. Three separate regions will be investigated. The first is when $M_0 = M_i = 0$, in which $g = 0$. The second is when $M_i = 0$ and $M_0 > 0$, so that ripples at peak gain enter the amplifier and oscillate, but then grow exponentially as the intensity increases. Using $\mu_c = \theta^2/4$, $\theta_c = 2(\mu_0)^{\frac{1}{2}}$, we rewrite (7.58) as

$$\ln\left(\frac{R}{R_0}\right) = \frac{k\theta}{\alpha}\left[\left(4\mu_0-\theta^2\right)^{\frac{1}{2}}-\theta\cos^{-1}\left(\frac{\theta}{2\mu_0^{\frac{1}{2}}}\right)\right] \quad . \quad (7.59)$$

To find the angle at which the maximum gain occurs, we must evaluate

$$\frac{d}{d\theta}\left\{\theta\left[\left(4\mu_0-\theta^2\right)^{\frac{1}{2}}-\theta\cos^{-1}\left(\frac{\theta}{2\mu_0^{\frac{1}{2}}}\right)\right]\right\} = 0 \quad , \quad (7.60)$$

obtaining

$$2\theta\cos^{-1}\left(\frac{\theta}{2\mu_0^{\frac{1}{2}}}\right) = (\mu_0)^{\frac{1}{2}}\left[1-\left(\frac{\theta}{2\mu_0^{\frac{1}{2}}}\right)^2\right]^{\frac{1}{2}} \quad . \quad (7.61)$$

If we define

$$\nu = \frac{\theta}{\theta_c} \quad , \tag{7.62}$$

then (7.61) can be written

$$2\nu \cos^{-1}(\nu) = (1-\nu^2)^{\frac{1}{2}} \quad . \tag{7.63}$$

Numerical evaluation of this transcendental equation results in

$$\nu = \frac{\theta}{\theta_c} \simeq 0.394 \quad . \tag{7.64}$$

In a passive medium ($\alpha=0$), from (7.38) we find that

$$\frac{\theta}{\theta_c} = \frac{1}{\sqrt{2}} \simeq 0.71 \quad . \tag{7.65}$$

Thus, in a high-gain medium, the peak-gain ripples are a factor of $\simeq 1.8$ greater than in a passive medium. Evaluating (7.59) at the maximum angle $\theta_m$, we obtain

$$\ln\left(\frac{R}{R_0}\right)_{\theta=\theta_m} = 2\nu(1-\nu^2)^{\frac{1}{2}} \frac{k\mu_0}{\alpha} \simeq 0.7246 \frac{k\mu_0}{\alpha} \quad , \tag{7.66}$$

which shows that, for fixed output intensity, the ripple gain decreases as the gain coefficient $\alpha$ increases. These results may be depicted graphically as in Fig.7.2 where the ripple gain g is shown as a function of the weak-beam angle $\theta$ for a 30-cm-long rod of ED-2 laser glass with an output flux of 20 GW/cm$^2$. The unpumped rod, calculated from (7.37), can be seen to display a peak gain for $\theta \simeq 0.0034$ rad. As the gain is increased, however, the peak angle becomes smaller, and for very great gain approaches 0.0019 rad. The

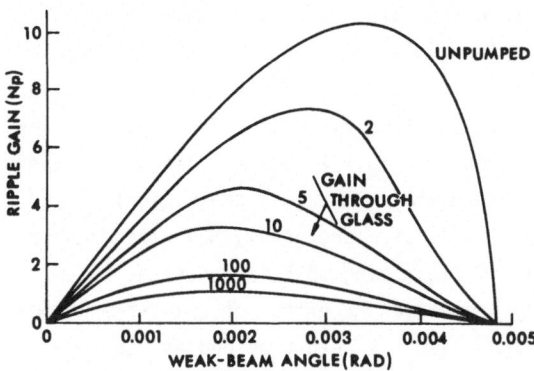

Fig. 7.2. Small-scale ripple growth as a function of weak-beam angle (inverse ripple size) for 30 cm ED-2 glass (output flux 20 GW/cm$^2$) [7.10]

peak ripple gain at high gain is thus $\approx 1.9$ times that in the passive rod, in agreement with the previous results.

For the third, low-gain case, we have $M_i > 0$ and once again we must evaluate d $\ln(R/R_0)d\theta = 0$. TRENHOLME [7.10] has calculated the result,

$$
\left[ \frac{1-2\nu^2}{(1-\nu^2)^{\frac{1}{2}}} \right] - \frac{[1/G-2\nu^2]}{[1/G-\nu^2]^{\frac{1}{2}}} - 2\nu \cos^{-1}\left[ G^{\frac{1}{2}}\left\{ \nu^2 + [(1-\nu^2)(1/G-\nu^2)^{\frac{1}{2}}]^{\frac{1}{2}} \right\} \right]
$$
$$
+ G^{\frac{1}{2}}\nu^3 \frac{2 - [1+(1/G-2\nu^2)]/[(1-\nu^2)(1/G-\nu^2)]^{\frac{1}{2}}}{\left[ \nu^2\left( (1+G) - 2G\left\{ \nu^2 + [(1-\nu^2)(1/G-\nu^2)]^{\frac{1}{2}} \right\} \right) \right]^{\frac{1}{2}}} = 0 \quad . \tag{7.67}
$$

which shows that $\theta_m$ is a function of the gain G.

We are interested in finding the gain $G_T$ when the transition occurs between the second high-gain and third low-gain regions examined above. It occurs when $M_i = 0$ is at $\theta_m$ or, from (7.52)

$$
\theta_m = 2(\mu_i)^{\frac{1}{2}} = 2(\mu_0/G)^{\frac{1}{2}} = \frac{\theta_c}{G^{\frac{1}{2}}} \quad . \tag{7.68}
$$

The maximum angle is $\theta_m = \nu\theta_c$; thus

$$
\nu = G^{-\frac{1}{2}} \quad . \tag{7.69}
$$

Substitution in (7.66) gives

$$
G_T = \frac{2}{1 + \cos(G_T - 1)^{\frac{1}{2}}} \quad , \tag{7.70}
$$

or, after numerical evaluation

$$
G_T \simeq 6.434 \quad . \tag{7.71}
$$

The conclusions of this analysis are clear: amplifiers that have gains greater than 6.434 have a maximum-gain ripple angle independent of the gain given by (7.70). For amplifiers with gain less than that value, the maximum gain angle increases to the passive value given by (7.65). These results are in agreement with the calculations shown in Fig.7.2. It is obvious from the above analysis that it is desirable to maximize the gain coefficient of an amplifier in a system, because the effects of $\alpha$ are great. Maximizing amplifier gain is the most effective way of minimizing SSSF in a system.

In the above, we have completely neglected the effect of gain saturation in ripple growth. The analysis of that case has been provided by ELLIOT [7.16]

who combined the analysis of SUYDAM [7.12] with the classical picture of
gain saturation (see Chap.1).

In the foregoing analysis, we have completely neglected the effect of
boundaries which of course occur in real laser systems, because beams propagate often from glass to air and air to glass. The effect of the boundary
can be drastic, as shown by the work of AURIE et al. [7.17] and TRENHOLME
[7.10,11]. We follow here a similar treatment by LEE [7.18]. The following
analysis is particularly important because it is easy to handle computation-
ally as it involves a matrix formulation.

In the foregoing, we have shown how a strong plane wave is unstable under
linear phase or amplitude perturbations. Furthermore, it was seen that spatial frequencies could be associated with a direction of propagation different from the main beam, at the so-called weak-beam angle $\theta$. TRENHOLME [7.10]
has considered the case of weak beams interfering with a strong beam; such
interference gives rise to intensity irregularities that spatially modulate
the refractive index of the medium, giving rise to the "self-induced diffraction grating". This variation of the refractive index phase modulates the
strong beam, transferring power to the weak beam and leading to the growth
of intensity irregularities. We begin the analysis with the paraxial equation (7.9) and assume that the field A is described by a linear superposition
of a strong beam together with a finite number of small-scale ripple modes,
where

$$A = A_0 \left[ 1 + \sum_i a_i(z) e_i(x,y) \right] \quad . \tag{7.72}$$

Here the eigenfunctions $e_i$ are real and satisfy the equation

$$\nabla^2 e_i = -\sigma_i^2 e_i \tag{7.73}$$

and the $\sigma_i$ are the corresponding eigenvalues. When the ripples are small
($\sum_i a_i e_i \ll 1$), we assume that the coefficients $a_i$ are independent (i.e., no
coupling between the small-scale ripple modes) and by substitution of (7.72)
in (7.9) we find

$$\frac{\partial a_i}{\partial z} = \frac{i\sigma_i^2}{2k} a_i - 2ik\mu \, \text{Re}\{a_i\} + \frac{\alpha(I)}{2} a_i \quad . \tag{7.74}$$

Because the $a_i$ are independent of each other, we can drop the index i in all
further calculations. The coefficient a(z) is a complex function of z and can
be written

$$a(z) = u(z) + iv(z) \quad . \tag{7.75}$$

Substitution in (7.74) and separation into real and imaginary parts results in two coupled first-order differential equations

$$\frac{du}{dz} = - \frac{\sigma^2}{2k} v + \frac{\alpha}{2} u \tag{7.76}$$

and

$$\frac{dv}{dz} = \left( \frac{\sigma^2}{2k} - 2k\mu \right) u + \frac{\alpha}{2} v \quad . \tag{7.77}$$

If we differentiate (7.76) and substitute from (7.77) we obtain

$$\frac{d^2u}{dz^2} = \alpha \frac{du}{dz} + \left( \sigma^2 \mu - \frac{\alpha^2}{4} - \frac{\sigma^4}{4k^2} + \frac{1}{2} \frac{d\alpha}{dz} \right) u \quad . \tag{7.78}$$

Assuming that u grows according to $u = e^{\beta z}$ where $\beta$ is the growth coefficient, we obtain the characteristic equation

$$\beta^2 - \alpha\beta + \left( \frac{\sigma^4}{4k^2} + \frac{\alpha^2}{4} - \sigma^2\mu - \frac{1}{2} \frac{d\alpha}{du} \right) = 0 \quad , \tag{7.79}$$

whose roots are

$$\beta_{1,2} = \frac{\alpha}{2} \pm \sqrt{\sigma^2\mu + \frac{1}{2} \frac{d\alpha}{dz} - \frac{\sigma^4}{4k^2}} \quad , \tag{7.80}$$

which is the generalized Bespalov-Talanov gain coefficient for small-scale ripple growth. The general solution to (7.78) is

$$u = c_1 e^{\beta_1 z} + c_2 e^{\beta_2 z} \tag{7.81}$$

or

$$u = e^{\alpha z/2} \left( c_1 e^{\beta z} + c_2 e^{-\beta z} \right) \quad , \tag{7.82}$$

where

$$\beta = \left( \sigma^2\mu + \frac{1}{2} \frac{d\alpha}{dz} - \frac{\sigma^4}{4k^2} \right)^{\frac{1}{2}} \quad . \tag{7.83}$$

From (7.76) we obtain the solution of v as

$$v = \frac{2k\beta}{\sigma^2} e^{\alpha z/2} \left( -c_1 e^{\beta z} + c_2 e^{\beta z} \right) \quad . \tag{7.84}$$

If, for $z = 0$, we assume the initial conditions $u = u_0$ and $v = v_0$, using (7.82, 84) we find

$$c_1 = \frac{1}{2} \left( u_0 - \frac{v_0}{S} \right) \tag{7.85}$$

and

$$c_2 = \frac{1}{2} \left( u_0 + \frac{v_0}{S} \right) \quad , \tag{7.86}$$

where $S$ is

$$S = \frac{2k\beta}{\sigma^2} \quad . \tag{7.87}$$

The complete solutions for $u$ and $v$ can thus be written

$$u = e^{\alpha z/2} \left( u_0 \cosh \beta z - \frac{v_0}{S} \sinh \beta z \right) \tag{7.88}$$

and

$$v = e^{\alpha z/2} (-S u_0 \sinh \beta z + v_0 \cosh \beta z) \quad . \tag{7.89}$$

We can write the real and imaginary parts of a small-scale ripple mode as the vector

$$a = \begin{pmatrix} u \\ v \end{pmatrix} \quad . \tag{7.90}$$

Then it is easy to see that the initial amplitude $a_0$ is transformed in a medium of length L to an amplitude a given by

$$a = e^{\alpha L/2} M a_0 \quad , \tag{7.91}$$

where M is a $2 \times 2$ transformation matrix

$$M = \begin{bmatrix} \cosh \beta L & -\sinh(\beta L)/S \\ -S \sinh \beta L & \cosh \beta L \end{bmatrix} \quad . \tag{7.92}$$

The power gain of a ripple mode propagating through a medium of length L is

$$G = \frac{\|a\|^2}{\|a_0\|^2} = \frac{a_0^T M^T M a_0}{\|a_0\|^2} \quad , \tag{7.93}$$

where T denotes the matrix transpose. The gain G is the quantity we are interested in calculating for a stage in a laser system.

Before discussing specific applications of (7.93), we mention some properties of the matrix M. Although M is not symmetric, $\bar{M} = M^T M$ is, because

$$(\bar{M})^T = (M^T M)^T = M^T M = \bar{M} \quad . \tag{7.94}$$

also, because the determinant

$$\det M = \cosh^2\beta L - \sinh^2\beta L = 1 \quad , \tag{7.95}$$

then det $\bar{M} = 1$. The product of the eigenvalues of a square matrix is well known to be equal to its determinant, thus the two eigenvalues are inverse to each other. Because $\bar{M}$ is symmetric, the two eigenvalues of $\bar{M}$ must also be distinct and by (7.93), the maximum and minimum power gains are equal to the two distinct eigenvalues of $\bar{M}$. The trace T of a square matrix is known to be equal to the sum of its eigenvalues; for a $2 \times 2$ matrix M

$$M = \begin{bmatrix} M_{11} & M_{12} \\ M_{21} & M_{22} \end{bmatrix} \quad , \tag{7.96}$$

we can find T from

$$T = \text{trace}\{\bar{M}\} = \text{trace}\{M^T M\} \tag{7.97}$$

or

$$T = M_{11}^2 + M_{12}^2 + M_{21}^2 + M_{22}^2 \quad . \tag{7.98}$$

From (7.92) we see that

$$T = 2 + (S^2+1)^2 \sinh(\beta L)/S^2 \quad . \tag{7.99}$$

We are interested in solving the eigenvalue equation for $\bar{M}$

$$(M^T M - \lambda I) = \lambda^2 - \left( M_{12}^2 + M_{22}^2 + M_{11}^2 + M_{21}^2 \right)$$
$$+ \left[ \left( M_{11}^2 + M_{21}^2 \right)\left( M_{12}^2 + M_{22}^2 \right) - \left( M_{11}M_{12} + M_{21}M_{22} \right)^2 \right] = 0 \quad . \tag{7.100}$$

Using (7.98), we can find the roots $\lambda_\pm$ or power gains $G_\pm$

$$\lambda_\pm = \frac{T}{2} \pm \sqrt{\left(\frac{T}{2}\right)^2 - 1} = G_\pm \quad .$$

(7.101)

If $P_\pm$ are eigenvectors of $\bar{M}$ with corresponding eigenvalues $\lambda_\pm$, then

$$\bar{M}P_\pm = \lambda_\pm P_\pm \quad .$$

(7.102)

If we transpose this equation and recall that $\bar{M}$ is symmetric, we obtain

$$P_+^T\bar{M}P_- = \lambda_+ P_+^T P_-$$

(7.103)

and by the eigenvalue equation (7.102),

$$P_+^T\bar{M}P_- = \lambda_- P_+^T P_- \quad ,$$

(7.104)

we have finally

$$(\lambda_- - \lambda_+)P_+^T P_- = 0 \quad .$$

(7.105)

But because $\lambda_- \neq \lambda_+$, $P_+$ and $P_-$ are orthogonal. The main conclusion here is that small-scale ripples that strike a surface with phases that differ by $90°$ suffer maximum or minimum gain.

TRENHOLME [7.11] has shown that for a passive medium it is possible to rewrite the transformation matrix (7.92) as

$$M = \begin{bmatrix} \cosh\theta S & -\frac{\sinh\theta S}{S} \\ -S\sinh\theta S & \cosh\theta S \end{bmatrix} \quad ,$$

(7.106)

where $\theta$, the so-called mode vacuum diffractive angle is

$$\theta = \frac{\sigma^2 L}{2k} = \frac{1}{2} kL\theta^2 \quad ,$$

(7.107)

where as before the eigenvalue $\sigma = k\theta$ where $\theta$ is the weak-beam angle, and the quantity S, given by (7.87), is

$$S = \frac{2k\beta}{\sigma^2} = \left(\frac{2B}{\theta} - 1\right)^{\frac{1}{2}} \quad .$$

(7.108)

In vacuum, B = 0 and S = i. It is easy to show that the trace T can be rewritten

$$T = 2 + \left(\frac{2B \sinh \theta S}{\theta S}\right)^2 \quad . \tag{7.109}$$

The angles of maximum and minimum gain $\phi_\pm$ are given by

$$\phi_\pm = \frac{G_\pm - (M_{11}^2 + M_{12}^2)}{M_{11}M_{12} + M_{21}M_{22}} \quad , \tag{7.110}$$

where the $M_{ij}$ are matrix elements of (7.106). A mode entering the slab at an angle $\gamma$ with respect to the angle of maximum gain will then have a gain G where

$$G = \frac{(G_+^2 - 1)\cos^2\gamma + 1}{G_+} \quad . \tag{7.111}$$

If the modes arrive at the slab with random phase, the average gain $\bar{G}$ is

$$\bar{G} = \frac{2}{\pi} \int_0^{\pi/2} G \, d\phi \quad , \tag{7.112}$$

which, upon substitution of (7.111) gives

$$\bar{G} = \frac{G_+^2 + 1}{2G_+} \quad . \tag{7.113}$$

Substitution of (7.101,109) results in

$$\bar{G} = 1 + 2B^2 \frac{\sinh^2\beta}{\beta^2} \quad , \tag{7.114}$$

where $\beta = \theta S$. We rewrite the value $\beta$ in terms of the eigenvalue ratio r, where

$$r = \frac{\sigma}{\sigma_c} = \frac{\theta}{\theta_c} \tag{7.115}$$

and $\theta_c$, the critical weak-beam angle, is given as before by $\theta_c = 2\sqrt{\mu}$. Then it is an easy matter to show that

$$\beta = \theta S = 2Br \sqrt{1-r^2} \quad . \tag{7.116}$$

There are a number of interesting limits to the growth of ripples, the gain of which is given by (7.114). Very large ripples have small angles $\theta$ associated with them; the limit for large ripples is $r \to 0$. Since by (7.114) $\beta \to 0$ also, and

$$\lim_{\beta \to 0} \left(\frac{\sinh \beta}{\beta}\right) = 1 \quad , \tag{7.117}$$

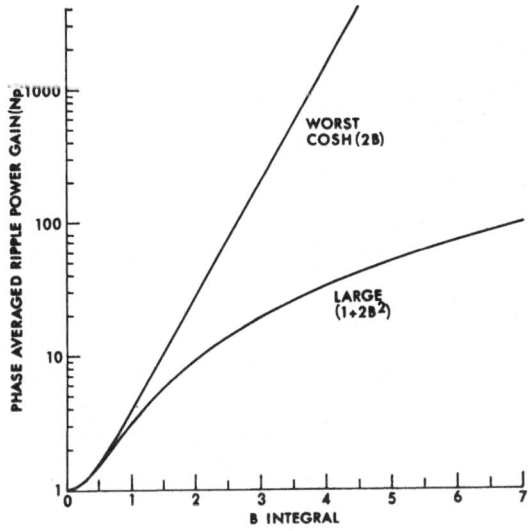

Fig. 7.3. Phase averaged ripple power gain as a function of B integral for large $(1+2B^2)$ and most rapidly growing [cosh 2B] ripples

from (7.114) we arrive at

$$\bar{G} = 1 + 2B^2 \quad \text{(large ripples)} \quad . \tag{7.118}$$

For the most rapidly growing ripples, $\theta = \theta_m = \sqrt{2\mu}$ and thus $r = 1/\sqrt{2}$. Then $\beta = B$ and

$$\bar{G} = 1 + 2 \sinh^2 B = \cosh(2B) \quad \text{(fastest growing ripples)} \quad . \tag{7.119}$$

The average ripple-power gain $\bar{G}$ can be plotted for large and fastest-growing ripples as a function of the B integral, as shown in Fig.7.2; this clearly demonstrates that the most rapidly growing ripples are indeed a source of great concern to the laser designer. As will be seen in Sect.7.5, the smallest ripples are filterable in a spatial filter and may be forced to re-expo-nentiate at strategic locations in a laser system. The large ripples, which also have high growth rates and are only partially or not at all filterable, are of most concern in present laser systems.

Various computer codes have been written to calculate ripple gain through a given stage of a laser system. These include RIPPER at LLL and BTGAIN at LLE. The method is to compute the transfer matrix $M_i$ for each element in the stage and to combine them to calculate (7.93). The results of such a calcula-tion are shown in Fig.7.4, which displays the natural log of the ripple gain as a function of weak-beam angle $\theta$ for the final stage of the GDL (Glass

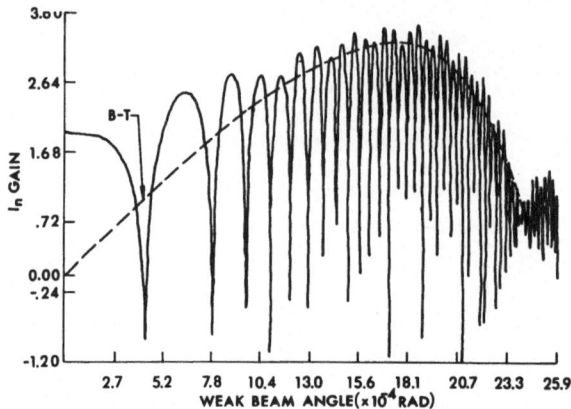

Fig. 7.4. Log of the ripple power gain as a function of weak-beam angle for a 90 mm GDL rod amplifier stage [7.18]

Development Laser) at LLE. Superimposed upon that graph is the Bespalov-Talanov gain, as calculated by use of (7.37). A crucial difference between the original Bespalov-Talanov theory and the matrix theory described in the foregoing is that the matrix theory predicts finite gain for small weak-beam angles (large ripples). The effect of the boundary conditions in producing the modulations in Fig.7.4 is obvious.

Comparison of such results with experimental measurements is qualitative at best, as may be seen in the work of FLECK et al. [7.4] who compared, with fair agreement, the computed and measured gains of ripples in the Cyclops laser system at LLL. The difficulty is that to find the spectral density of ripples in a laser system, it is necessary to have a knowledge of not only the growth rate for various ripple sizes, but also the density of initial-noise sources and their distribution. Such measurements are exceedingly difficult and the computation of such effects is prohibitive, hence semi-phenomenological treatments are normally used, which are normalized to experimental systems. Such models will be discussed in Sect.7.7. Simple computer results such as those shown in Fig.7.4 are nevertheless useful because they indicate at least a first-order guess at the pinhole diameter needed to cut off the most-damaging ripples in a laser system, as discussed in Sect.7.5.

## 7.3 Whole Beam Self-Focusing

The second major cause of beam degradation in Nd:glass laser systems is that
of WBSF. The treatment we follow here is that of ray optics, that is, diffraction is not taken into account. Diffraction effects are, of course, important in high-peak-power laser systems; a number of techniques available
for reducing such effects will be discussed in Chap.8. A first-order, ray-
optics treatment is, however, valuable in elucidating WBSF effects and showing what consequences are in store for propagating high-peak-power beams.
The following analysis is due to HUNT et al. [7.19] and BROWN et al. [7.20].
We begin by writing the equation that describes the angular deviation of a
ray that travels through an index gradient $\vec{\nabla} n$ as

$$\frac{d}{ds}\left(n\,\frac{d\vec{r}}{ds}\right) = \vec{\nabla} n(\vec{r}) \quad , \tag{7.120}$$

where s is the path length, $\vec{\nabla}$ the vector-gradient operator, and $\vec{r}$ the vector
tangent to any point along the path. The phase accumulation along the path
is given by

$$\phi = \frac{2\pi n_0}{\lambda}\int ds + \Delta\phi_{NL} + K \quad . \tag{7.121}$$

Here K is a constant, and $\Delta\phi_{NL}$ the phase accumulation due to the nonlinear
index $n_2$ or nonlinear coefficient $\gamma$, given by

$$\Delta\phi_{NL}(r_T,z) = \frac{2\pi\gamma}{\lambda}\int I(r_T,z)dz \quad , \tag{7.122}$$

where the intensity I is seen to vary with the transverse radial coordinate
$r_T$ and the propagation distance z. In the so-called constant-shape approximation,

$$I(r_T,z) = F(r_T/a)I(z) \quad . \tag{7.123}$$

Therefore, we assume that, at any point z, the beam profile can be described
by a shape factor $F(r_T/a)$, where a is the beam radius. Using (7.120-123) the
angular deviation $\delta\theta_{NL}$ and nonlinear phase distortion $\Delta\phi_{NL}$ can be derived,

$$\delta\theta_{NL} = \frac{\gamma}{n_0}\left|\vec{\nabla}_T F(r_T/a)\right|\int I(z)dz \tag{7.124}$$

$$\Delta\phi_{NL} = \frac{2\pi\gamma}{\lambda}F(r_T/a)\int I(z)dz \quad . \tag{7.125}$$

Fundamentally, the transverse gradient $\vec{\nabla}_T$ associated with the transverse intensity profile deviates rays through the small angle $\delta\theta_{NL}$. After passing through a nonlinear medium and being brought to a foxus by means of a lens, rays will thus change their focal position, depending upon the magnitude of the nonlinearity and the intensity. Because the intensity of a pulse is time dependent, at various times during the pulse duration, rays will focus at different locations (z), and thus the best focus will "zoom" giving rise to the phenomenon of "focal zooming" in spatial filters, which we will discuss in Sects.7.4,5. Equation (7.124) can be rewritten

$$\delta\theta_{NL} = \frac{B}{k_0 a} \frac{dF(\mu)}{d\mu} \quad , \tag{7.126}$$

where $k_0$ is the vacuum wavenumber, $\mu$ the normalized radius ($\mu = r_T/a$), and B the B-integral. Elementary optics and (7.126) allows us to find the change of focal position $\Delta z(t)$ from that of a low-intensity pulse for a Gaussian of 1/e width $t_0$

$$\Delta z(t) = \frac{4(f^\#)^2 B}{k_0 \mu} \left(\frac{dF(\mu)}{d\mu}\right) \exp\left[-(t/t_0)^2\right] \quad , \tag{7.127}$$

where $f^\# = f/2a$ is the f number of the lens, and f the focal length. Of note here is that for any shape factor $F(\mu)$ the zooming $\Delta z$ is directly proportional to the B integral. Various shape factors have been used to calculate $\Delta z(t)$ [7.19]; the most interesting is the so-called $n^{th}$ order super-Gaussian given by

$$F(\mu) = \exp(-c\mu^n) \quad , \tag{7.128}$$

where c is a constant and n the integer order. Using (7.127,128) we find for a super-Gaussian

$$\Delta z_p = -4B(f^\#)^2 \, cn\mu^{n-2} \, e^{-c\mu^n} \quad , \tag{7.129}$$

where the peak zoom $\Delta z_p$ obviously occurs at the peak of the pulse. Equation (7.129) may be used to estimate, in a first-order-ray optics sense, the zoom expected in spatial filters, and thus the translation required for a pinhole so that most of the main-beam energy is transmitted.

From (7.126), it can be seen that rays that travel through a nonlinear media are deviated from their original propagation direction by an amount $\delta\theta_{NL}$ proportional to the B integral. After propagation through a distance z, changes of the relative positions of all the rays will result in changes of the final intensity distribution and will lead, in particular, to a sharpening of the intensity gradient of an initially smooth profile. Thus beams develop "horns" at their edges, which are very dangerous because local SF can occur. The amelioration of this effect through the use of image relaying will be thoroughly discussed in Sect. 7.5 below.

The phenomenon of WBSF has serious consequences for laser system design, not only in the operation of spatial filters but also in the way in which a laser beam interacts with a fusion target. Uniform illumination of targets has become an important goal in laser-fusion research; its attainment is difficult in the presence of whole-beam effects because focal zooming and modifications of the intensity profile in the region of best focus are time dependent. It is obvious that the beam shape $F(\mu)$ can have a profound influence on the magnitude of zooming as well as on the intensity distribution. To see this, we show in Fig. 7.5 a plot of the relative angular variation $\delta\theta_{NL}$ given by (7.124) for three different beam shapes [7.21], including the super-Gaussian

$$F(\mu) = \exp[-6.91(\mu)^5] \quad , \qquad\qquad (7.130)$$

the modified quadratic

$$\left.\begin{aligned} F(\mu) &= 1 - 1.3(\mu)^2 \quad (0<\mu<0.79) \\ F(\mu) &= 0.30\ \exp\left(\frac{\mu-0.71}{0.124}\right)^2 \quad (0.79<\mu<1.0) \end{aligned}\right\} \quad , \qquad (7.131)$$

and the Airy disk

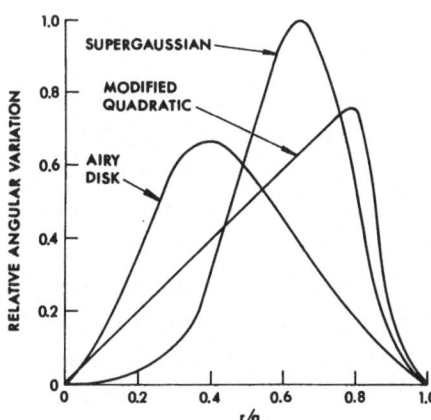

Fig. 7.5. Relative angular variation as a function of normalized radius for super-Gaussian, modified-quadratic, and Airy-disk beam profiles [7.21]

$$F(\mu) = 4\left(\frac{J_1(3.81\mu)}{3.81\mu}\right)^2 , \qquad\qquad (7.132)$$

where $J_1$ is the first-order Bessel function. The constants in (7.130-132) have been chosen such that $F(1) \simeq 10^{-3}$, to minimize diffraction effects. The angular deviation $\delta\theta_{NL}$ is rather complicated for the super-Gaussian and Airy disk as a function of $\mu = r/a$; however, the modified quadratic gives a linear variation over most of the rays. Because of this it is expected that the modified quadratic may be brought to a common focus along the distance z, whereas the super-Gaussian and Airy disk yield complex intensity and phase variations in the region of the focus. The focal zoom is also best for the modified quadratic. Since we want the maximum fill factor at the output stage of a system to maximize energy or power output, the Airy disk is essentially useless, having a fill factor of only 23%. The modified quadratic and fifth-order super-Gaussian have fill factors of 41%, which is low by modern standards (typically 70-75%). Beam shapes emitted by most laser systems are generally fitted by much-higher-order super-Gaussians (typically tenth order); consequently, the details of the focal-region distribution are very complicated. Although ray-trace calculations such as those mentioned in the foregoing are illuminating and lead to a physical understanding of what is a very complex problem, the details must be investigated by use of sophisticated computer codes, which include the effects of WBSF and classical diffraction theory as embodied in the paraxial equation discussed in Sect.7.1. Results of such calculations have been described in detail [7.22].

## 7.4 System Consequences of Self-Focusing

We are particularly interested in the far-field consequences of SF in high-peak-power laser systems. In particular, effects associated with spatial-filter pinholes and beams focused on laser-fusion targets. We describe an experiment performed at LLL [7.23,24], which dramatically demonstrated the now well-known effect of loss of beam power due to self-focusing effects. A high-power beam emitted by the ILS laser (0.75J/100 ps) was reduced in diameter and split into a reference beam and a second beam that propagated through a glass rod. Both beams were then focused, by use of the same lens, into the far field, where the data were recorded on film. After processing, the ratio

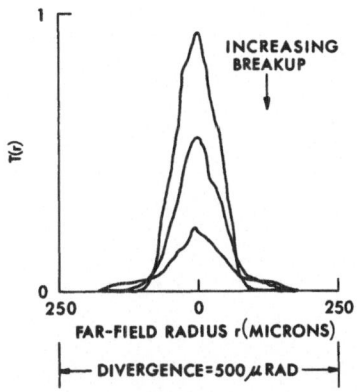

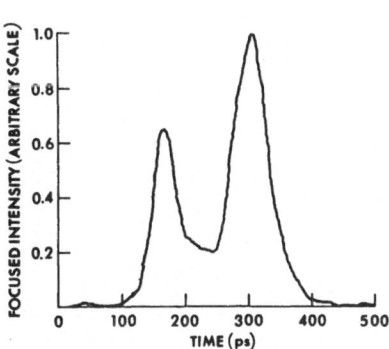

Fig. 7.6. Ratio T(r) of energy of the transmitted and reference beams within a radius r for increasing beam intensities (breakup) [7.23]

Fig. 7.7. Temporal distortion due to small-scale beam breakup [7.23]

$$T(r=0) = \frac{\varepsilon_d(0)_t}{\varepsilon_d(0)_a} \tag{7.133}$$

of the on axis (r=0) energy density $\varepsilon_d(0)_t$ for the transmitted and reference beam $\varepsilon_d(0)_a$ was calculated for various laser intensities. It was found that T(r=0) decreased as laser intensity increased, as did the ratio

$$T(r) = \frac{\int_0^r \varepsilon_d(r)_t dA}{\int_0^r \varepsilon_d(r)_a dA} \tag{7.134}$$

for all values of r up to a full-angle divergence of 500 μrad. The effect can be seen in Fig.7.6, where T(r) is shown as a function of r for three increasing intensities. Thus, as the intensity, or equivalently the B integral increases, T(r) decreases and broadens. The total energy within the 500 μrad angle is obviously decreased; in fact, some energy is found beyond the aperture, at angles greater than 500 μrad. Physically, as the beam intensity becomes greater and greater, more and more main-beam energy is coupled into small-scale ripple modes, in particular those with high growth rates which are also associated with larger angles. The effect can become so severe that it may be seen in the near field as well. The familiar "halo", present around the main beam in near-field photographs of high-power beams, is also due to SSSF.

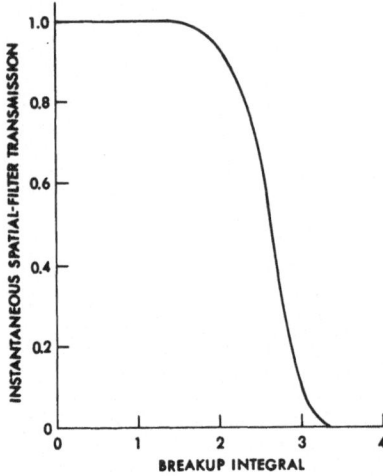

Fig. 7.8. Instantaneous spatial-filter transmission as a function of B integral [7.24]

It is obvious that the patterns recorded in such experiments are time integrated. When a finite aperture, such as the pinhole in a spatial filter, is placed at the focus, temporal modulation is observed, as shown in Fig.7.7. If the system is driven harder and harder, the modulation may become 100%; of course the loss of beam energy is severe and the pulse is essentially useless to drive a laser-fusion target. It is also an experimental fact that before temporal modulation of a beam occurs, there is considerable pulse broadening, this effect being predicted by simulation codes. The transmission of spatial-filter pinholes has been studied extensively and, as might be expected, parametrized as a function of the B integral. Various phenomenological functions have been used by various investigators and are extremely important in simulation codes that model system performance, as will be discussed in Sect.7.7. In Fig.7.8, the power-transmission function is shown as a function of the B integral, as derived from experimental data obtained at LLL [7.24]. Transmission drops dramatically for a B integral between 2 and 3; the B limit for a given system depends upon many factors, including the types of components, spacing, intensity level, and noise sources. The value of B at which a spatial filter begins to intercept a substantial portion of the beam energy is a critical system parameter and, as will be seen in Chap.8, determines the system staging for short pulses. Specific spatial-filter power-transmission functions will be discussed in Sect.7.7 in connection with system simulation and modeling.

As has been seen in the foregoing, the far-field consequences of nonlinear effects in high-peak-power laser systems are profound and include modifications of the far-field spatial profile, temporal modulation of the main beam, and severe loss of beam energy to SSSF effects. It should also be obvious that beam zooming in spatial filters can also affect the transmission and must be included in any analysis of beam-system performance. It must also be included in any analysis of illumination uniformity on laser-fusion targets.

## 7.5 Spatial Filters and Imaging

Disregarding amplification stages, the most important elements in any high-peak-power laser system are spatial filters. Indeed, the common use of them in such systems has made the routine operation of multi-terawatt systems possible. We treat here three important properties of spatial filters. They are spatial-filtering removal of unwanted ripples, imaging, and magnification, and are discussed in that order.

To understand how a spatial filter works, recall that spatial frequencies K in a beam have a propagation angle $\theta$ associated with them, given by (7.36). From Fig.7.9, it can be seen that spatial frequencies contained in a certain angle $\theta_c$ will be passed by a pinhole (aperture) placed in the far field (focal plane of a lens). We may, therefore, by choosing the pinhole diameter d, control which spatial frequencies are passed. In analogy with electrical engineering, what we have is a low-pass filter through which all spatial frequencies up to a cutoff $K_c$ are passed and all others are rejected. It is an easy matter to show that the cutoff spatial frequency $K_c = k\theta_c$ is

$$K_c = \frac{kd}{2Df^{\#}} = \frac{kd}{2f} \quad , \tag{7.135}$$

where D is the spatial-filter input-lens diameter and $f^{\#} = f/D$ is the f number of the spatial filter.

Recall the exponential growth of ripples in high-peak-power laser systems, as discussed in Sect.7.1. If left uncontrolled, such ripples induced by dust,

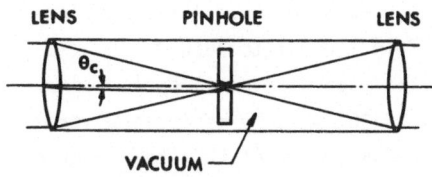

Fig. 7.9. A typical spatial filter showing the cutoff or bandpass angle $\theta_c$

scattering particles, etc., grow to disastrous proportions and seriously damage expensive optical components as well as lead to severe energy loss and beam degradation. The spatial filter is designed to remove the most-damaging ripples while passing most of the beam energy; therefore, such ripples are forced to re-exponentiate at strategic locations in a laser system.

The use of spatial filters in a high-peak-power laser system leads to the requirement that they be operated in vacuum, because the high intensities reached in the vicinity of the focal spot lead to optical breakdown if gas or air at much more than $10^{-5}$-$10^{-6}$ Torr pressure are present. We will not discuss the construction of spatial filters here. The interested reader is referred to the work of SIMMONS et al. [7.25]. It should be obvious that it is virtually impossible to describe accurately from first principles the power transmission of a spatial filter because that would require a knowledge of the ripple-noise spectrum in a laser system, the transverse and longitudinal distribution, the nonlinear properties of the media and boundary conditions and the spatial filter properties. We are forced to resort to a phenomenological description of the transmission function; early attempts to do so are described in [7.25]. Others may be found in various LLL Annual Reports, for example, [7.26]. By far the most successful formulation is that due to TRENHOLME [7.27] who guessed that a power-transmission function T(B) should depend upon the B integral, according to

$$T(B) = 1 - \epsilon[\beta(1+2B^2)+(1-\beta)\cosh 2B] \quad , \tag{7.136}$$

where $\beta$ is the fraction of noise in the large $(1+2B^2)$ ripples and $(1-\beta)$ the fraction in the fastest (cosh 2B) growing ones, as previously described by (7.118,119), respectively. Usually, the noise spectrum, phenomenologically described by the noise parameter $\epsilon$, is assumed to be divided equally between the large and fastest-growing ripples ($\beta=\frac{1}{2}$). The noise parameter $\epsilon$ is adjustable and must be determined by fitting to actual experimental data on a system. This has been done for a number of large laser systems. For the CYCLOPS and early ARGUS systems at LLL, values of $\epsilon \approx 5 \times 10^{-3}$ were obtained [7.28]. The value for the GDL system at LLE is $\epsilon \sim 1.4 \times 10^{-4}$ [7.29]. The approximate factor-of-ten difference between the early disk-amplifier system used at LLL and rod-amplifier systems at LLE is thought to result from the larger number of surfaces in disk systems. That speculation is supported by the more recent work of HUNT and associates [7.30] who, by meticulous attention to the cleanliness of surfaces in the ARGUS system at LLL, have consid-

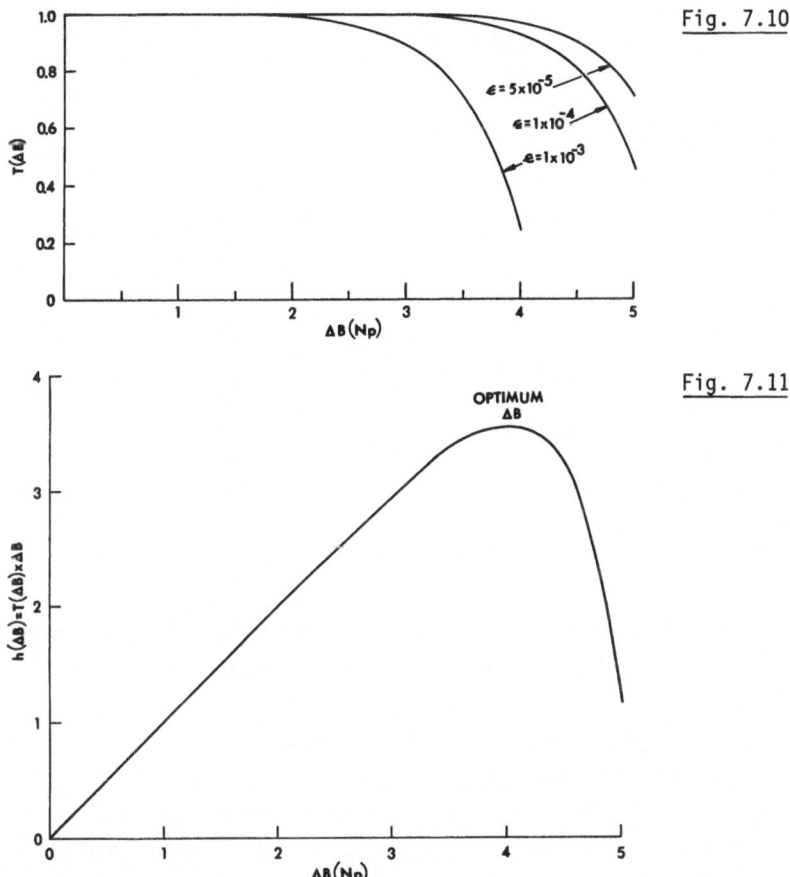

Fig. 7.10

Fig. 7.11

Fig. 7.10. Spatial-filter power transmission function as a function of B in tegral for three values of the noise parameter $\epsilon$

Fig. 7.11. $T(B) \times \Delta B$ as a function of accumulated B integral ($\Delta B$) in a stage, for $\beta = 1/2$ and $\epsilon = 1.4 \times 10^{-4}$.

erably reduced the value of $\epsilon$. The form of $T(B)$ is shown in Fig.7.10 for a number of values of $\epsilon$. The magnitude of $\epsilon$ is obviously related to the maximum allowable $\Delta B$ between stages (pinholes), as may be seen directly from Fig.7.10. That quantity, in fact, determined the staging of a high-peak-power laser system as will be discussed in Sect.7.6. As discussed there, the optimum value of $\Delta B$, which results in the maximum power output from a spatial filter, is proportional to $T(B) \times \Delta B$, an example of which is shown in Fig.7.11 for (7.136), $\epsilon = 1.4 \times 10^{-4}$, and $\beta = 1/2$ which gives an optimum $\Delta B \simeq 4$ nepers. It is easy to construct a plot, as shown in Fig.7.12, which gives the maximum allowable $\Delta B_{max}$ between stages as a function of $\epsilon$. For the original ARGUS system,

218

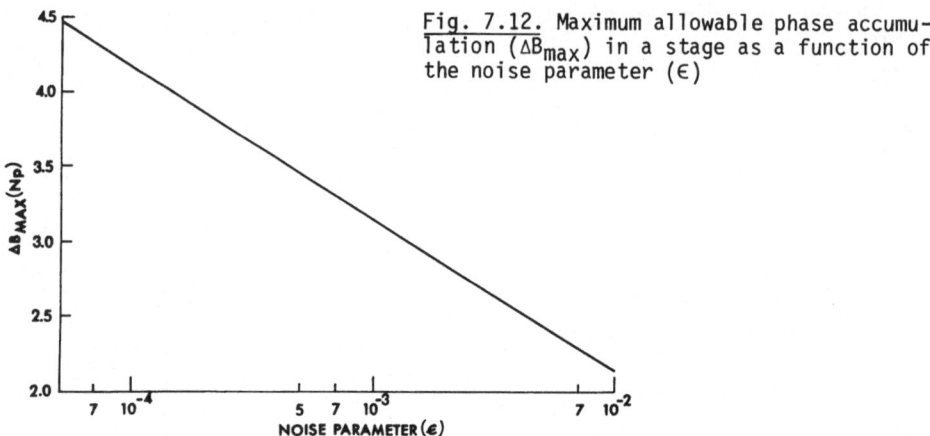

Fig. 7.12. Maximum allowable phase accumulation ($\Delta B_{max}$) in a stage as a function of the noise parameter ($\epsilon$)

$\epsilon \sim 5 \times 10^{-3}$ resulted in $\Delta B_{max} \approx 2.5$ nepers. For the GDL system, $\epsilon \sim 1.4 \times 10^{-4}$ gives $\Delta B_{max} \simeq 4$ nepers. Recently, $\Delta B_{max} \simeq 3\text{-}3.5$ nepers have been achieved on the ARGUS system [7.30], thus $\epsilon \sim 4 \times 10^{-4}\text{-}1.5 \times 10^{-3}$, a significantly smaller number that is apparently due to the system cleanliness [7.30]. For the OMEGA system at LLE and the GDL system boosted by four active-mirror amplifiers [7.31], the value $\epsilon \sim 1.4 \times 10^{-4}$ has been found from consistent results obtained by both experiments and simulation codes. From the foregoing, it should be obvious that in designing a high-peak-power laser system, it is important to use the correct value of $\epsilon$ for the type of system chosen because its value and thus $\Delta B_{max}$ determines the staging, a subject to be discussed in Chap.8.

We now turn our attention to an important treatment of spatial filtering, which shows explicitly the dependence of spatial-filter transmission and focal-spot size on the B integral, due originally to HUNT et al. [7.19]. We begin by assuming that the whole beam is perturbed by a soft Gaussian scatterer whose dimensions are small compared to the main beam. The scattering center is located at a distance $z_1$ in front of a spatial-filter lens of focal length f. The total electric field $\psi$ at the location of the lens is given by

$$\psi = \psi_0 - \psi_s , \qquad (7.137)$$

where $\psi_0$, the unperturbed electric field, is

$$\psi_0 = A_0 F(r)^{\frac{1}{2}} \qquad (7.138)$$

and, as before, F(r) is the normalized intensity distribution of the unperturbed wave and $\psi_s$, the electric field of the scatterer, is

$$\psi_S = A_0 \exp\left(-\frac{1}{2}\left[r/\omega(z_1)\right]^2\right) \exp\left[-ikr^2/2R(z_1)\right] \quad . \tag{7.139}$$

Here $\omega(z_1)$, the normal Gaussian waist size, is given by the usual expression

$$\omega^2(z_1) = \omega_0^2\left[1+\left(\frac{z_1}{k\omega_0^2}\right)^2\right] \quad , \tag{7.140}$$

and $R(z_1)$, the radius of curvature, is found from

$$R(z_1) = z_1\left[1+\left(\frac{k\omega_0^2}{z_1}\right)^2\right] \quad . \tag{7.141}$$

Use of the ABCD law of KOGELNIK [7.32] enables us to find the electric field $\psi_S(f)$ at the focal plane of a thin lens

$$\psi_S(f) = \frac{A_0\omega_0}{\omega(f)} \exp\left(-\frac{1}{2}\left[r/\omega(f)\right]^2\right) \exp\left[-ikr^2/2R(f)\right] \quad . \tag{7.142}$$

Now $\omega(f)$ and $R(f)$ are found to be

$$\omega(f) = f/k\omega_0 \tag{7.143a}$$

and

$$R(f) = \frac{f}{1-(z_1/f)} \quad . \tag{7.143b}$$

If we assume that the size of the scattered beam in the focal plane is significantly more than the pinhole diameter d, then uniform illumination of the pinhole results in the classical Airy pattern at the output lens. Then, after filtering, the beam amplitude will be of the form

$$\psi_S = A_0\omega_0^2\left(\frac{\pi d}{\lambda f}\right)^2\left[\frac{J_1(x)}{x}\right] \quad , \tag{7.144}$$

where $\lambda$ is the light wavelength, $J_1(x)$ the first-order Bessel function, and x the transverse parameter, given by

$$x = \left(\frac{\pi d}{\lambda f}\right)r \quad . \tag{7.145}$$

It is obvious that all scatterers that produce a beam waist in excess of the pinhole diameter d will result in a modulation of the beam that passes through the spatial filter. Thus, if the scattering-center dimension $\delta$ satisfies

$$\delta > \frac{f}{k\omega_0} \quad , \tag{7.146}$$

modulation results, is independent of the size of the scattering center, and has an amplitude determined by the square of its diameter.

Another important source of modulation in laser systems is the pinhole itself. It is a well-known fact that the truncation of a Fourier spectrum, for example, by an aperture placed at the focus of a lens, results in a modulation at the cutoff frequency $K_c$ of the lens-pinhole combination, where $K_c$ is given by (7.135). If the diameter d of the pinhole is significantly larger than that of the central lobe due to the main beam, then GLAZE [7.33] has shown that the rms amplitude modulation $\Delta\psi_{rms}$ can be written

$$\Delta\psi_{rms} = (1-T)^{\frac{1}{2}}\psi_i \quad , \tag{7.147}$$

where $\psi_i$ is the beam amplitude before filtering. Calculation [7.19] shows that the modulation due to millimeter scatterers in a system is comparable to that produced by a typical pinhole. The foregoing analysis has shown that one scatterer can produce significant modulation of a beam, that an optimal pinhole diameter exists for any real system, and that use of an optimal pinhole will always result in a modulation, usually by not more than a few percent rms.

We now investigate the effect of a harmonic perturbation of amplitude on a main beam as it propagates through nonlinear elements. The initial wave function $\psi_i$ of the beam can be written

$$\psi_i = A_0[F(r)]^{\frac{1}{2}}(1+\delta\cos\vec{K}\cdot\vec{r}) \quad , \tag{7.148}$$

where K, the wave vector of the perturbation is comparable to the spatial filter cutoff frequency $(K \sim K_c)$. At the input lens, the field is

$$\psi_f(r) = A_0[F(r)]^{\frac{1}{2}}(1+\delta\cos\vec{K}\cdot\vec{r})\exp\left\{-[iBF(r)(1+\cos\delta\vec{K}\cdot\vec{r})^2]\right\} \quad , \tag{7.149}$$

where B is the total accumulated B integral. Equation (7.149) can be expanded for $\delta \ll 1$ to [7.34]

$$\psi_f(r) = A_0[F(r)]^{\frac{1}{2}}e^{-iBF(r)}\left[\sum_j \alpha_j\cos(j\vec{K}\cdot\vec{r})\right] \quad , \tag{7.150}$$

where, for example, the first three coefficients $\alpha_j$ are

$$\left. \begin{aligned} \alpha_0 &= J_0(x) + i\delta J_1(x) \\ \alpha_1 &= \delta + 2iJ_1(x) - \delta J_2(x) \\ \alpha_2 &= i\delta J_1(x) - 2J_2(x) - iJ_3(x) \end{aligned} \right\} \quad , \tag{7.151}$$

and the argument x is

$$x = 2B\delta F(r) \quad . \tag{7.152}$$

The far-field diffraction pattern in the focal plane of the lens is

$$I(r) = |\psi_f(r)|^2 \quad . \tag{7.153}$$

The resulting pattern is shown in Fig.7.13 and is seen to consist of a central lobe of intensity $\alpha_0^2$ and a number of side bands of intensity $\alpha_1^2$, $\alpha_2^2$, etc. The centers of the side lobes are separated from each other by integral multiples of the wave vector $|\vec{K}|$, and have the same shape and width as the central lobe. It can be seen that as B increases, the power in the side lobes does also, leading to a significant power loss. For $2B\delta < 1$, negligible loss occurs and we may, by using (7.151), obtain

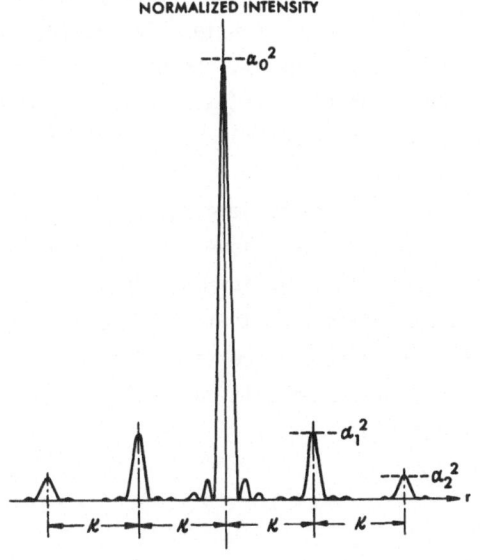

Fig. 7.13. Power spectrum of a high-power beam with a harmonic perturbation [7.19]

$$\frac{I_1}{I_0} = \frac{|\alpha_1|^2}{|\alpha_0|^2} = (1+2B^2)\delta^2 \tag{7.154}$$

for the ratio of the intensities $I_1$ and $I_0$ in the first side lobe and main lobe, respectively. Thus, the ratio increases as $(1+2B^2)$, the growth rate for large ripples derived previously (7.118). Obviously, when $B \to 0$, the diffraction-limited case occurs and most of the beam energy is found in the central lobe.

In Sect.7.3, WBSF was explored and it was shown that the maximum change of focus $\Delta z_p$, given by (7.129) was proportional to B. It is an easy matter to show that the diameter g of the smallest focused spot is given by

$$g = \left(\frac{r_0}{f}\right)\Delta z_p \quad , \tag{7.155}$$

where $r_0$ is the radial position of the ray. From (7.129) it is easy to see that $g \sim B$; that is, the diameter of the best focus spot increases linearly with B.

The foregoing analysis has given us two important results for the design of a high-peak power laser system, the first that the side-lobe power increases according to $(1+2B^2)$, whereas the beam-spot size varies linearly with B, for spatial frequencies within the bandpass of a spatial filter. We could envision choosing successive spatial filters that have smaller and smaller pinholes, each successive one filtering out the bandpass frequency of the preceding one. The difficulty is that, as B is accumulated in a system, the focal-spot size increases linearly with B and results in increased modulation. Calculation of optimum pinhole diameter is extremely difficult. Usually, the best procedure is to determine the diameter experimentally for each system by examining how near- and far-field patterns change with different pinhole sizes. Typical values of d are in the range 0.5-1.5 mm.

The second important property of spatial filters is imaging, a technique now widely known to alleviate significantly the deleterious effects of whole-beam self-focusing. In essence, the method leads to achievement of a much more constant beam profile and larger fill factor than that obtainable without spatial filters. We again follow here a treatment due to HUNT et al. [7.19]. In Fig.7.14 we show a typical optical system with confocal lens pairs. The transfer matrix $t_k$ for the $k^{th}$ element is

$$t_k = \begin{bmatrix} -m_k & -m_k d_k^1 - (m_k)^{-1} \; d_k^2 + f_k^1 + f_k^2 \\ 0 & -1/m_k \end{bmatrix} \quad , \tag{7.156}$$

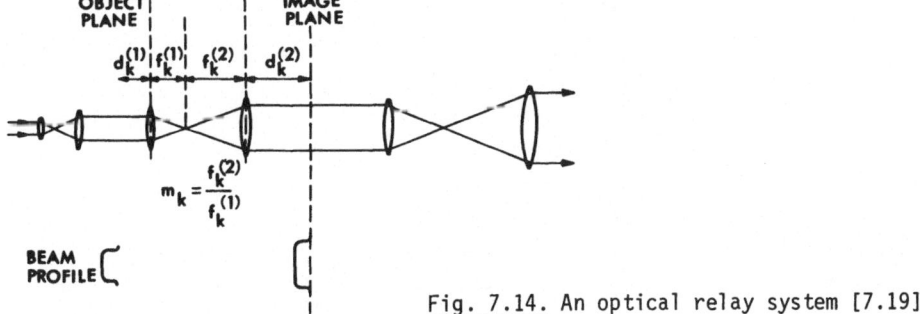

<u>Fig. 7.14.</u> An optical relay system [7.19]

where the magnification $m_k$ is given by

$$m_k = \left(\frac{f_k^2}{f_k^1}\right) \ , \tag{7.157}$$

and $d_k^1$ and $d_k^2$ are the object and image distances, respectively, of the lens pairs. For a system that consists of N lens pairs, the total system matrix $T_N$ is

$$T_N = \prod_{k=1}^{N} t_k \ . \tag{7.158}$$

By use of geometrical optics, the beam profile F(r) in the plane located at $d_k^1$ can be shown to be imaged at the plane located at $d_{k+1}^1$ with magnification $m_k$, if,

$$m_k d_k^1 + (m_k)^{-1} d_k^2 - f_k^1 - f_k^2 = 0 \ . \tag{7.159}$$

If (7.159) is satisfied for all k, the beam will be imaged, or relayed, through the entire system. Modern laser systems image an apodized or hard aperture through the entire system to near or at the location of the target chamber. It has been seen previously (Sect.7.3) that the existence of an intensity-dependent index of refraction leads to steepening of gradients of an initially smooth beam profile after propagation through a nonlinear element, leading to the development of "horns" on the beam, as can be seen schematically in Fig.7.15. That effect is reduced by imaging as may be seen from the following. For a nonrelayed beam, the radial position $r_N$ and angle $\theta_N$ after transversing N nonlinear elements is given by

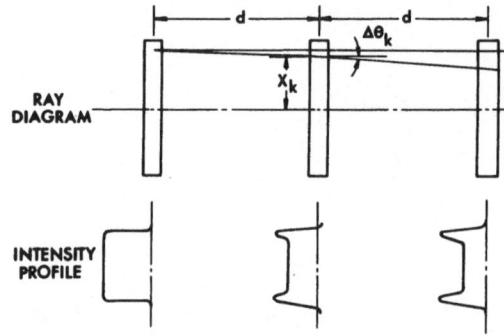

RAY
DIAGRAM

INTENSITY
PROFILE

Fig. 7.15. Beam distortion from self-focusing in a nonimaged system [7.19]

$$r_N = r_1 + d \sum_{k=1}^{N-1} (N-k)\Delta\theta_k \qquad (7.160)$$

and

$$\theta_N = \sum_{k=1}^{N} \Delta\theta_k \quad , \qquad (7.161)$$

where the angular deviation $\Delta\theta_k$ of the $k^{th}$ element was given previously as (7.124) and $r_1$ is the initial ray position. The change of intensity of the beam upon passing through an element is assumed to be negligible; the major change is assumed to be the wavefront phase, formally given by the B integral. If a confocal lens pair is placed between each of the elements in the system, then the radial position $\overline{r_N}$ and angle $\overline{\theta_N}$ become

$$\overline{r_N} = (-1)^N \left[ r_1 - (4f-d) \sum_{\ell=1}^{N} (N-\ell)\overline{\Delta\theta}_{\ell+1} \right]$$

$$\overline{\theta_N} = (-1)^N \sum_{\ell=1}^{N} \overline{\Delta\theta}_\ell \quad . \qquad (7.162)$$

The effective propagation length $z_e = 4f-d$ is zero or very small near image planes. Thus $\overline{r_N}$ is

$$\overline{r_N} = (-1)^N r_1 \quad , \qquad (7.163)$$

while $\overline{\theta_N}$ is again given by (7.162), and thus, although rays are deviated from their initial directions, their positions are unchanged and a smooth intensity profile results, identical to that of the original beam. In real laser systems, amplifiers and other nonlinear components extend over a distance that is usually greater than the depth of field of the optical system, resulting in less than ideal imaging. Nevertheless, modulation is considerably reduced when compared to that of a nonimaged system, and larger system output occurs because the technique allows maximization of the fill factor.

It is obvious that a spatial filter can also act as a beam expander at strategic locations in a system. The staging of a system, to be discussed in Sect.7.6 and Chap.8, requires frequent changes of beam diameter for optimal performance. Thus, spatial filters are important not only for filtering unwanted ripples and reducing whole-beam effects but also because they are a necessary and critical system component for implementing a given staging philosophy.

## 7.6 X-Factor Analysis

One of the most powerful techniques for staging a high-peak-power laser system is X-factor analysis, a subject that we now take up. It has become a standard tool of the laser designer. The power of the technique has been proved in a number of cases, including the design of ARGUS and SHIVA at LLL, and the GDL and OMEGA systems at LLE. The analysis we follow here is due to TRENHOLME [7.35]; it was used by BROWN and BURNHAM [7.36] and BROWN et al. [7.37] in the analysis of the ZETA and GDL systems, respectively, at LLE. In Sect.7.5, it was seen that the output power $P_o$ from a spatial filter is related to the input power $P_i$ through the relationship

$$P_o = T(B)P_i G \quad , \tag{7.164}$$

where $T(B)$ is the transmission function, $G$ the total gain of the stage and $P_i$ the output of the proceeding filter. This may be rewritten as

$$P_o = \left(\frac{P_i}{B}\right)BT(B) = X \, BT(B) \quad , \tag{7.165}$$

where

$$X = \frac{P_o}{B} = \frac{P_i G}{B} \tag{7.166}$$

is the so-called X factor or power-handling factor of the group of components in a stage. Note that because in the absence of saturation, B is linearly proportional to the intensity and since intensity is directly proportional to the power output, X is a measure of the power-handling capability of a stage. We rewrite (7.165) as

$$P_o = X \, h(B) \quad . \tag{7.167}$$

For a given group of components in a stage, X is a constant and thus the power output is linearly proportional to h(B). Recall that in Sect.7.5 it was shown that $h(B) = BT(B)$ displays a maximum as shown in Fig.7.11. Then, the ultimate power output $P_u$ from a stage is

$$P_u = X \, h_m(B_m) \quad , \tag{7.168}$$

where $h_m$ is the maximum value of h(B), and $B = B_m$ is the maximum B value, determined by the noise content in a stage, as discussed in connection with Fig.7.12.

A group in a stage would normally consist of n components, including spatial-filter lenses, amplifiers, isolation devices, etc., with each component having a gain $G_n$ (note that we use the general definition of gain here, which in some cases may mean a loss). Because

$$B = \frac{P_o}{X} = \frac{GP_i}{X} \quad , \tag{7.169}$$

we can rewrite (7.167) as

$$P_o = X \, h(B) = X \, h\!\left(\frac{GP_i}{X}\right) \quad , \tag{7.170}$$

which is an important equation in the design of high-peak power laser systems.

We are interested in how the individual X factors, $X_n$, in a stage combine. For n components, it is easy to see that the total X factor $X_T$ is

$$X_T = \frac{P_o}{B_T} = \frac{P_n}{\sum\limits_{j=1}^{n} B_j} \quad , \tag{7.171}$$

where $B_T$ is the total B integral. Some algebra and manipulation along with the relationship

$$P_{j+1} = G_{j+1} P_j \quad , \tag{7.172}$$

yield

$$X_T = \left[\frac{1}{X_n} + \frac{1}{G_n X_{n-1}} + \frac{1}{G_n G_{n-1} X_{n-2}} + \dots\right]^{-1} , \tag{7.173}$$

which shows that, for n elements with equal gains the X factors add like resistors in parallel. The effect of the gain is to make the final components in a stage more important, which of course is true if amplifiers are increasing the beam intensity. How do we calculate the X factor for components in a laser system? The $X_R$ factor was calculated previously in Chap.2, (2.34), for a rod amplifier. A similar calculation for a disk amplifier gives

$$X_D = \frac{K \alpha n_d \sqrt{n_d^2+1}\; D^2}{n_2^T \lambda_p} \left(1 - \frac{1}{G_0}\right)^{-1} . \tag{7.174}$$

The only difference between (7.174) and (2.34) is the factor $(n_d^2+1)$, which arises from the difference of area between a disk orthogonal to a beam and one at Brewster's angle in the case of a disk amplifier. We can rewrite (7.174) and (2.34) as

$$X_D = X_{\infty D}\left(1 - \frac{1}{G_0}\right)^{-1} \tag{7.175}$$

$$X_R = X_{\infty R}\left(1 - \frac{1}{G_0}\right)^{-1} , \tag{7.176}$$

where the factors $X_{\infty D}$ and $X_{\infty R}$ are often referred to as the ultimate X factors, which depend only upon the type of glass, size of the element and pumping. Then $G_0$ is a function of the number of disks in the stage. Owing to the overlap of beams inside an active-mirror amplifier, it is not possible to obtain an analytical expression for its X factor, as with the rod and disk, except in the case of very short pulses. Instead, it must be evaluated by numerical analysis, or by defining an "equivalent" rod amplifier. The B integral for an active mirror of thickness $\ell$ is the same as for a rod amplifier of twice the thickness only in the limit of a vanishingly small pulse width, when no overlap occurs. For all other cases, the B integral accumulated in an active mirror is significantly greater than in a rod amplifier of twice the thickness.

We now consider a single disk amplifier located in a stage, as analyzed previously by TRENHOLME [7.35]. If the only elements are spatial-filter lenses with X and factors $X_1$ and $X_2$ and a disk amplifier with X factor $X_D$, the total X factor $X_T$ is

$$X_T = \left|\frac{1}{X_1} + \frac{1}{X_D} + \frac{1}{X_2 G_0}\right|^{-1} . \tag{7.177}$$

The output power is given by (7.170), whereas the maximum possible power is given by (7.168) or

$$P_u = X_T h_m(B_m) \simeq X_\infty D h_m \left(1 - \frac{1}{G_0}\right)^{-1} \quad , \tag{7.178}$$

where we have assumed that $X_1, X_2 \gg X_D$, which is usually the case. Note that (7.178) predicts that fewer disks (less gain $G_0$) results in more output power. This is TRENHOLME's "less is more" principle [7.35]; the power required to drive the disks, of course, increases according to

$$P_i \simeq X_\infty D B_m / (G_0 - 1) = \frac{P_0}{G_0} \quad . \tag{7.179}$$

These conclusions can be shown by simple analysis to be valid even though other components are left in the group.

To find the X factor of passive components such as beam splitters, Pockels cells and Faraday rotations, we evaluate the B integral as in Chap.2 and find the passive X factor $X_p$ as

$$X_p = \frac{K n_d D^2}{n_2^T \lambda_p L} \quad , \tag{7.180}$$

where L is the total length of the element. For Brewster polarizers,

$$X_{BP} = \frac{K n_d \sqrt{n_d^2 + 1}\ D^2}{n_2^T \lambda_p L} \quad , \tag{7.181}$$

in analogy with disk amplifiers. For lenses, (7.180) is a good first approximation, but due to the curved surfaces characteristic of low f number lenses, a more detailed analysis which we shall not repeat here, is needed [7.38] for exact results.

Rod amplifiers present a rather special case because the entire length is not pumped. Some unpumped portion is needed for mechanical mounting of the rod. Thus there are essentially passive elements on either side of the active length. The separate X factors are given by (2.34) for the active length and by (7.180) for the end regions of length $\ell_1$ at the input and $\ell_2$ at the output. By (7.173), it is easy to see that the effect is to reduce the power-handling capability of a rod amplifier. Typically, for OMEGA rod amplifiers, a 20% reduction of the power-handling capability occurs for $\ell_1 = \ell_2$. By making $\ell_2 < \ell_1$, as far as possible within the mechanical constraints, the power-handling capability is reduced by only 6%.

X-factor analysis forms the backbone of short-pulse, high-peak-power laser design. Its application to the design of laser systems will be discussed in detail in Chap.8.

## 7.7 Simulation and Modeling of Small-Scale Self-Focusing Effects

We end this chapter with a discussion of how SSSF is simulated by use of sophisticated computer codes. Two computer codes have been developed, one at LLL (VORPAL) and the other (RAINBOW) at the LLE. The latter code will be treated here because of the author's intimate experience with it. Both codes have been found to describe accurately, to within 10%, the performance of real laser systems once normalization was accomplished on an initial system, as will be discussed below. Furthermore, because VORPAL and RAINBOW are essentially ray-trace codes involving no diffraction, it should be realized that they are first-order tools in any design. If diffraction is well controlled as it can be in a collimated, spatially filtered, imaged system, then RAINBOW is a powerful tool for doing system analysis.

We first list the general features of RAINBOW and then discuss each in order in which they appear in Table 7.1 below.

That diffraction is ignored in RAINBOW has already been discussed. The second feature arises because rod and active-mirror amplifiers display variation of gain with radius, in rod amplifiers, owing to the radial attenuation of pump radiation, and in active-mirror amplifiers because of vignetting of the pump array. Disk amplifiers show a small ($\approx$10%) variation of gain across the aperture and are treated as having flat profiles. The third feature of RAINBOW is the use of the classical Frantz-Nodvik gain equations, as we de-

Table 7.1. RAINBOW features

---
1) Ray trace code
2) Radial-gain dependence
3) Frantz-Nodvik gain equations
4) Linear and two-photon absorption
5) Saturation
6) Phenomenological simulation of nonlinear effects
7) Forward and backward propagation
8) Flexible pulse shapes (input or output)
9) Active-mirror amplifiers
10) Damage threshold as a function of pulse width for standard coatings
11) Damage, $\Delta B$ flag system

---

scribed in Chap.1 (1.20,21). Because these equations are unsolvable analytically, they are written in finite-difference form. The fourth feature, involving losses due to linear absorption and two-photon absorption, is incorporated
into RAINBOW by adding another loss term to (1.20,21) proportional to the
square of the intensity. Note that linear absorption has already been included. The two-photon absorption coefficient in the added term has been
tabulated for a number of materials by SMITH [6.11]. For propagation at
1.06 μm, two-photon absorption is negligible but, in frequency-tripled systems
such absorption can be substantial. Saturation is a well-known phenomenon in
laser physics; it is fairly easy to incorporate it into the code by using
Frantz-Nodvik equations for a 3- or 4-level system. The real situation is
not usually that simple, however; for long pulses on the GDL system it was
found that a short (≃200 ps) terminal-level lifetime was required to fit
RAINBOW to the experimental results [7.31]. That result contradicted previous
measurements of the terminal-level lifetime measurements based upon gain recovery which found a value of ≃1.25 ns for ED-2 laser glass [1.6] and even
longer for phosphates. Based on gain-recovery measurements, a 3-level system
was predicted; in fact, the experimental data showed behavior close to a 4-
level system [7.31]. This conclusion has been verified recently by workers
at LLL [1.19] who measured the saturation fluence for a number of laser
glasses. The results show (see Fig.1.7) that the saturation fluence is in
fact dependent upon the fluence incident upon the amplifier, and that for
the conditions in GDL a 4-level system is a good approximation. The fluence-
dependent saturation fluence has been incorporated into RAINBOW by using
functions that were nonlinear least-square fitted to the data provided by
MARTIN [1.19]. At present, two versions of RAINBOW exist; one uses the terminal lifetime approach to saturation and the other a saturation fluence.
Of course, the two are equivalent. One of the most important features of
RAINBOW is simulation of SSSF by the use of a power or B integral-dependent
spatial-filter transmission function (Feature 6). The function is, in fact,
(7.136), with suitable adjustment of the noise parameter $\epsilon$. For RAINBOW,
initial code normalization was performed by use of GDL data [7.31] which
yielded the value of $\epsilon \sim 1.4 \times 10^{-4}$. As has been seen, that value of $\epsilon$ has also
been found to be appropriate to other systems constructed at LLE. The next
feature of RAINBOW (7) is unique. Normal propagation proceeds from the oscillator to the system output. It has been found by BROWN [7.39] that it is
possible to invert (1.20,21) so that a pulse may propagate backwards. Although
the procedure substantially increases the computer running time, it is valuable because systems should, in fact, be staged from the output to the input,

as will be discussed in Chap.8. Backward propagation, in addition to considerably simplifying system design, permits predetermination of the initial
pulse shape, temporally and spatially, required to achieve a specified beam
profile and temporal shape at the output. The scheme at present is applicable
only to rod and disk amplifiers, but active-mirror amplifiers may be simulated using "equivalent" rod amplifiers. The next feature (8) is that a number
of different pulse shapes are available in RAINBOW, including flat-top,
Gaussian, or $n^{th}$-order rising pulses, such as linear ramps or cubic pulses.
The pulse shape at the output is an important consideration in systems designed to drive ablative laser-fusion targets. A unique feature (9) of RAIN-
BOW is the treatment of active-mirror amplifiers. Unlike single-pass rod or
disk amplifiers, in active mirrors it is necessary to keep track of the stored-
energy density and local-beam flux, because of the overlap of two beams. RAIN-
BOW follows the instantaneous values of the stored-energy density, beam flux,
and B integral in a correct way; of course, it is considerably more complicated and involves a longer running time than codes for disk or rod amplifiers.
A recent addition to RAINBOW is the incorporation of a flag system (11), which
stops computer execution when a limit has been reached in the system, either
the damage limit at a surface or coating, or when the accumulated $\Delta B$ in a
stage exceeds the maximum permissible value. The pulse-width dependence of
the damage threshold (10) has also been incorporated in RAINBOW based upon
data provided by LOWDERMILK and MILAM [6.59].

RAINBOW is a powerful tool for analyzing existing systems or predicting
the performance of new ones. Typical output from RAINBOW at any stage is shown
in Fig.7.16. It consists of a two-dimensional (radial and temporal) profile,
time-integrated energy density as a function of radius, maximum total B integral accumulated in the system to the location as a function of radius, and
the total integrated output power as a function of time. If the system input
is driven harder and harder, pulse broadening is observed, and finally modulation, as shown in Fig.7.17, in agreement with experimental results discussed
in Sect.7.4. In Fig.7.18, we show the focusable power output from the GDL system (shown in Fig.7.19) as a function of beamline input [7.31]. Experimental
measurements are compared with RAINBOW prediction; a best fit yielded
$\epsilon \sim 1.4 \times 10^{-4}$, and served as code normalization. In Fig.7.20, GDL output,
output with linear polarization and circular polarization as predicted by
RAINBOW, is seen to agree well with the measured values; recall that circular
polarization results in a lower nonlinear index than linear. Further verifi-
cations of RAINBOW predictions have been obtained on the OMEGA system, and
on GDL driving four 17-cm-aperture active-mirror amplifiers as shown in

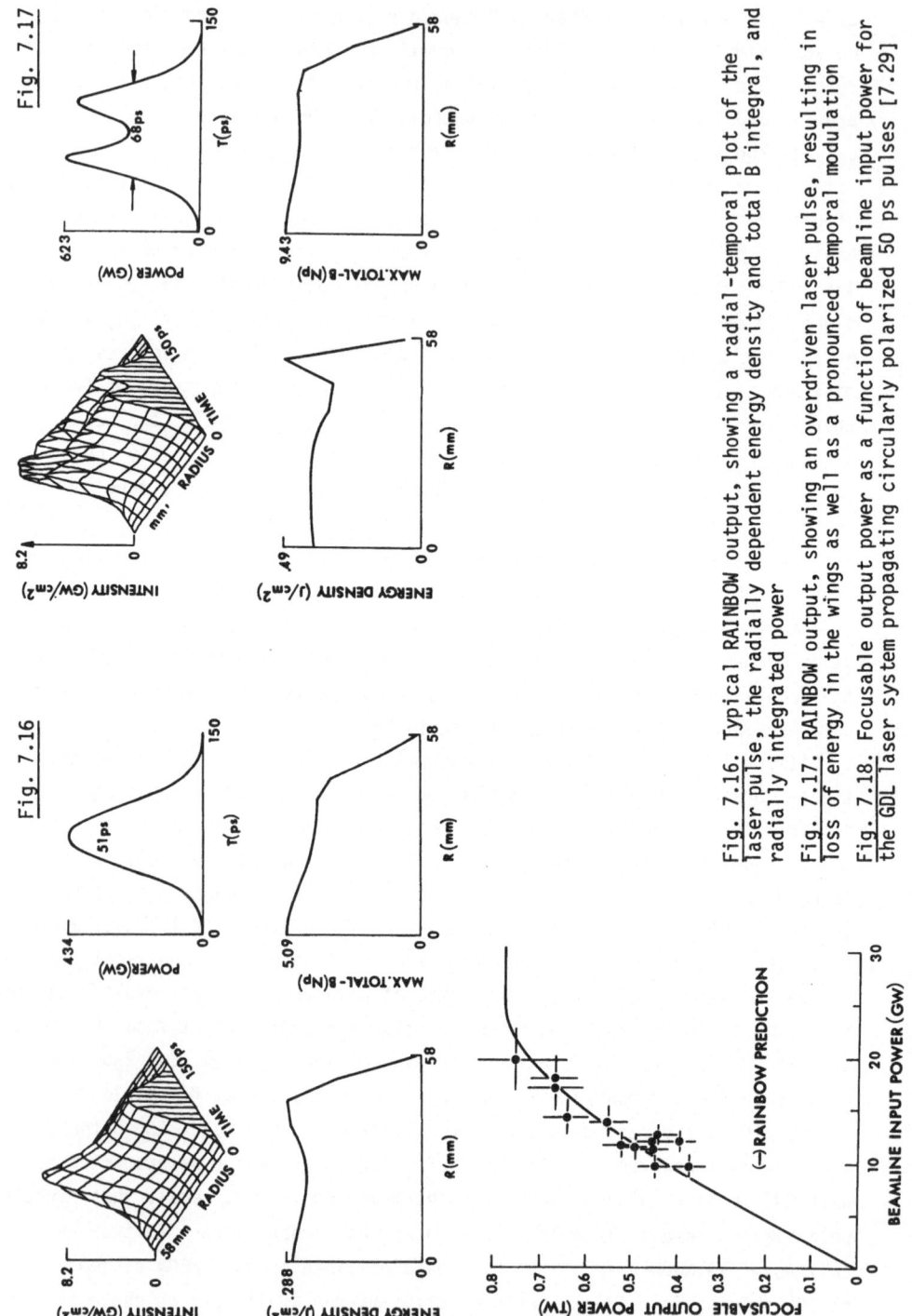

Fig. 7.16

Fig. 7.17

Fig. 7.16. Typical RAINBOW output, showing a radial-temporal plot of the laser pulse, the radially dependent energy density and total B integral, and radially integrated power

Fig. 7.17. RAINBOW output, showing an overdriven laser pulse, resulting in loss of energy in the wings as well as a pronounced temporal modulation

Fig. 7.18. Focusable output power as a function of beamline input power for the GDL laser system propagating circularly polarized 50 ps pulses [7.29]

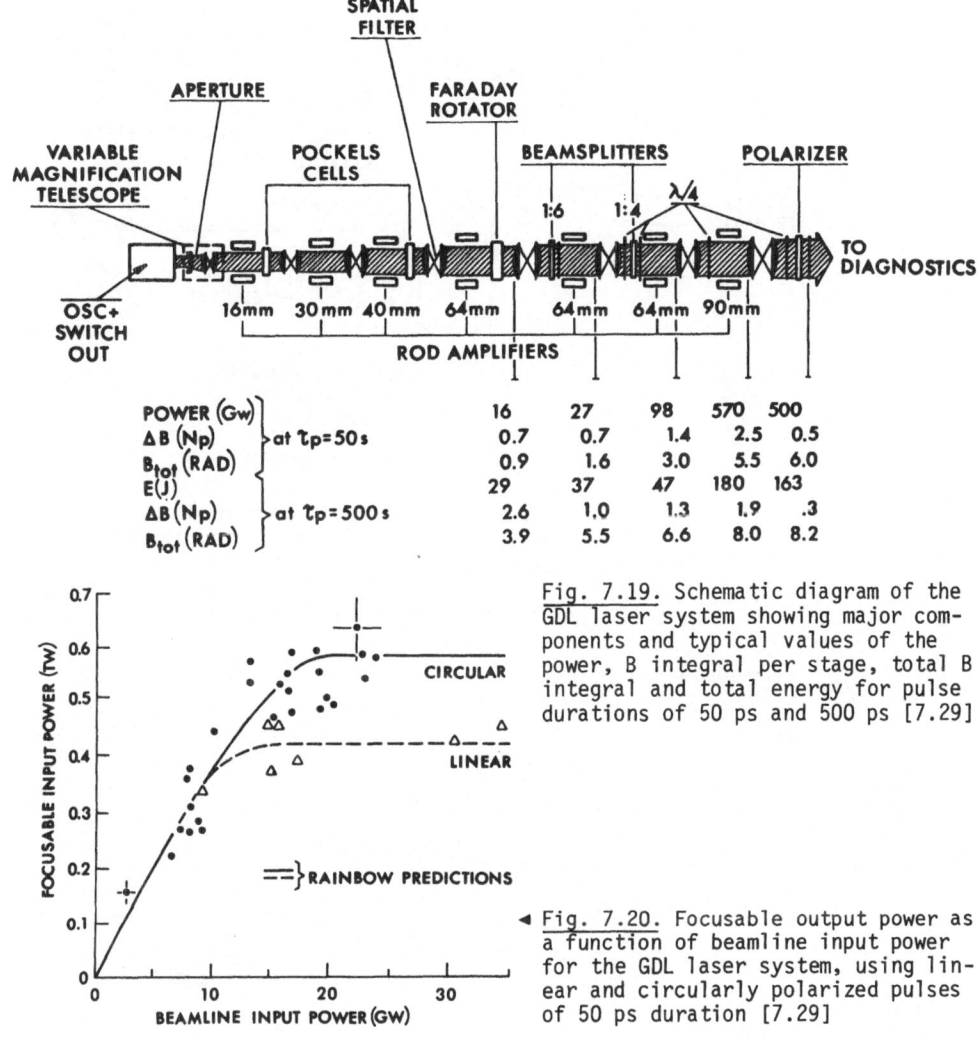

| | | | | POWER (Gw) | | 16 | 27 | 98 | 570 | 500 |
|---|---|---|---|---|---|---|---|---|---|---|
| | | | } at $\tau_p$=50 s | ΔB (Np) | | 0.7 | 0.7 | 1.4 | 2.5 | 0.5 |
| | | | | B$_{tot}$ (RAD) | | 0.9 | 1.6 | 3.0 | 5.5 | 6.0 |
| | | | | E (J) | | 29 | 37 | 47 | 180 | 163 |
| | | | } at $\tau_p$=500 s | ΔB (Np) | | 2.6 | 1.0 | 1.3 | 1.9 | .3 |
| | | | | B$_{tot}$ (RAD) | | 3.9 | 5.5 | 6.6 | 8.0 | 8.2 |

**Fig. 7.19.** Schematic diagram of the GDL laser system showing major components and typical values of the power, B integral per stage, total B integral and total energy for pulse durations of 50 ps and 500 ps [7.29]

◄ **Fig. 7.20.** Focusable output power as a function of beamline input power for the GDL laser system, using linear and circularly polarized pulses of 50 ps duration [7.29]

Fig.7.21. The RAINBOW predictions for short (≈50 ps) and long (≈750 ps) pulses, and the corresponding data are shown in Figs.7.22, 7.23, respectively; they can be seen to agree very well. Having verified the predictions of RAINBOW, we have gained confidence in the code and use it often to predict the performance of future systems, including the one shown in Fig.7.24 in which the GDL system has been used to drive four 21-cm aperture active-mirror amplifiers

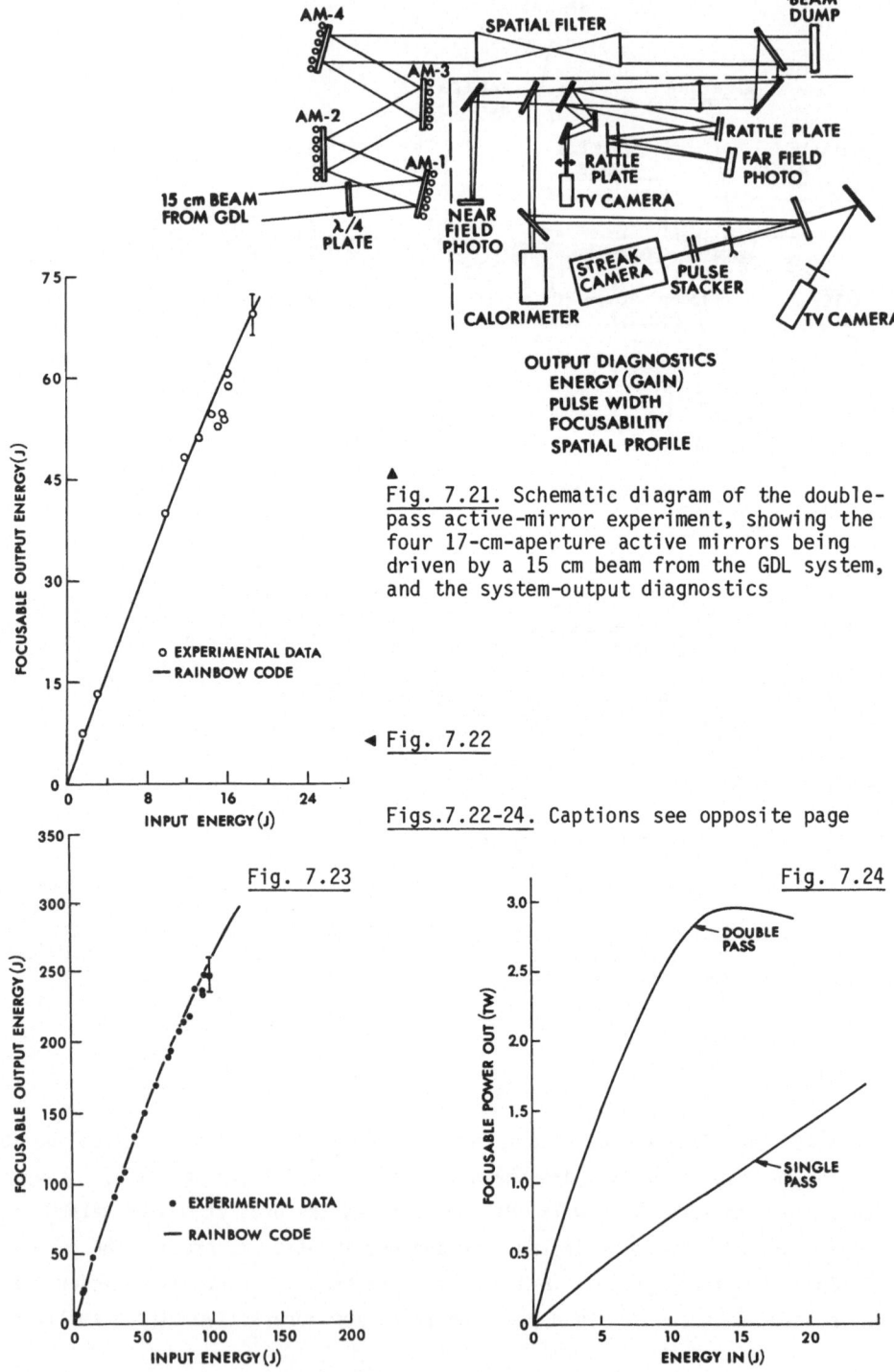

Fig. 7.21. Schematic diagram of the double-pass active-mirror experiment, showing the four 17-cm-aperture active mirrors being driven by a 15 cm beam from the GDL system, and the system-output diagnostics

Fig. 7.22

Figs.7.22-24. Captions see opposite page

Fig. 7.23

Fig. 7.24

using the double-pass scheme described in Chap.5 (Fig.5.16). Based upon our experience with RAINBOW, we confidently expect that the GDL$^+$ system presently being constructed at LLE will conform closely with the RAINBOW predictions shown in Fig.7.24.

Fig. 7.22. Short-pulse ($\approx 50$ ps) performance of the double-pass active-mirror experiment. Four 17-cm-aperture active-mirror amplifiers were used, two containing LHG-8 and two Q-88 laser glass

Fig. 7.23. Long-pulse ($\approx 750$ ps) performance of the double-pass active-mirror experiment. Four 17-cm-aperture active-mirror amplifiers were used, two containing LHG-8 and two Q-88 laser glass

Fig. 7.24. Predicted (RAINBOW) 50 ps performance of the single- and double-pass, four 21-cm-aperture active-mirror system (GDL$^+$) being constructed at LLE

# 8. The Design of High-Peak-Power Nd : Glass Laser Systems

In the previous seven chapters we have treated, in as much detail as possible in a book such as this, the important physics relevant to the design of Nd: glass laser systems. This final chapter will concentrate on applying what we have learned to the actual design of real laser systems operating in the short-pulse, high-peak-power regime when output is limited by nonlinear effects. Although it will become apparent that the subject matter preceding this chapter will be very relevant to the task at hand, we here introduce another variable to an already complicated problem — namely, cost. In recent years, the cost associated with building large laser devices for laser-fusion experiments has assumed great importance and must be factored into the overall laser-design process. Indeed, methods have been developed that optimize system performance and cost for both the short-pulse (high-peak power) and the long-pulse (high-energy) regimes. Although we will treat here only the design of the high-peak-power systems, the optimization procedure for high-energy systems is straightforward and, in fact, much simpler.

In Sect.8.1, we first review and discuss the major design issues relevant to short-pulse operation. In Sect.8.2, we present the general design methodology used to arrive at a system configuration and apply it to the general design of B-limited laser systems, with cost optimization. The separate stages of laser design will be shown and methods for arriving at a final system configuration discussed.

## 8.1 Design Issues in High-Peak-Power Nd:Glass Lasers

We begin this section by listing in Table 8.1 the major issues relevant to the design of high-peak-power Nd:glass lasers.

To design a laser system it is first necessary to know a great deal about expected system performance; the designer normally begins by listing under

Table 8.1

1) System-Output Specifications
2) Radial-Gain Dependence
3) Saturation
4) Pulse Shape
5) Amplifier Performance
6) Cost
7) Glass Type(s) - Material Selection
8) Frequency Conversion
9) Noise - B limit
10) Isolation
11) Repetition Rate
12) Damage
13) System Reliability and Maintenance

(1) the specifications, including desired total energy, peak power, pulse
shape and pulse width, wavelength, polarization, and the total number of beams.
Although it is possible to cost optimize the number of beams needed for a
fixed output, that variable is usually fixed by other constraints, such as
the expected degree of illumination uniformity on a target or the number of
available ports on a target chamber. While it is desirable to maximize the
fill factors of a stage to obtain the peak power or energy output the system
is capable of, requirements of uniform illumination on a target often determine the beam shape and result in less than maximum-fill factors. It is important to know ahead of time whether a system will be frequency converted
to the second, third or fourth harmonic, because the staging can be affected.
An effect that must be taken into account is the radial-gain dependence of
amplifiers (2) which is extremely important in rod amplifiers and, to a lesser
extent, in disk and active-mirror amplifiers. The reason, of course, is that
the presence of a radial-gain dependence modifies the propagation of pulses;
if left uncontrolled, it would lead to an undesirable output profile. In rod
amplifiers, a large variation of radial gain occurs and is most pronounced
in the smaller-diameter heads. Disk amplifiers show little variation of gain
across the aperture, as do active mirrors except near the edges, where vignetting occurs. It is obvious that the presence of ASE or parasitics can also
modify the radial-gain profile of amplifiers. Although short-pulse laser systems are normally operated at fluences only small fractions of the saturation
fluence, it is nevertheless still important to take saturation (3) into account because profile modifications and slight pulse distortions are found
when such pulses are propagated through a laser chain. The pulse shape (4)
is also a design issue because owing to the combined effects of saturation
and, if driven too hard, nonlinear effects can produce a pulse shape at the

system output different from that at the input. If a desired pulse shape and spatial profile is desired at the output, we will see how it is possible, using backward propagation, to find the pulse that must be produced at the input, as long as nonlinear effects have not occurred. The next issue, amplifier performance (5) is extremely important. We have seen in Chap.5 how, by normalizing sophisticated computer codes to real amplifier performance, it is possible to predict with good accuracy how amplifier performance is related to such parameters as the bank energy, disk thickness, diameter, etc. More importantly, it was shown how, by spectroscopic analysis of small laboratory samples of laser glass, we can predict how performance is related to glass type. This predictive capability is one of the main ingredients in the laser-design process to be described.

The cost of components (5) is also a very important design issue, particularly for amplifiers because a major part of the cost of a system is invested in them. Cost equations must be developed for each component in the chain; for amplifiers it is necessary to know how the cost scales with diameter, disk thickness, bank energy, etc. That information, when combined with amplifier performance data and X-factor analysis allows us to design cost-optimized laser chains. Materials selection, particularly laser glass (7) is another design issue. The selection of a certain type of glass in a laser system can optimize performance and cost. It may be desirable to utilize one type of glass in one section of a system and another in a different section. For example, in ASE or parasitic-limited amplifiers, a glass with a low cross section but possessing other desirable properties such as a low $n_2$ may be desirable. If frequency conversion (8) is to be used in the system, material properties become critical because of two-photon absorption and enhanced nonlinear effects. If frequency tripling to 0.35 μm is considered, then any transmitting optical components that follow the conversion must possess a band gap below about 0.175 μm because of the onset of two-photon absorption. An additional consideration is that the nonlinear index is generally greater the smaller the wavelength, the B integral is wavelength dependent as $1/\lambda$ and, if a two-photon resonance is approached, further enhancement of the nonlinear index is expected. Because conversion to the second or third harmonic can be accomplished with high efficiency, third-harmonic radiation is much more strongly absorbed than the fundamental by a target, and the use of short wavelengths may help solve the fast-electron problem associated with current laser-fusion targets, there is an impetus at the time of this writing to frequency double or triple major systems at LLL and LLE. SEKA et al. [8.1] have recently demonstrated conversion efficiencies up to 80% in two Type II KDP crystals, using a novel scheme developed by CRAXTON [8.2]. KMS Fusion Inc. has, for some time, done experiments at the second harmonic where nonlinear and two-photon problems are not severe. At present, no measurements of $n_2$ exist at the second and third harmonic and measurements of the two-photon absorption coefficient at the second and third harmonic are sparse, although such a program is currently being pursued by SMITH at LLL. The next design issue, the noise level (9), which determines the B limit of a system, has been described in Chap.7. We mention here only that the B limit of a given system is determined by the appropriate noise level and has been found to be different for rod amplifier, active-mirror or disk amplifier systems. A rather important consideration is isolation (10), of which there are three distinct types. The first is isolation from back reflection or opposing beams, if a target is not in place during a system firing. The designer always designs for the worst case, which is 100% back reflection: if no isolation were in place, such a pulse would eventually damage or destroy costly components in the system. Recall that, after passage of a short pulse, much of the amplifier gain still remains. Isolation is normally provided by a Faraday rotator placed in an appropriate location. Such devices normally contribute a large B integral to a stage because of their thickness and usually large nonlinear index. A new device, known as a liquid-crystal isolator [8.3] has recently been developed which has very low insertion loss, negligible B integral contribution, a rejection ratio of greater than 100:1 and low cost. The liquid-crystal layer is designed so that, for example, right-handed circularly polarized light which is passed by a reflected, left-handed polarized beam, is rejected. The second type of isolation required is from ASE prepulse on a target. ASE energy, resulting from the amplification of spontaneous emission emitted along the direction of the optical axis of the system, arrives at the target during hundreds of microseconds in advance of the main laser pulse and has been found to interfere with the compression of laser-fusion targets. The acceptable level of ASE for the worst-case target must be specified in advance and isolation devices, usually Pockels cells, must be placed in appropriate locations to reduce the ASE to an acceptable level. The difficulties with Pockels cells are that they normally have low transmittance due to intrinsic absorption and associated birefringence losses, they add a large B-integral contribution to any stage, and are costly and hard to fabricate in large diameters. A new device, essentially a narrow-band filter [8.4] may lead to a passive device with acceptable rejection rates, low cost, and much smaller B-integral contributions. If systems are to be frequency converted, ASE prepulse energy is no longer important. The third type of isolation, between amplifier stages, may be necessary for high-gain stages. Oscillation between stages may be avoided by wedging amplifier faces and paying close attention to secondary reflections. ASE, however, leading to energy loss and profile modification as well as inordinately high ASE levels along the system optical axis, may require the use of Pockels-cells isolators in the early, low-diameter high-gain stages. The next design issue, repetition rate (11), may or may not be important, depending upon the end use of the system. For the National Users Facility located at LLE at the University of Rochester, a repetition rate of one firing per half hour was deemed necessary. It forced the early selection of rod amplifiers for the OMEGA system. Subsequently, the development of the active-mirror amplifier has been stimulated by the decision to use an amplifier with a high repetition rate. At LLL, however, where a few shots per day are sufficient, the disk amplifier has been chosen. Normally, the repetition rate is only a secondary consideration after the system performance and cost have been evaluated. Damage phenomena (12), normally associated with long-pulse (>200 ps) laser systems, should also be taken into account in any designs. The designer should calculate the damage level appropriate to a given component, estimate the expected beam modulation for a given pulse length, and determine that the damage limit is not reached for any component. Generally, for high-peak-power laser systems, a B limit is reached before the damage limit. The final design issue (13) is rather vague; it involves the problems associated with system reliability, and maintenance. As for reliability, for example, the system designer is often faced with tradeoffs that are difficult to quantify. Xe flashlamps, known to have a finite lifetime and associated with explosive failures, may be jacketed in either $N_2$ or $H_2O$. Although the lifetime exponent and explosion constant are well known for lamps fired in a $N_2$ atmosphere, they are not known for $H_2O$-jacketed flashlamps; the OMEGA system has had virtually no explosive flashlamp failures using this arrangement. $H_2O$-jacketed Xe flashlamps have about 10% loss of pumping efficiency, but the reliability of such lamps has dictated their use in the OMEGA system. The choice of rod amplifiers in the OMEGA system was influenced strongly by the low maintenance requirements of such devices. In considering the cost of a system, the designer must also enter into the final decision fixed cost and maintenance cost. For disk amplifiers, for example, it is necessary to assess the cost of an elaborate clean-room facility for their assembly and maintenance and, in a large system, the costs associated with personnel needed to keep the system operating near its full performance. Disk amplifiers, having a far larger number of surfaces than an equivalent rod or active-mirror system, as well as surfaces that are particularly sensitive to particulate accumulation and damage due to their direct exposure to flahlamp radiation, require more elaborate handling and assembly facilities.

Having reviewed the major design issues involved in arriving at a system configuration conforming to a given set of output parameters, we now, in Sect.8.2, show the methodology that is generally followed in the process.


## 8.2 A Design Methodology for High-Peak-Power Nd:Glass Laser Systems

How does one go about designing a high-peak power Nd:glass laser system? The method of course will vary from one facility or designer to another but, in general, the process consists of three distinct stages:

*Stage 1* First-Order Design
*Stage 2* Design Exploration
*Stage 3* Full-System Simulation

Methods now exist for cost optimizing a system design; the details will be discussed in Sect.8.2.1. Generally, the first Stage is a so-called first-order design in which by use of X-factor analysis and cost data for components, an initial cost-optimized design is obtained. This design is only preliminary, however, because the designer must then investigate the system by use of auxiliary codes to calculate, for example, the ASE levels expected to reach a target after making initial guesses about what isolation may be required. If the initial guess was wrong, the process must be repeated until a satisfactory arrangement is found. This second stage, referred to above as design exploration, would also include, for example, the prediction of system performance at other pulse durations. If the system is optimized for 50 ps pulses, it may be necessary, because of secondary performance requirements, to find out how the system behaves, for example, at 200 ps or 1 ns. The second stage will be covered in Sect.8.2.2. The final stage is very important because it will insure that the first-order design, when fine tuned according to the results of this stage, will perform as predicted. Up to this point, we have totally ignored the effects of diffraction or WBSF in the system. Such phenomena should be included, using sophisticated (and complicated) propagation codes to simulate system performance fully. If the effects of diffraction and WBSF are taken into account by the use of a full simulation code, then the system will perform closely to performance predictions generated in

the first and second design stages. In Sect.8.2.3 we explore the use of such
codes to effect the third stage of system design.

The methodology followed is not unique to Nd:glass laser systems, it may
be applied to the design of any high-peak-power laser system, ignoring or
including the several considerations according to their relevance to the system under investigation.

### 8.2.1 *Stage 1* First-Order Design

After a set of design parameters has been specified for a system, the designer
is faced with the task of assembling a system that will meet them in an optimal way. That is, using the known (or predicted) performance of amplifiers
and other system components, combined with the relevant cost data, he has to
stage the system to deliver a given performance at minimum cost. To do this,
he/she must first determine the cost formulas for the system components. From
the laser designer's point of view, it is the *scaling* of cost with certain
critical parameters that is important. Based on the scaling, he can arrive at
a first-order system configuration. The absolute cost data can be provided at
a later date, to compare the relative merits of different system configurations as well as to determine the system cost. The determination of cost formulas is complicated; it usually involves solicitation of quotations from
manufacturers for various values of the relevant parameters, the use of actual
purchase data for prior systems if any have been built, and educated guesses
at the expected scaling. One such set of cost equations is shown in Table 8.2;
they were used in the design of the NOVA system at LLL [8.5]. For disk amplifiers, for example, the critical parameters in Table 8.2 are the disk thickness, diameter, glass cost [dollars/cm$^3$] and energy-storage bank-related costs.
Important considerations are finishing cost, volume of glass, and costs of the
edge-cladding, which is used for suppression of PO. The efficiency of an amplifier at the determined operating point (storage-bank energy and disk thickness) is obviously an important consideration, because use of a more efficient
amplifier would translate directly into a smaller energy-storage bank and
therefore cost. Active-mirror amplifiers have costs associated with multi-
layer dielectric coatings applied to both faces, whereas the storage-bank
costs are usually less owing to the efficiency of active-mirror amplifiers,
which are higher than that of a disk amplifier. Other considerations in determining cost data are discounts granted for quantity purchases, contingency
funds, and escalation due to inflation.

Table 8.2. Laser chain cost formulae

| Component | Mechanical cost | Optical cost | Electrical cost |
|---|---|---|---|
| Disk amplifier | $\frac{7.8\ D + 0.53\ D^2}{60}\ N_D$ | $(0.00312\ t + 0.0032)A$ | $2.3\ N_C + 0.12\ E_B$ |
| Spatial filter (f/10) | $7 + 0.35\ \bar{D}$ | $0.5 + 0.005\ D^2$ | 0 |
| Polarizer | $1 + 0.004\ D^2$ | $0.02\ D^2$ | 0 |
| Faraday rotator (FR-5) | $6 + 0.007\ D^2$ | $0.03\ D^2\ t$ | $0.0032\ D^3/t^2$ |
| Focusing Optics<br>Nova I: (Single window & shield) | $0.9\ D$ | $0.032\ D^2 + 0.55\ D^3 \times 10^{-3}$ | 0 |
| Nova II: (Double window & shield) | $0.9\ D$ | $0.039\ D^2 + 0.65\ D^3 \times 10^{-3}$ | 0 |
| Preamplifier | 80 | 50 | 20 |

Cost in thousands of dollars in FY 76 at same prices as paid for Shiva materials.

Alignment, control, and diagnostics costs = $250 K total per chain

$D$ = clear aperture in cm; $N_D$ = number of disks

$t$ = thickness in cm

$A = \frac{\pi}{4}\ LD^*$ = surface area of disk, where $L = \sqrt{1+n^2}D + \frac{t}{n} + 0.6$ cm = major axis of disk, and $D^* = 1.04\ D + 0.6$ cm = minor axis of disk

$N_C$ = number of flashlamp circuits with two flashlamps per circuit

$E_B$ = energy storage bank in kJ

In Chap.5, we investigated characteristics of the major types of amplifiers now in common use in most Nd:glass laser systems. It was shown that predictive pumping models now exist for all disk and active-mirror amplifiers. Such models, when combined with the cost formulas in Table 8.2 for disk amplifiers and those for active-mirror amplifiers, allow first-order chain designs to be obtained. We begin the process by first examining the figure of merit $Z_S$ for a laser system. We have seen in Chap.7 (7.168) that the appropriate measure of performance in a short-pulse, B-limited laser system is the X factor, because the maximum focusable output power is proportional to X. The figure of merit $Z_S$ (also known as the cost effectiveness) is defined as the maximum output power $P_S$ per unit cost $\$_S$, or [8.6]

$$Z_S = \frac{P_S}{B_A\ \$_S}\ ,\qquad (8.1)$$

where the $B_A$ integral is that appropriate to one chain, for fixed $\$_S$ or fixed $P_S$. If the number of chains is N, and the output per chain is $P_A$, then

$$P_S = N\ P_A\ ,\qquad (8.2)$$

and

244

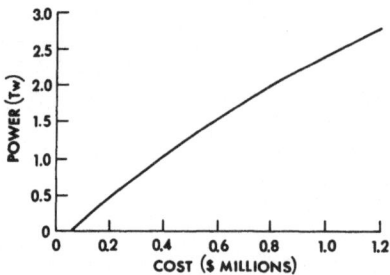

Fig. 8.1. Focusable power of a single Z-optimized laser chain as a function of cost [8.6]

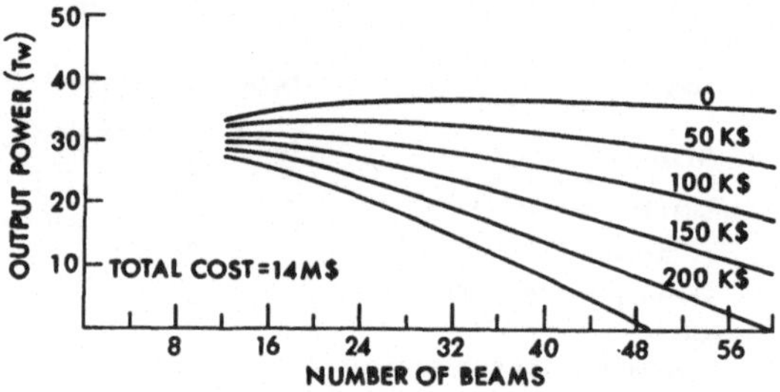

Fig. 8.2. Shiva output power as a function of number of beams (Z-optimization) [8.6]

$$\$_S = N \$_A \quad , \tag{8.3}$$

where $\$_A$ is the cost of an individual chain. Then, maximizing $Z_S$, given by

$$Z_S = \frac{P_A}{B_A \$_A} \quad , \tag{8.4}$$

implies maximizing the performance of individual chains. An optimized chain does not necessarily consist of individually optimized components; the way in which the designer arrives at such a chain will now be described. In Fig. 8.1 we show the performance (focusable power output) from a chain as a function of the chain cost, for a fixed B limit of 7. Each point on the curve represents a system chain that, for fixed total dollars, maximized $P_A/B_A\$_A$. To find the system performance as a function of the number of arms, we consider the total number of dollars $\$_T$ that are available for fixed system costs $\$_F$, such as building, target chamber, computer control, etc., fixed costs per arm $\$_{FA}$ independent of arm size (e.g. input mirrors), and the variable cost per arm, $\$$. Then $\$_T$ is

$$\$_T = \$_F + \$_{FA}N + \$N \quad . \tag{8.5}$$

Assuming fixed values for $\$_F$ and $\$_{FA}$ and varying the cost per arm according to Fig.8.1, we obtain Fig.8.2 which shows $P_S$ as a function of the number of arms N, for six values of $\$_{FA}$. In Fig.8.2 the total system cost was assumed to be 17 million and a fixed system cost of 3 million. As N increases, the performance decreases because the fixed costs per chain become an ever larger portion of the total dollars available. Although such performance curves are useful and will become very important when large laser-fusion demonstration

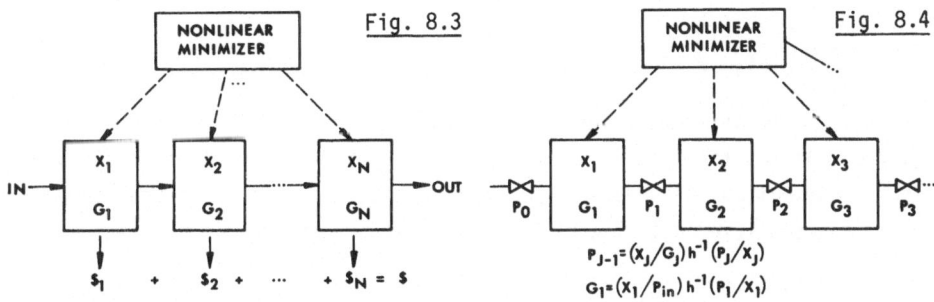

Fig. 8.3. The chain to be optimized is divided into N stages, each specified by the P/B ratio X and its gain G. The costs of the lowest priced amplifiers meeting these specifications are found, and thus the total cost. The values of X and G are varied to find the chain of lowest total cost. Overall chain X and G are maintained during the process [8.6]

Fig. 8.4. Optimal design of a spatially filtered laser chain: The values of X and G (that yield a given $P_{out}$ and $P_{in}$) are apportioned among groups so as to minimize the dollar cost of the laser. Each group in turn must be optimally designed to yield its X and G at the lowest possible cost [8.6]

facilities are constructed, in fact the number of chains N is not normally determined by such considerations but rather by a desire to achieve uniform illumination on targets. Nevertheless, given a fixed N, it is of obvious interest to build Z-optimized chains under the constraints of either fixed-system output or fixed dollars available.

How do we design a Z-optimized chain? For simplicity, we first consider a chain without spatial filters, as illustrated in Fig.8.3. We first specify the input and output power, and thus the gain and overall X factor for the chain, given a total B-integral constraint. We assign values of the $X_n$ factor and gain $G_n$ to the $n^{th}$ component in such a way that the total X and G remain constant, because they determine the performance of the chain as a whole. Using cost formulas $\$_s$ similar to those in Table 8.2, we then calculate the total cost $\$$ for the chain and thus obtain the overall chain Z. By varying the gain and X factors of the individual stages using well-known nonlinear minimization methods, we find the distribution of $X_n$ and $G_n$ that minimizes the total system cost. The optimization of individual components in the chain will be covered later; another variable that must be considered and optimized is the number of amplifiers n. Because by necessity, we must have spatial filters for imaging the small-scale-focusing suppression in real systems, we consider the optimization of such chains here and illustrate them in Fig.8.4. Recall from Chap.7 that the output power of a $j^{th}$ stage is given by (7.170)

$$P_j = X_j h \left( \frac{G_j P_{j-1}}{X_j} \right) \quad , \tag{8.6}$$

where h is a function that describes the spatial-filter transmittance. We again divide the total X and G among the groups and attempt to find the optimum distribution. We choose the number of stages n, and by fixing the output at the specified value, work our way back through the chain, using

$$P_{j-1} = \left( \frac{X_j}{G_j} \right) h^{-1} (P_j/X_j) \quad . \tag{8.7}$$

The gain of the first group is not varied but is found by use of

$$G_1 = (X_1/P_{in}) h^{-1} (P_1/X_1) \quad , \tag{8.8}$$

where $P_{in}$ is the specified input power. The values of $X_i$ and $G_i$ are varied, again by use of a nonlinear minimization program, until the system is found that produces the maximum Z, or minimum cost. The designer must also vary the total number n of stages to find the optimum system. This procedure, used to design the SHIVA system at LLL, a short-pulse-laser system, has been discussed in detail [8.6]. The procedure just described may have to be modified if additional constraints are placed on the system, for example, if we specify that the accumulated B integral in a stage $\Delta B_j$ must satisfy the criterion $\Delta B_j \lesssim \Delta B_m$, where $\Delta B_m$ is the maximum permissible B limit allowable for near-zero pulse distortion.

The next problem to be solved is the design of individual amplifiers with specified $X_j$ and $G_j$, so as to minimize cost. An optimum amplifier will obviously consist of optimum disks; therefore we must develop procedures for arriving at an optimum disk. In Chap.7, we showed in (7.174) the X factor for a disk amplifier, which we consider here. It is an easy matter to show that the X factor for m disks in series is given by

$$X_m = X_\infty \left( 1 - \frac{1}{G_m} \right)^{-1} \quad , \tag{8.9}$$

where $X_\infty$, the ultimate performance factor defined in Chap.7, gives the performance of m disks independently of how many there are, and $G_m$ is

$$G_m = (G_0)^n \quad , \tag{8.10}$$

where $G_0$ is the small signal gain of an individual disk and n the number of disks in the $j^{th}$ stage. If we are interested in cost-effective designs, the appropriate parameter in such a case is the so-called gain-specific cost S, given by [8.7]

$$S = \frac{\$}{\ln(G_0)} \quad .$$
(8.11)

If the designer doubles the number of disks, clearly S will be unchanged. Optimum amplifiers will then maximize $X_\infty$, because $X_m$ is also maximized and leads to the best performance. Minimizing S for a fixed $G_0$ leads to the lowest cost. Thus, in optimizing amplifiers for a given stage, the X factor and gain-specific cost are the important parameters. In Chap.5, we saw how by using an amplifier-simulation program, it was possible to predict amplifier performance accurately. For a fixed disk diameter, thickness and storage-bank energy, we also saw that the designer could find the optimum glass $Nd_2O_3$ doping and flashlamp-drive pulse width to achieve maximum stored-energy density or gain, as in Fig.5.4. It is, of course, true that because the factor $X_\infty$ is proportional to the gain coefficient, $X_\infty$ will be maximized there also. We also saw in Fig.5.5, how the designer can arrive at amplifier-gain coefficient contours for which the storage-bank energy and disk thickness are the variables, and for which each point on the plot has been optimized for doping and flashlamp-drive pulse width. We can also find contours of the so-called short-pulse figure of merit $M_S$ formally given by

$$M_S = \frac{X_\infty \ln(G)}{\$} = \frac{X_\infty}{S}$$
(8.12)

and illustrated in Fig.8.5. We can see that for an amplifier of diameter D there is a disk thickness and storage-bank energy that optimizes the figure of merit $M_S$. Alternatively, for any disk diameter the designer may find a disk thickness and storage-bank energy that maximizes the performance while minimizing gain-specific cost. Returning to Fig.5.5, we may assign to each point on the contours a performance $X_\infty$, and using cost formulas for flashlamps and circuits, hardware, disk costs, etc., find the appropriate S value. The designer can then obtain a set of curves, such as those shown in Fig.8.6, that show S contours as functions of storage-bank energy and performance factor. Each set of such contour reaches a maximum value for a set of four parameters (storage-bank energy, disk thickness, wt-% doping, flashlamp-drive pulse width) and represents the best performance that can be obtained for a given S. We can, from Fig.8.6, determine the lowest specific-cost storage-bank

248

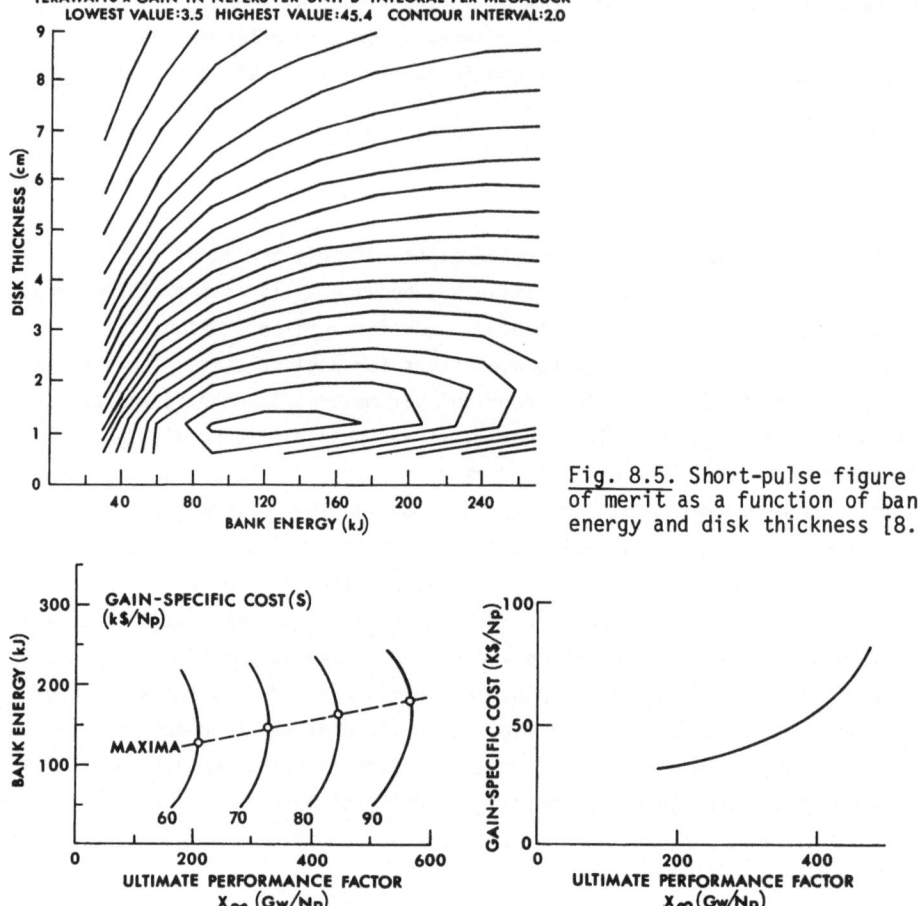

**TERAWATTS x GAIN IN NEPERS PER UNIT B-INTEGRAL PER MEGABUCK**
**LOWEST VALUE:3.5  HIGHEST VALUE:45.4  CONTOUR INTERVAL:2.0**

DISK THICKNESS (cm)

BANK ENERGY (kJ)

Fig. 8.5. Short-pulse figure of merit as a function of bank energy and disk thickness [8.5]

GAIN-SPECIFIC COST(S) (k$/Np)

BANK ENERGY (kJ)

MAXIMA

ULTIMATE PERFORMANCE FACTOR
$X_\infty$ (Gw/Np)

GAIN-SPECIFIC COST (k$/Np)

ULTIMATE PERFORMANCE FACTOR
$X_\infty$ (Gw/Np)

Fig. 8.6. Curves of gain-specific cost(s) as a function of ultimate performance factor and bank energy. The most-cost-effective amplifier is the one giving the largest performance factor at any fixed S. These optimal values are indicated by dots on the curves [8.6]

Fig. 8.7. Minimum cost required to produce a desired performance for a given amplifier diameter. As the performance increases, the disks become thinner and the band larger [8.6]

for any $X_\infty$ to obtain Fig.8.7, which shows the minimum cost to produce a specified performance with a fixed diameter. For higher performance, the designer is forced to thinner disks and larger storage-bank energy. If the outlined process is repeated for a number of disk diameters, a series of curves, such as those shown in Fig.8.8, is obtained. Therefore, for a given performance

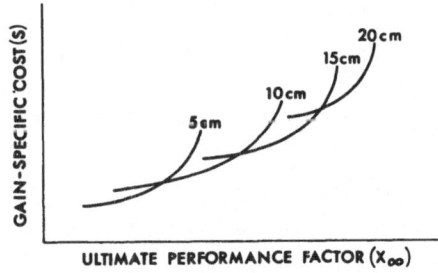

Fig. 8.8. A set of curves like those in Fig.8.7, but for different beam diameters, allowing the designer to choose the best diameter for any performance level [8.6]

level, we can choose an amplifier that has the lowest specific cost at the performance level in which we are interested.

The process just described, which results in Z-optimized amplifiers of overall performance $X_m$ given by (8.9) and total gain $G_m$, includes the creation of extensive files that describe the expected performance of amplifiers of various diameters, and should be repeated for each available glass type. The overall design procedure is to use the nonlinear minimization process described previously to arrive at a given array of $X_m$ and $G_m$. The best amplifier to achieve the desired performance $X_{m\infty}$ is then found from a chart, such as that in Fig.8.8, and yields the desired diameter, thickness, storage-bank energy, disk doping and pump pulse width. Knowing the gain $G_0$ of one disk, we then find the number of disks $N_m$ in the stage, from

$$N_m = \frac{\ln(G_m)}{\ln(G_0)} \quad , \tag{8.13}$$

and the cost of the $m^{th}$ stage $\$_m$ is given by

$$\$_m = S_m \ln(G_m) \quad . \tag{8.14}$$

The designer then determines the total system cost and arrives at the cost effectiveness of the chain. The process is repeated until the minimum chain cost for a given overall chain performance has been achieved. The designer must also vary the number of stages in the chain to find the absolute minimum.

The first-order design procedure described has a number of shortcomings, including the exclusion of important elements in a laser chain, such as spatial-filter lenses and isolators. After an initial configuration has been arrived at, the designer must include such elements and iterate the design until a given performance is achieved. Because the costs of isolators and spatial-filter lenses are normally small compared to amplifier costs, the initial guess at system cost will not be too far from the final value. Inclusion of other elements does, however, degrade performance owing to their finite contribution to the B integral. The designer may in a number of ways include in the initial calculations their expected effects in degrading system performance, for example, by overdesigning the system, degrading the performance by a fixed amount proportional to ln(G), etc.

After a cost-effective design has been arrived at by the process described, the designer should investigate its performance in detail by using a system-simulation code such as RAINBOW described in Chap.7, including all elements. Up until now, the designer has, for example, ignored isolation, making only educated guesses at placement of isolators and beam-transport optics to the target chamber. At this point, the designer begins a somewhat more sophisticated stage of laser design, as described in the next section.

### 8.2.2 Design Exploration

By using amplifier-pumping models and cost formulas combined with the powerful X-factor analysis approach, we may, through the process described in Sect.8.2.1 arrive at a system configuration that we expect will come close to the specified performance levels. We must now include isolation for ASE and back reflection, spatial-filter lenses, and any other elements needed to achieve the expected system-performance levels. The suppression of ASE in a laser system requires the addition of Pockels cells, devices that normally introduce large losses due to linear absorption or birefringence, as well as large B-integral contributions due to the type and length of material used (KD*P) and associated polarizers. We must also propagate linear polarization in such devices; thus, we cannot take advantage of a reduced nonlinear index due to circular polarization. The placement of Pockels cells is determined by the use of an auxiliary ASE code, such as those described briefly in Chap.4. Knowing the stored-energy density in amplifiers, the spectral distribution around 1.06 $\mu$m, the angular acceptance angle of spatial filters, the placement of Pockels cells, and the time history of the inversion density in each amplifier, the designer can accurately determine the amount of ASE that will reach the target before the arrival of the main laser pulse. It is then a fairly simple matter to determine the number and placement of Pockels cells or other devices in a system to limit ASE to an acceptable level. If initial guesses were included for Pockels-cell placement in the original first-order design, the designer should modify the system to include the new information. In certain cases, Pockels-cell deployment may so limit the performance of the system from B-integral considerations as to warrant the inclusion in the system of

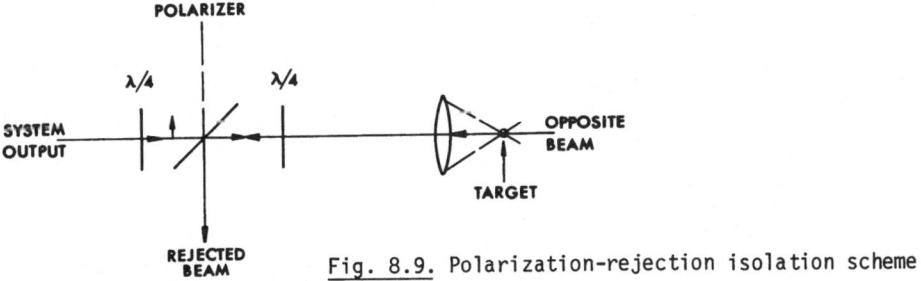

**Fig. 8.9.** Polarization-rejection isolation scheme

an entirely new, additional stage that may be used for isolation only. An
advantage of that scheme is that it may also be used for frequency conversion,
if such is desired. Frequency-doubling or -tripling crystals act as natural
isolators against the effects of ASE. We must also insure that back reflections or opposite beams in a target illumination scheme are properly suppressed by the use of a Faraday rotator, liquid-crystal isolator, or a polar-
ization-sensitive device. Because Faraday rotators, heretofore used almost
exclusively for this application, are costly in large apertures and may contribute a large B integral to a stage, it has been the practice to seek other
alternative means, one of which is illustrated in Fig.8.9. The system output,
if it is linearly polarized, is passed through an output polarizer and converted by the use of a quarter-waveplate to circular polarization. If the
beam is totally reflected from the target, circularly polarized light is
changed from one handedness to the other, converted back to linear polarization that is rotated to 90° with respect to the polarization of the initial
beam, and rejected at the polarizer. Opposite beams should be arranged so
that they have reverse handedness and will also be rejected. Although, for
opposite beams, the scheme has the amount of contrast required to isolate the
system against damage, in the case of real targets some depolarization of the
reflected beam occurs, resulting in significant leakage through the polarizer.
Accurate measurements of the depolarization ratio are lacking, at present.
Another means of blocking back reflections is the use of the plasma-closed
spatial filter in which some portion of the forward beam is used to create a
plasma in the vicinity of the pinhole. This scheme has been investigated recently [8.8] and results in zero B-integral contribution to the system and
scalability to large sizes. Regardless of what particular scheme is used, it
is easy to perform an analysis of what isolation is required at various locations in the system, to suppress back reflections. A simple analysis that as-
sumes the worst case (small-signal gain conditions) may be used, or a more-
sophisticated model, such as RAINBOW, to obtain the answers.

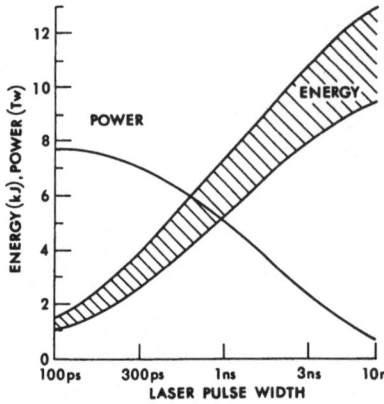

<u>Fig. 8.10.</u> Performance of a NOVA chain as a function of pulse width. The shaded areas extend from a conservative lower limit to the region that may be achieved with expected or demonstrated improvements [8.7]

After the initial design configuration has been obtained including isolation and all other components, it is necessary to verify that the system will perform as expected. This is normally accomplished by the use of a full simulation code, such as RAINBOW, to propagate pulses at the design pulse width through the system and verify that the expected power output will be achieved. If it is found that the performance deviates significantly from that expected, the designer must modify the initial configuration, changing the first-order staging until it is achieved. Usually, after a system configuration has been iterated until the desired system performance at a specified pulse width is reached, it is also of interest to see what performance may be achieved at other pulse widths. Therefore, a code such as RAINBOW, which has limits built into it based on damage at surfaces or coatings as functions of pulse width and which correctly treats saturation, is employed to calculate the expected performance. For the NOVA laser system at LLL, an early design yielded the data shown in Fig.8.10, which shows expected beamline power or beamline energy as a function of pulse width. The system was staged for a pulse width of 100 ps and would yield close to 8TW/beamline; that power drops precipituously beyond 200 ps to about 2TW/beamline at 3 ns. For short pulses ($\lesssim 200$ ps) little energy is extracted from the system, but as the pulse width increases so does the energy output which is limited in the 1-3 ns regime by damage to coatings, and beyond that by saturation. This kind of curve can be obtained for each cost-optimized system/glass type used to find the glass that yields the desired performance limit at a given pulse width while also maximizing performance at other pulse widths of interest.

The techniques described in this and Sect.8.2.1 are the general rules that must be followed to arrive at a cost-optimized system configuration that will deliver a given system performance. It has been found that simulation codes, such as RAINBOW, will predict the performance of a system to within acceptable accuracy (10-15%); that accuracy has been verified in numerous short-pulse laser systems at LLL, LLE and other facilities. It should be clear, however, that such predictions are accurate only in the sense that a ray-trace code such as RAINBOW, is suitable; in other words, we must insure that diffraction effects, whole-beam effects, the choice of appropriate pinhole diameters, etc., are such as to minimize their effect on system performance. We will investigate this in Sect.8.2.3. At this point, after a final system configuration is determined that seems to meet all of the desired performance criteria, it is necessary to produce a detailed system layout that will image the initial hard or soft aperture of the system to the location of the target chamber. The importance of imaging was discussed in Chap.7; imaging must now be considered because the spacing of components as well as their nonlinear properties determines the diffraction behavior of a system, a subject to which we now turn in Sect.8.2.3.

## 8.2.3  Full-System Simulation

The design and analysis of laser systems so far presented has one major short-coming; although it accounts in a phenomenological way for SSSF effects, it totally ignores the phenomena of linear diffraction and nonlinear phase effects associated with the beam as a whole. Linear diffraction is important because of hard apertures located in the system and the induction of ripples, which, if uncontrolled, propagate through a system and result in catastrophic damage. Hard apertures are elements such as amplifiers, lenses, etc., that have finite (sharp) edges. The designer normally wants to propagate a smooth beam that fills as much of an amplifier aperture as possible. The associated fill factor F is important because the output of an amplifier is directly proportional to it. This is particularly important in the final stage, where the ultimate power of the system is determined. The desire to achieve a large fill factor is complicated, however, by alignment difficulties if the beam is too close to the edge of an amplifier, and the necessity to keep the beam intensity at the edge very low, typically down by $10^{-3}$ of the center intensity, so that linear diffraction is minimized. If that condition is realized, modulation is virtually undetectable. To arrive at a high fill factor in the final stage, it would be desirable to propagate a high-order super-Gaussian through it, because such a profile has associated with it a high fill factor and can also be made to decrease very quickly at the edges. To minimize diffraction

and WBSF effects, the system is designed to image an initial aperture through
the entire system to the target chamber, as discussed in Chap.7. The type of
aperture desired may be determined by backward propagation through a system
or by choosing an apodized aperture that, when convoluted with the Gaussian
spatial profile of the oscillator, results in a large-fill-factor profile of
the output of the final stage. In between, we must contend with changes of
the beam profile due to diffraction, WBSF and SSSF effects, saturation, and
possibly radial-gain profiles of amplifiers. In fact, in practice, it has
proven very difficult to synthesize the desired apodized apertures and to
contend with damage problems associated with their application. The result
is that most major systems now use hard apertures at the system's input; the
resulting modulation, due to the truncation of the Gaussian input, is then
propagated through the entire system but has been found to result in acceptable modulation and fill factor at the output. The way in which an initial
Gaussian is converted to a high fill factor in the final stage may be illustrated as in Fig.8.11 where the beam profile at various locations in the GDL
system (see Fig.7.19) at LLE is shown after each amplifier. The radial gain
found in rod amplifiers has the effect of raising the edges of the beam profile because radial locations near the edge show higher gain. The effect was
simulated using a linear diffraction code (BEAMPROP), which ignores the effects of whole-beam self-focusing but includes saturation and radial gain.
The designer must, by varying the size of the initial hard aperture or apodized aperture and the Gaussian full width at half maximum of the oscillator
pulse and using fixed or variable spatial-filter magnifications, arrive at a
configuration that results in a high fill factor at the system output. To
achieve this, the designer must also insure that the intensity at the amplifier edges is very low, even in the presence of large radial gain variations.

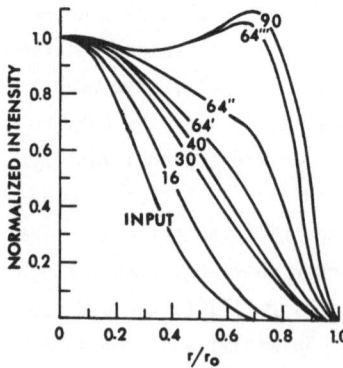

Fig. 8.11. Normalized intensity as a function
of normalized radius showing the change in
beamshape as it propagates through amplifiers
of various diameter (16, 30, 40, 64', 64",
64''', 90φ) in the GDL system at LLE

To account properly for all of the effects known to affect beam propagation
in a high-power Nd:glass laser system, it is necessary to use a full simulation code that includes diffraction, saturation effects, radial gain, WBSF
and SSSF effects. Although a number of such codes are presently available,
such as FLAC discussed briefly in Chap.7, the only one that includes most of
the important effects and which has been successfully tested on a number of
systems at LLL is MALAPROP, written and implemented by GLASS and WARREN [8.9].
Because of its significance, we discuss the application of this code here.
MALAPROP enables the study of beam propagation in a laser system in the presence of surface noise, which is represented as stochastic perturbations of
amplitude and phase. It has been written to be available in either one- or
two-dimensional forms. Normally, owing to the large number of grid points
needed in a two-dimensional (x-y) calculation, the code is run primarily in
the one-dimensional (r) axially symmetric form. Propagation is treated by the
usual paraxial equation encountered previously in Chap.7 (7.9). Although in
MALAPROP only nonlinear phase distortion due to the medium nonlinear index $n_2$
is treated, it is an easy matter to include other whole-beam phase distortions,
such as those due to thermal lensing in rod amplifiers (see Chap.2). The code
distinguishes between propagation in nonlinear media or in free space. For
free-space propagation, it can be shown that propagation may be accomplished
by the use of fast-Fourier-transform (FFT) techniques. If nonlinear media are
encountered, then (7.9) is integrated using finite-difference techniques.
MALAPROP will treat nonlinear media such as amplifiers, Pockels cells, Faraday
rotators, apertures, spatial filters, beam expanders, beam splitters and
free-space propagation. A feature of MALAPROP is that it does not require a
temporal history of the pulse, but integrates (7.9) at a particular position
in the incident pulse. It is then not necessary to keep track of the local
stored-energy density at each axial and radial position in an amplifier, free-
dom from which necessity substantially reduces the storage (core) requirements
for execution on a computer, and thus the cost of running the code. The details of how saturation is treated, as well as how individual elements are
handled, can be found in the description by GLASS and WARREN [8.9]. A unique
feature of MALAPROP is the simulation of noise, which is assumed to occur
only on surfaces and to arise from dust particles, inhomogeneities, scratches,
or other sources. The noise added at any surface is specified by a mean-square
amplitude $\epsilon$ for the fluctuation and a correlation length $\ell$. It is further assumed that the fluctuation may be modeled as a Gaussian random process in the
frequency domain, which reproduces the desired power spectrum; the exponential
distribution F, given by

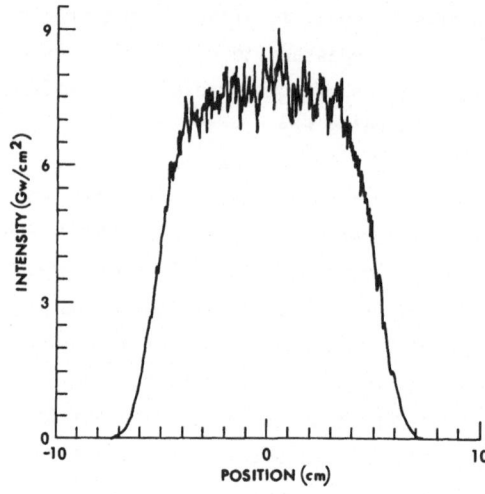

Fig. 8.12. Amplifier output intensity as a function of radial
position, with 2.5% amplitude
noise added at entrance surface,
generated using MALAPROP [8.9]

$$F(x_1 - x_2) = \epsilon^2 \exp[|x_1 - x_2|/\ell] \quad , \qquad\qquad (8.15)$$

was chosen from a number of possible candidates; for $\simeq$1-2% noise/surface it
was found to give excellent agreement with experiment. No measurements of
surface noise have been performed to date; thus a one percent noise level and
the assumed exponential distribution are, at best, guesses. Nevertheless, the
excellent agreement with experiment and qualitative reproduction of features
found at the output of real laser systems gives us confidence in the use of
such a code and its predictive capabilities. Typical output from MALAPROP is
shown in Fig.8.12, which shows intensity at the output of an amplifier as a
function of radial distance; a $\simeq$5% modulation may be observed from the assumed 2.5% noise on the input surface of the amplifier. The use of MALAPROP
in showing the advantages of spatial filtering over long-path (near-field)
filtering has also been described [8.9].

A code such as MALAPROP gives the laser designer a powerful tool for investigating the details of all the important effects presently known in high-
power Nd:glass laser systems. After the initial design procedure described
above, it is a necessity to explore the fine details by using MALAPROP or
another appropriate code.

We have seen in Chap.7 two critical parameters that must be adjusted to
achieve maximum system performance. The first is the diameter of the pinhole
in all spatial filters. Although we have discussed some methods by which initial guesses may be generated, in practice, it is very difficult to find the
optimum diameter that results in effective filtering for the most damaging

ripples but which also minimizes band-pass modulation on the beam. The second effect is WBSF zooming in spatial filters. Again, although approximate methods exist for estimating the focal zoom and corresponding pinhole displacement for maximum pinhole transmittance, it is difficult to determine whether the optimum has been found. Although an experimental verification is often possible, it is extremely time-consuming and expensive on a large system and requires optimization at each power-output level. MALAPROP, in its present or modified version, may be used to determine both parameters. Although considerable computer time is required to arrive at an optimum solution, it is usually far less expensive than finding one experimentally. Another important phenomenon that may be studied is the intensity loading experienced by a given pinhole; the designer may then design the spatial filter so that the peak intensity needed for plasma breakdown is not exceeded. When a plasma has been formed, it may distort a pulse that passes through the pinhole; this phenomenon is particularly important for long pulses.

At this stage in the evaluation of Nd:glass laser systems, fairly complete methods exist for design and performance predictions. Most laboratories that have large laser systems rely upon a combination of experience and modeling to arrive at final configurations. The development of in-house modeling and computational capability is a time-consuming and often expensive process. The investment is usually justified, however, when it is realized that such methods may, from minimum experimental measurements, predict the output of a system usually to within 10-15%. Furthermore, later changes in system configuration, glass type, etc., are easily and cost-effectively evaluated without actually having to build the system.

# Acronyms

| | | | |
|---|---|---|---|
| ASE | Amplified spontaneous emission | LLE | Laboratory for Laser Energetics |
| E-D | Electric dipole | LLL | Lawrence Livermore Laboratory |
| FLN | Fluorescence line narrowing | PFN | Pulse-forming network |
| FOM | Figure of merit | PO | Parasitic oscillation |
| GDL | Glass-Development Laser | SF | Self-Focusing |
| GLAMP | Glass laser and materials program | SSSF | Small-scale self-focusing |
| | | TIR | Total internal reflection |
| GW | Gigawatt | TW | Terawatt |
| J-O | Judd Ofelt | WBSF | Whole-beam self-focusing |

# References

*Chapter 1*

1.1  G.H. Dieke, H.M. Crosswhite: Appl. Opt. *2*, 675 (1963)
1.2  D.K. Duston: Ph.D. Thesis, Rensselaer Polytechnic Institute, May (1969)
1.3  A.A. Mak, D.S. Prilezhaev, V.A. Serehryakov, A.J. Starikov: Opt. Spectrosc. (USSR) *33*, 381 (1972)
1.4  Yu.P. Rudnitskii, R.V. Smirnov, V.M. Chernyak: Sov. J. Quantum Electron. *6*, 1107 (1976)
1.5  R. Dumanchin, J.C. Farcy, M. Michon, P. Vincent: IEEE J. QE-*7*, 53 (1971)
1.6  W.E. Martin, D. Milam: Appl. Phys. Lett. *32*, 816 (1978)
1.7  Spectra generated by S. Jacobs, J. Rinefierd: Laboratory for Laser Energetics, University of Rochester
1.8  W. Koechner: *Solid-State Laser Engineering*, Springer Series in Optical Sciences, Vol.1 (Springer, Berlin, Heidelberg, New York 1976)
1.9  Lawrence Livermore Laboratory: Laser Program Semi-Annual Report, July-December 1973, UCRL-50021-73-2 (1973)
1.10 Lawrence Livermore Laboratory: Laser Program Annual Report, UCRL-5-21-74 (1974)
1.11 C.B. Layne: Ph.D. Thesis, University of California, Lawrence Livermore Laboratory, UCRL-51862 (1975)
1.12 T. Forster: Z. Naturforsch. *4a*, 321 (1949)
1.13 I.A. Bondar', B.I. Denker, A.I. Domanskii, T.G. Mamedov, L.P. Mezentseva, V.V. Osiko, I.A. Shcherbakov: Sov. J. Quantum Electron. *7*, 167 (1977)
1.14 C. Brecher, L.A. Riseberg, M.J. Weber: Appl. Phys. Lett. *30*, 475 (1977)
1.15 C. Brecher, L.A. Riseberg, M.J. Weber: Preprint UCRL-81181, Lawrence Livermore Laboratory (May 1978)
1.16 E.E. Fill, K.G.V. Finckenstein: IEEE J. QE-*8*, 24 (1972)
1.17 O. Lewis, W. Seka: unpublished data, Laboratory for Laser Energetics, University of Rochester (1978)
1.18 D. Milam, W. Martin: private communication, Lawrence Livermore Laboratory
1.19 W. Martin, D. Milam: private communication, Lawrence Livermore Laboratory
1.20 M. Weber: private communication, Lawrence Livermore Laboratory
1.21 R. Reisfeld, C.K. Jørgensen: *Lasers and Excited States of Rare Earths* (Springer, Berlin, Heidelberg, New York 1977)
1.22 L.A. Riseberg, M.J. Weber: *Progress in Optics*, Vol.XIV (North-Holland, Amsterdam 1976)
1.23 B.R. Judd: Phys. Rev. *127*, 750 (1962)
1.24 G.S. Ofelt: J. Chem. Phys. *37*, 511 (1962)
1.25 W.F. Krupke: IEEE J. QE-*10*, 450 (1974)
1.26 W.T. Carnall, P.R. Fields, K. Rajnak: J. Chem. Phys. *49*, 4424 (1968)
1.27 R.R. Jacobs, M.J. Weber: IEEE J. QE-*11*, 846 (1975)
1.28 R.R. Jacobs, M.J. Weber: IEEE J. QE-*12*, 102 (1976)
1.29 P.H. Sarkies, J.N. Sandoe, S. Parke: Brit. J. Appl. Phys. *4*, 1642 (1971)
1.30 C. Hirayama, D.W. Lewis: Phys. Chem. Glasses *5*, 44 (1964)

1.31 S. Stokowski, R. Saroyan, M.J. Weber: "Nd:Laser Glasses-Data Sheets", Lawrence Livermore Laboratory (1977)
1.32 O.K.E. Deutschbein: IEEE J. QE-*12*, 551 (1976)
1.33 P.H. Sarkies: Ph.D. Dissertation, University of Sheffield (1971)
1.34 G.H. Dieke: In *Paramagnetic Resonance*, Vol.1, ed. by W. Low (Academic Press, New York 1963) p.237
1.35 L.A. Riseberg, H.W. Moos: Phys. Rev. *174*, 429 (1968)
1.36 C.B. Layne, W.H. Lowdermilk, M.J. Weber: Phys. Rev. B *16*, 10 (1977)
1.37 A. Kiel: Ph.D. Thesis, Johns Hopkins University, Baltimore (1962)
1.38 C.B. Layne, W.H. Lowdermilk, M.J. Weber: IEEE J. QE-*11*, 798 (1975)
1.39 R. Risefeld: Struct. Bonding (Berlin) *22*, 129 (1975)
1.40 R.W. Hellwarth: Prog. Quantum Electron., Vol.5 (Pergamon Press, Oxford, New York 1977)
1.41 O. Svelto: *Progress in Optics*, Vol.XII (North-Holland, Amsterdam 1974)
1.42 W. Seka, J. Soures, O. Lewis, J. Bunkenburg, D. Brown, S. Jacobs, G. Mourou, J. Zimmerman: Appl. Opt. *19*, 409 (1980)
1.43 U.L. Boling, A.J. Glass, A. Owyoung: IEEE J. QE-*14*, 601 (1978)
1.44 A. Owyoung: IEEE J. QE-*9*, 1064 (1973)
1.45 Lawrence Livermore Laboratory: Laser Program Annual Report, UCRL 50021-75 (1975)
1.46 D. Milam, M.J. Weber: J. Appl. Phys. *47*, 2497 (1976)
1.47 D. Milam, M.J. Weber, A. Glass: Appl. Phys. Lett. *27*, 822 (1977)
1.48 M.J. Weber, D. Milam, W.L. Smith: Opt. Eng. *17*, 463 (1978)
1.49 M.J. Weber, C.F. Cline, W.L. Smith, D. Milam, D. Heiman, R.W. Hellwarth: Preprint UCRL-80434, Lawrence Livermore Laboratory (1977)
1.50 S.D. Jacobs: unpublished results, Laboratory for Laser Energetics (1978)
1.51 M.J. Moran, C-Y. She, R.L. Carman: IEEE J. QE-*11*, 259 (1975)
1.52 E.S. Bliss, D.R. Speck, W.W. Simmons: Appl. Phys. Lett. *25*, 728 (1974)
1.53 D.C. Brown: unpublished results, Laboratory for Laser Energetics (1976)
1.54 J. Rinefierd, S.D. Jacobs, D.C. Brown, J. Abate, O. Lewis, H. Appelbaum: Paper presented at Laser Induced Damage in Optical Materials, Boulder, CO 1978
1.55 D.C. Brown, J.M. Rinefierd, S.D. Jacobs, J.A. Abate: Laser Induced Damage in Optical Materials, Boulder, CO 1979
1.56 R.H. Lehmberg, J. Reintjes, R.C. Eckardt: Phys. Rev. A *13*, 1095 (1976)
1.57 M.J. Weber: private communication, Lawrence Livermore Laboratory
1.58 M. Sparks, C.J. Duthler: Xonics, Inc., Fifth Techn. Rpt, Contract PAHC-15-73-C-0127, ARPA (June 1975)


*Chapter 2*

2.1  J.B. Trenholme: Laser Program Semi-Annual Report, UCRL-50021-73-1, Lawrence Livermore Laboratory (1973)
2.2  L.G. De Shazer, L.G. Komai: J. Opt. Soc. Am. *55*, 940 (1965)
2.3  J.M. Soures, L.M. Goldman, M.J. Lubin: Appl. Opt. *12*, 927 (1973)
2.4  D.C. Brown, S.D. Jacobs, N. Nee: Appl. Opt. *17*, 211 (1978)
2.5  J.F. Holtzrichter, T.R. Donich: UCID-16860 (Nov.1973)
2.6  J.A. Abate, D.C. Brown, C. Cromer, S.D. Jacobs, J. Kelly, J. Rinefierd: Proc. of the Boulder Conf. on Laser Induced Damage in Optical Materials, NBS Special Publications 509 (1977)
2.7  J.M. McMahon: private communication and paper V5 presented at 10th Int. Quantum Electronics Conf., Amsterdam (1976)
2.8  S. Stokowski, R. Saroyan, M.J. Weber: "Nd:Laser Glass-Data Sheets", Lawrence Livermore Laboratory (1977)
2.9  W.F. Krupke: IEEE J. QE-*10*, 450 (1974)
2.10 D.C. Brown, S.D. Jacobs, J.A. Abate, O. Lewis, J. Rinefierd: Proc. of the Boulder Conf. on Laser Induced Damage in Optical Materials, NBS Special Publication 509 (1977)

2.11 S.D. Jacobs, J. Rinefierd: *Properties of Laser Glasses*, private communication
2.12 D.C. Brown: unpublished results presented at ERDA review of Laboratory for Laser Energetics Beta Laser System (Jan.1977)
2.13 CP&D Interim Report, Shiva Nova, Mis.107, Lawrence Livermore Laboratory (1977)
2.14 Laser Program Annual Report, UCRL-50021-74, Lawrence Livermore Laboratory (1974)
2.15 S.D. Jacobs: paper presented at 10th Annual Electro-Optics:Laser Conf., Boston, MA (1978)
2.16 S.D. Jacobs: private communication
2.17 C.B. Layne: Ph.D. Thesis, University of California, Lawrence Livermore Laboratory, UCRL-51862 (1975)


*Chapter 3*

3.1 J.H. Goncz, P.B. Newell: J. Opt. Soc. Am. *56*, 87 (1966)
3.2 American Institute of Physics Handbook, 3rd. ed. (McGraw-Hill, New York 1972)
3.3 J.L. Emmett, A.L. Schawlow, E.H. Weinberg: J. Appl. Phys. *35*, 2601 (1964)
3.4 Laser Program Annual Report, Lawrence Livermore Laboratory, UCRL-50021-77 (1977)
3.5 W. Finkelnburg: J. Opt. Soc. Am. *39*, 185 (1949)
3.6 J.F. Holzrichter, T.R. Donich: UCID-16860, Lawrence Livermore Laboratory (Nov.1973)
3.7 C.H. Church: Arc Discharge Sources, Final Report, ARPA Contract 4647(00) (March 1967)
3.8 J.H. Trenholme, J.L. Emmett: Proc. 9th Intern. Congress on High Speed Photography, ed. by W.G. Hyzer and W.G. Chace, SMPIE, 299 (1970)
3.9 J.H. Kelly, D.C. Brown, K. Teegarden: Appl. Opt. *19*, 3817 (1980)
3.10 J.H. Goncz: J. Appl. Phys. *36*, 742 (1965)
3.11 P.B. Newell: EG&G Technical Report B-4420 (Jan.1976)
3.12 L. Noble, C.B. Kretschmer: Tri-Annual Report #1, ECOM-0239-1, Contract DAAB07-71-C-0239 (March 1972)
3.13 J.P. Markiewicz, J.L. Emmett: IEEE J. QE-*2*, 707 (1966)
3.14 D.C. Brown, N. Nee: IEEE Trans. ED-*24*, 1285 (1977)
3.15 J.F. Holzrichter, J.L. Emmett: Appl. Opt. *8*, 1459 (1969)
3.16 R.H. Dishington: SPIE Vol.69, Laser Systems, 135 (1975)
3.17 J.B. Trenholme: In Laser Program Annual Report, Lawrence Livermore Laboratory, UCRL-50021-75
3.18 J.B. Trenholme: Naval Research Laboratory Report #2480 (1972)
3.19 J. Kelly, D.C. Brown, J. Abate, K. Teegarden: Appl. Opt. *20*, Apr. 15 (1981)
3.20 S. Singh, R.G. Smith, L.G. Van Uitert: Phys. Rev. B *10*, 2566 (1974)
3.21 H.E. Edgerton: *Electronic Flash, Strobe* (McGraw-Hill, New York 1920)
3.22 P.B. Newell, A.P. Benson: EG&G Technical Report B-4426 (March 1975)
3.23 A.P. Benson: EG&G Technical Report, 2nd Report (Sept.1974)
3.24 D.C. Brown: Laboratory for Laser Energetics Annual Report 1977, Vol.1 (March 1978)
3.25 J.A. Abate, D.C. Brown, J.H. Kelly: Laboratory for Laser Energetics Report #76 (April 1978)
3.26 R.A. Hill: Appl. Opt. *7*, 2184 (1968)
3.27 D.J. Baker, A.J. Steed: Appl. Opt. *7*, 2190 (1968)
3.28 C.H. Church, L. Gompel: Appl. Opt. *5*, 241 (1966)
3.29 I. Liberman, C.H. Church, J.A. Asars: Appl. Opt. *6*, 779 (1967)
3.30 P. Bochhave, C.H. Church: Appl. Opt. *7*, 2200 (1968)

3.31 C.H. Church, R.D. Schlecht, I. Liberman: J. Quant. Spectr. Radiat. Transfer *8*, 403 (1968)
3.32 Manufactured by EG&G/Princeton Applied Research Corp., Princeton, NJ


*Chapter 4*

4.1  L. Tonks: J. Appl. Phys. *35*, 1134 (1963)
4.2  E. Sibert, F. Tittel: J. Appl. Phys. *40*, 4434 (1969)
4.3  J.B. Trenholme: Naval Research Laboratory Report 2480 (July 1972)
4.4  J.M. Soures, L.M. Goldman, M.J. Lubin: Appl. Opt. *12*, 927 (1973)
4.5  D.C. Brown, S.D. Jacobs, N. Nee: Appl. Opt. *17*, 211 (1978)
4.6  V.N. Alekseev, A.A. Mak, E.G. Pininskii, B.M. Sedov, A.D. Starikov, A.D. Tsvotkov: Sov. J. QE-*16*, 126 (1976)
4.7  W.D. Fountain, W.W. Simmons, J.B. Trenholme, J.S. Chin: Laser Program Annual Report UCRL-50021-75, 249 (1976)
4.8  R. Sampath, O. Lewis: unpublished results (1978)
4.9  G.P. Kostometov, N.N. Rozanov: Sov. J. QE-*6*, 696 (1976)
4.10 J.E. Swain, R.E. Kidder, K. Pettipiece, F. Rainer, E.L. Baird, B. Loth: J. Appl. Phys. *40*, 3973 (1969)
4.11 J.M. McMahon, J.L. Emmett, J.F. Holzrichter, J.B. Trenholme: IEEE J. QE-*9*, 992 (1972)
4.12 J.M. McMahon: NRL Report 7838, 9 (1974)
4.13 Laser Program Annual Report, UCRL-50021-74 (1974)
4.14 G. Dubé, N.L. Boling: Appl. Opt. *13*, 699 (1974)
4.15 S. Guch Jr.: UCRL-76874 (May 1975), and [4.13]
4.16 D.C. Brown: Appl. Opt. *12*, 2215 (1973)
4.17 J.A. Glaze, S. Guch, J.B. Trenholme: Appl. Opt. *13*, 2808 (1974)
4.18 J.M. McMahon: private communication and paper presented at 10th Int. Quantum Electronics Conf., Amsterdam (June 1976)
4.19 J.B. Trenholme: Laser Program Annual Report, UCRL-50021-74, Lawrence Livermore Laboratory (1974)
4.20 J.B. Trenholme: CPD Interim Report, Shiva Nova, Misc.107, Lawrence Livermore Laboratory (1977)
4.21 P. Labudde, W. Seka, H.P. Weber: Appl. Phys. Lett. *29*, 732 (1976)
4.22 J.M. Rinefierd, S.D. Jacobs, D.C. Brown, J.A. Abate, O. Lewis, H. Appelbaum: Laser Induced Damage in Optical Materials, Boulder, CO (1978)
4.23 Corning Glass Works, Corning, NY
4.24 S.D. Jacobs, J.A. Abate: Laboratory for Laser Energetics, private communication


*Chapter 5*

5.1  This program is described in Naval Research Laboratory Memorandum Report 2480, J.B. Trenholme (1972)
5.2  J.B. Trenholme: Shiva Nova CP&D Interim Report, Vol.I, Laser Fusion Program (Oct.1977)
5.3  J.B. Trenholme: Naval Research Laboratory Memorandum Report 2480 (1972)
5.4  G.J. Linford, R.A. Saroyan, J.B. Trenholme, M.J. Weber: IEEE J. QE-*15*, 510 (1979)
5.5  Laser Program Annual Report URCL-50021-77, Lawrence Livermore Laboratory (1977)
5.6  Laser Program Annual Report UCRL-50021-74, Lawrence Livermore Laboratory (1974)
5.7  *Face Pumped Laser*, General Electric Co., Final Technical Report, ONR Contract No. NONR-4659(00), Project Code 4730, ARPA Order #306 (1 June 1966 - 30 Nov.1966)

5.8  J. Kelly: Ph.D. Thesis, Institute of Optics, University of Rochester
     (1980);
     J. Kelly, D.C. Brown, J.A. Abate, K. Teegarden: Appl. Opt. *20*, Apr. 15
     (1981)
5.9  S.D. Jacobs: paper presented at 10th Annual Electro Optics/Laser Conf.,
     Boston (Sept.1978)
5.10 J. Kelly, D.C. Brown, J.A. Abate: unpublished results, Laboratory for
     Laser Energetics (1980)
5.11 O. Lewis, J. Soures, J. Jacobs, S. Refermat, T. Almguist: Technical
     Digest, Topical Meeting on Inertial Confinement Fusion, San Diego, CA
     (1978)
5.12 W.S. Martin: *Face Pumped Lasers*, Report #68-C-285, General Electric
     Research and Development Center, Schenectady, NY (August 1968)
5.13 R. Sampath: OMEGA Technical Note #77, Laboratory for Laser Energetics
     (August 1978)
5.14 H.S. Carlson, J.C. Jaeger: *Conduction of Heat in Solids* (Oxford Uni-
     versity Press, London 1948)
5.15 P. Manzo: Science Applications Inc., private communication
5.16 D.C. Brown: University of Rochester, patent docket UR-0027 (1979)
5.17 D.C. Brown, J.A. Abate, L. Lund, J. Waldbillig: Appl. Opt. *20*, Apr. 15
     (1981)
5.18 Yu.A. Kalinin, A.A. Mak: Opt. Technol. *37*, 129 (1970)
5.19 C.H. Cooke, J. McKenna, J.C. Skinner: Appl. Opt. *3*, 957 (1964)
5.20 C.E. Devlin, J. McKenna, A.D. May, A.L. Schawlow: Appl. Opt. *1*, 11 (1962)
5.21 O. Lewis, S. Jacobs, L. Lund: Technical Digest, Topical Meeting on
     Inertial Confinement Fusion, San Diego, CA (1978)
5.22 Laboratory for Laser Energetics, 1977 Ann. Rpt., Vol.I
5.23 F.W. Quelle, Jr.: Appl. Opt. *5*, 633 (1966)
5.24 D.C. Brown, S.D. Jacobs: results presented at ERDA Review of Prototype
     Beam Performance, Laboratory for Laser Energetics, University of
     Rochester (Jan.6-7, 1977)


*Chapter 6*

6.1  Laser Induced Damage in Optical Materials: ASTM Special Publication 469
     (U.S. Department of Commerce, National Bureau of Standards, Washington,
     D.C. 1969)
6.2  Laser Induced Damage in Optical Materials: NBS Special Publication 341
     (U.S. Department of Commerce, National Bureau of Standards, Washington,
     D.C. 1970)
6.3  Laser Induced Damage in Optical Materials: NBS Special Publication 356
     (U.S. Department of Commerce, National Bureau of Standards, Washington,
     D.C. 1971)
6.4  Laser Induced Damage in Optical Materials: NBS Special Publication 372
     (U.S. Department of Commerce, National Bureau of Standards, Washington,
     D.C. 1972)
6.5  Laser Induced Damage in Optical Materials: NBS Special Publication 387
     (U.S. Department of Commerce, National Bureau of Standards, Washington,
     D.C. 1973)
6.6  Laser Induced Damage in Optical Materials: NBS Special Publication 414
     (U.S. Department of Commerce, National Bureau of Standards, Washington,
     D.C. 1974)
6.7  Laser Induced Damage in Optical Materials: NBS Special Publication 435
     (U.S. Department of Commerce, National Bureau of Standards, Washington,
     D.C. 1975)

6.8 Laser Induced Damage in Optical Materials: NBS Special Publication 462
(U.S. Department of Commerce, National Bureau of Standards, Washington,
D.C. 1976)
6.9 Laser Induced Damage in Optical Materials: NBS Special Publication 509
(U.S. Department of Commerce, National Bureau of Standards, Washington,
D.C. 1977)
6.10 Laser Induced Damage in Optical Materials: NBS Special Publication 541
(U.S. Department of Commerce, National Bureau of Standards, Washington,
D.C. 1978)
6.11 W.L. Smith: Opt. Eng. *17*, 490 (1978)
6.12 N. Bloembergen: IEEE J. QE-*10*, 375 (1974)
6.13 W.L. Smith, J.H. Bechtel, N. Bloembergen: Opt. Commun. *18*, 592 (1976)
6.14 E. Yablonovitch: Ph.D. Thesis, Harvard University (1972)
6.15 A. von Hippel: Z. Physik *75*, 145 (1932); *88*, 358 (1934)
6.16 J.P. Anthes, M. Bass: Appl. Phys. Lett. *31*, 412 (1977)
6.17 E. Yablonovitch, W. Bloembergen: Phys. Rev. Lett. *29*, 907 (1972)
6.18 A.A. Vorob'ev, G.A. Vorob'ev, L.T. Muraasko: Sov. Phys. Solid State *4*,
1441 (1963)
6.19 D.B. Watson, W. Heyes, K.C. Kao, J.H. Calderwood: IEEE Trans. Elect.
Insul. *1*, 30 (1965)
6.20 G.A. Vorob'ev, N.J. Lebedeva, G.S. Nadorova: Sov. Phys. Solid State *13*,
736 (1971)
6.21 D.W. Fradin, E. Yablonovitch, M. Bass: Appl. Opt. *12*, 700 (1973)
6.22 D.W. Fradin, N. Bloembergen, J.P. Lettelier: Appl. Phys. Lett. *22*,
635 (1973)
6.23 W.L. Smith, J.H. Bechtel, N. Bloembergen: Phys. Rev. R. *12*, 706 (1975)
6.24 N. Alyassini, J.H. Parks: In Laser Induced Damage in Optical Materials:
NBS Special Publication 435, 356 (1975)
6.25 J.R. Bettis, R.A. House II, A.H. Guenther: Laser Induced Damage in
Optical Materials: NBS Special Publication 462 (1976)
6.26 M. Sparks: Laser Induced Damage in Optical Materials: NBS Special Publication 435 (1975)
6.27 M.J. Soileau, M. Bass, W.W. Van Stryland: Laser Induced Damage in
Optical Materials: NBS Special Publication 541 (1978)
6.28 A. Schmid, P. Kelly, P. Bräulich: Phys. Rev. B *16*, 4569 (1977)
6.29 S.E. Stokowski, D. Milam, M.J. Weber: Laser Induced Damage in Optical
Materials: NBS Special Publication 541 (1978)
6.30 N.L. Boling, G. Dubê: Laser Induced Damage in Optical Materials: NBS
Special Publication 372 (1972)
6.31 N.L. Boling, M.D. Crisp, G. Dubê: Appl. Opt. *12*, 650 (1973)
6.32 N. Bloembergen: Appl. Opt. *12*, 661 (1973)
6.33 D.W. Fradin, M. Bass: Appl. Phys. Lett. *22*, 159 (1973)
6.34 W.L. Boling, J.A. Ringlien, G. Dubê: In Laser Induced Damage in Optical
Materials: NBS Special Publication 414 (1974)
6.35 R.A. House, J.R. Bettis, A.H. Guenther, R. Austin: In Laser Induced
Damage in Optical Materials: NBS Special Publication 435 (1975)
6.36 R.A. House, A.H. Guenther, J.M. Bennett: In Laser Induced Damage in
Optical Materials: NBS Special Publication 509 (1977)
6.37 D. Milam, W.L. Smith, M.J. Weber, A.H. Guenther, R.A. House, J.R. Bettis:
In Laser Induced Damage in Optical Materials: NBS Special Publication 509
(1977)
6.38 D. Milam: Appl. Opt. *16*, 1204 (1977)
6.39 D. Milam: In Laser Induced Damage in Optical Materials: NBS Special
Publication 541 (1978)
6.40 D.L. Burdick: In Laser Induced Damage in Optical Materials: NBS Special
Publication 541 (1978)
6.41 N.C. Fernelius, D.A. Walsh: In Laser Induced Damage in Optical Materials:
NBS Special Publication 541 (1978)

6.42 R.R. Austin, R.C. Michaud, A.H. Guenther, J.M. Potman, R. Harniman: In Laser Induced Damage in Optical Materials: NBS Special Publication 372 (1972)
6.43 L.G. De Shazer, B.E. Newnam, K.M. Leung: In Laser Induced Damage in Optical Materials: NBS Special Publication 387 (1973)
6.44 S.R. Scheele, J.W. Bergstrom: In Laser Induced Damage in Optical Materials: NBS Special Publication 509 (1977)
6.45 J.R. Bettis, R.A. House, A.H. Guenther, P. Austin: In Laser Induced Damage in Optical Materials: NBS Special Publication 435 (1975)
6.46 W.L. Smith, D. Milam, M.J. Weber, A.H. Guenther, J.R. Bettis, R.A. House: In Laser Induced Damage in Optical Materials: NBS Special Publication 509 (1977)
6.47 R.A. House, J.R. Bettis, A.H. Guenther: In Laser Induced Damage in Optical Materials: NBS Special Publication 462 (1976)
6.48 D. Milam: In Laser Induced Damage in Optical Materials: NBS Special Publication 462 (1976)
6.49 A.F. Turner: In Laser Induced Damage in Optical Materials: NBS Special Publication 356 (1971)
6.50 B.W. Newnam, D.H. Gill, G. Faulkner: In Laser Induced Damage in Optical Materials: NBS Special Publication 435 (1975)
6.51 J.H. Apfel, J.S. Matteveci, B.E. Newnam, D.H. Gill: In Laser Induced Damage in Optical Materials: NBS Special Publication 462 (1976)
6.52 J.H. Apfel: In Laser Induced Damage in Optical Materials: NBS Special Publication 509 (1977)
6.53 J.H. Apfel, E.A. Enemark, D. Milam, W.L. Smith, M.J. Weber: In Laser Induced Damage in Optical Materials: NBS Special Publication 509 (1977)
6.54 C.K. Carniglia, J.H. Apfel, C.B. Carrier, D. Milam: In Laser Induced Damage in Optical Materials: NBS Special Publication 541 (1978)
6.55 D.L. Burdick: In Laser Induced Damage in Optical Materials: NBS Special Publication 541 (1978)
6.56 P.A. Temple, D.L. Decker, J.M. Donovan, J.W. Bethke: In Laser Induced Damage in Optical Materials: NBS Special Publication 541 (1978)
6.57 W.C. Fernelius, D.A. Walsh: In Laser Induced Damage in Optical Materials: NBS Special Publication 541 (1978)
6.58 A.C. Tam, C.K.M. Datel, R.J. Kerl: Opt. Lett. 4, 81 (1979)
6.59 W.H. Lowdermilk, D. Milam: Technical Digest, Topical Meeting on Inertial Confinement Fusion, Paper THF6, San Diego, CA (1980), and Lawrence Livermore Laboratory Preprint 83920 (Jan.1980)


*Chapter 7*

7.1 J.H. Marburger: Prog. Quantum Electron. 4, 35 (1975)
7.2 O. Svelto: *Progress in Optics XII*, ed. by E. Wolf (North-Holland, Amsterdam 1974)
7.3 M.S. Sodha, A.K. Ghatak, V.K. Tripathi: *Self-Focusing of Laser Beams* (Tata McGraw-Hill, New Delhi 1974)
7.4 J.A. Fleck, Jr., J.R. Morris, E.S. Bliss: UCRL-80138 (1977)
7.5 J.A. Fleck, Jr., J.R. Morris, M.D. Feit: Appl. Phys. 10, 129 (1976)
7.6 F.P. Mattar: Ph.D. Dissertation, Polytechnic Institute of New York, Dec.1975; available from Ann Arbor, Michigan
7.7 V.I. Bespalov, V.I, Talanov: JETP Lett. 3, 307 (1966)
7.8 A.J. Glass: In Laser Induced Damage in Optical Materials: NBS Special Publication 387 (1973)
7.9 J. Marburger, R. Jokipii, A.J. Glass, J. Trenholme: In Laser Induced Damage in Optical Materials: NBS Special Publication 387 (1973)
7.10 J. Trenholme: Laser Program Annual Report, UCRL-50021-74, Lawrence Livermore Laboratory (1975)

7.11 J. Trenholme: Laser Program Annual Report, UCRL-50021-75, Lawrence Livermore Laboratory (1976)

7.12 B.R. Suydam: In Laser Induced Damage in Optical Materials: NBS Special Publication 387 (1973)

7.13 K.A. Brueckner, S. Jorna: Phys. Rev. Lett. *17*, 78 (1966)

7.14 A.J. Campillo, S.L. Shapiro, B.R. Suydam: Appl. Phys. Lett. *24*, 178 (1974)

7.15 E.S. Bliss, D.R. Speck, J.F. Holzrichter, J.H. Erkkila, A.J. Glass: Appl. Phys. Lett. *25*, 448 (1974)

7.16 C.J. Elliot: Appl. Phys. Lett. *24*, 91 (1974)

7.17 D. Aurie, A. Labadeus, J. Guyot: paper presented at 9th Intern. Quantum Electronics Conf., Amsterdam (1976)

7.18 K.K. Lee: OMEGA Technical Note No.74, Laboratory for Laser Energetics, University of Rochester (1977)

7.19 J.T. Hunt, J.A. Glaze, W.W. Simmons, P.A. Renard: Appl. Opt. *17*, 2053 (1978)

7.20 D.C. Brown, K. Lee, R. Sampath, W. Seka: Laboratory Report #78, Laboratory for Laser Energetics, University of Rochester (1978)

7.21 J. Glaze, W. Simmons, W. Hagan: Laser Program Annual Report UCRL-50021-74, Lawrence Livermore Laboratory (1974)

7.22 C. Fujikawa, A. Glass: Laser Program Annual Report UCRL-50021-74, Lawrence Livermore Laboratory (1974)

7.23 E.S. Bliss, J.T. Hunt, P.A. Renard, G.E. Sommargren, H.J. Weaver: IEEE I. QE-*12*, 402 (1976)

7.24 E. Bliss, G. Sommargren, H.J. Weaver, D.R. Speck, J. Holzrichter, J. Erkkila, A. Glass: Laser Program Annual Report UCRL-50021-74, Lawrence Livermore Laboratory (1974)

7.25 W.W. Simmons, S. Guch Jr., F. Rainer, J.E. Murray: UCRL-76873 (1975)

7.26 Laser Program Annual Report UCRL-50021-74, Lawrence Livermore Laboratory (1974)

7.27 J.B. Trenholme: Shiva Nova, CP&D Interim Report, Lawrence Livermore Laboratory, Misc.107 (1977)

7.28 J.B. Trenholme: private communication, Lawrence Livermore Laboratory

7.29 W. Seka, J. Soures, O. Lewis, J. Bunkenburg, D. Brown, S. Jacobs, C. Mourou, J. Zimmerman: Appl. Opt. *19*, 409 (1980)

7.30 J. Hunt: private communication

7.31 J.A. Abate, L. Lund, D. Brown, S. Jacobs, S. Refermat, J. Kelly, O. Lewis, M. Gavin, J. Waldbillig: Appl. Opt. *20*, 351 (1981)

7.32 A. Maitland, M.H. Dunn: *Laser Physics* (North-Holland, Amsterdam, London 1969)

7.33 J.A. Glaze: Opt. Eng. *15*, 136 (1976)

7.34 M. Abramowitz, J.A. Stegun (eds.): Handbook of Mathematical Functions, AMS 55 (U.S. Govt. Printing Office, Washington, DC 1965)

7.35 J.B. Trenholme: Laser Program Annual Report UCRL-50021-74, Lawrence Livermore Laboratory (1974)

7.36 D.C. Brown, G. Burnham: OMEGA Technical Report #67, Laboratory for Laser Energetics, University of Rochester (1977)

7.37 D.C. Brown, K.K. Lee, R. Sampath, W. Seka: OMEGA Technical Report #78, Laboratory for Laser Energetics, University of Rochester (1978)

7.38 A.J. Glass, J.B. Trenholme: Laser Program Annual Report UCRL-50021-75, Lawrence Livermore Laboratory (1975)

7.39 D.C. Brown: unpublished results, Laboratory for Laser Energetics, University of Rochester (1980)

*Chapter 8*

8.1 W. Seka, S.D. Jacobs, J.E. Rizzo, R. Boni, R.S. Craxton: Opt. Commun. *34*, 469 (1980)

8.2 R.S. Craxton: Opt. Commun. *34*, 474 (1980)

8.3 S.D. Jacobs, J.A. Abate, K.A. Bauer, R.P. Bossert, J.M. Rinefierd: Topical Meeting on Inertial Confinement Fusion, Paper THF10, San Diego, CA (1980)

8.4 D.C. Brown: Laboratory for Laser Energetics, University of Rochester; patent applied for

8.5 Shiva Nova CP&D Interim Report, Laser Fusion Program, Lawrence Livermore Laboratory, Misc.107 (1977)

8.6 Laser Program Annual Report, UCRL-50021-74, Lawrence Livermore Laboratory (1974)

8.7 Laser Program Annual Report, UCRL-50021-76, Lawrence Livermore Laboratory (1976)

8.8 J.F. Stowers, L.P. Bradley, C.B. McFann: Topical Meeting on Inertial Confinement Fusion, paper THF11, San Diego, CA (1980)

8.9 A.J. Glass, W.E. Warren: UCRL-80316 (1977)

# Subject Index

Abbe value  27-28,48-51,59,91

ABCD absorption function  65-66,149

ABCD law  219

Absorption coefficient  60

Absorption efficiency  66,70,82

Absorption-emission coefficient  101-106

Absorption into heat  60-62

Absorption into inversion  60-62

Active-mirror amplifier  *see* Amplifier

Amplified spontaneous emission  66, 119,121,128-145,150-154,157,160, 238-240,250

Amplifier  76-77,79-87,146-169,227-229

    active-mirror  61-62,76-77,79-87, 156-164,227,229

    box  154-155

    disk  62-63,76-77,79-87,146-156, 227-229

    rod  76-77,79-87,164-169,227,229

Avalanche ionization rate  173,175, 176

Average polarizability  46

Backward propagation  230-231,238

Bandgap energy  30-36,39,171,173,238

Bandpass angle  215,220,222

Beer's law  60

Bespalov-Talanov gain coefficient *see* Ripple

Bessel's equation  168

B integral  45,54-56,80,193,195-196, 205-207,209-210,213-214,216-217,218, 220-221,225-226,238-239,240,243-245

Birefringence  *see* Stress-induced birefringence or Passive birefringence

Blackbody distribution  101-106

Born-Oppenheimer approximation  43

Born-Oppenheimer coefficients  43, 52-53,69

Bose-Einstein relationship  33,38

Box amplifier  *see* Amplifier

Branching ratio  8-9,21,23-28,59, 69,82

Breakdown electric field  176-177, 181

Brewster's angle  146,157,228

Bulk breakdown intensity  177-179

Bulk damage  *see* Damage

Cavity transfer efficiency  *see* Transfer efficiency

Characteristic electron-collision time  173

Characteristic impedance  110,113-114

Claddings  *see* Edge claddings

Coefficient of ionization  175-176

Concentration quenching  5-7,150, 152,158,165

Constant-shape approximation  209-210

Cost effectiveness  *see* Figure of merit, laser system

W. Koechner
## Solid-State Laser Engineering
1976. 287 figures, 38 tables. XI, 620 pages
(Springer Series in Optical Sciences,
Volume 1)
ISBN 3-540-90167-1

**Contents:**
Optical Amplification. – Properties of Solid-State Materials. – Laser Oscillator. – Laser Amplifier. – Optical Resonator. – Optical Pump Systems. – Heat Removal. – Q-Switches and External Switching Devices. – Mode Locking. – Nonlinear Devices. – Design of Lasers Relevant to their Application. – Damage of Optical Elements. – Appendices: Laser Safety. Conversion Factors and Constants. – Subject Index.

## Laser Spectroscopy of Solids
Editors: W. M. Yen, P. M. Selzer

1981. 117 figures. Approx. 330 pages.
(Topics in Applied Physics, Volume 49)
ISBN 3-540-10638-3

**Contents:**
*G. F. Imbusch, R. Kopelman:* Optical Spectroscopy of Electronic Centers in Solids. – *T. Holstein, S. K. Lyo, R. Orbach:* Excitation Transfer in Disordered Systems. – *D. L. Huber:* Dynamics of Incoherent Transfer. – *P. M. Selzer:* General Techniques and Experiment Methods in Laser Spectroscopy of Solids. – *W. M. Yen, P. M. Selzer:* High Resolution Laser Spectroscopy of Ions in Crystals. – *M. J. Weber:* Laser Excited Fluorescence Spectroscopy in Glass. – *A. H. Francis, R. Kopelman:* Excitation Dynamics in Molecular Solids.

## Light Scattering in Solids
Editor: M. Cardona

1975. 111 figures, 3 tables. XIII, 339 pages
(Topics in Applied Physics, Volume 8)
ISBN 3-540-07354-X

**Contents:**
*M. Cardona:* Introduction. – *A. Pinczuk, E. Burstein:* Fundamentals of Inelastic Light Scattering in Semiconductors and Insulators. – *R. M. Martin, L. M. Falicov:* Resonant Raman Scattering. – *M. V. Klein:* Electronic Raman Scattering. – *M. H. Brodsky:* Raman Scattering in Amorphous Semiconductors. – *A. S. Pine:* Brillouin Scattering in Semiconductors. – *Y.-R. Shen:* Stimulated Raman Scattering.

## Relaxation of Elementary Excitations
Proceedings of the Taniguchi International Symposium, Susono-shi, Japan
October 12–16, 1979
Editors: R. Kubo, E. Hanamura

1980. 117 figures, 15 tables. XII, 285 pages
(Springer Series in Solid-State Sciences,
Volume 18)
ISBN 3-540-10129-2

**Contents:**
Lattice Relaxation. – Intermediate State Interaction. – Nonlinear Optical Phenomena. – Molecular Crystal and Biological System. – Molecular System. – Index. of Contributors.

# Springer-Verlag Berlin Heidelberg New York